Planned Change in Farming Systems: Progress in On-Farm Research

Edited by

Robert Tripp

JOHN WILEY & SONS

Chichester - New York - Brisbane - Toronto - Singapore

A Co-Publication with Sayce Publishing, Exeter, UK

Copyright © 1991 by R. Tripp

Published by John Wiley & Sons Ltd.
           Baffins Lane, Chichester
           West Sussex PO19 1UD, England

Other Wiley Editorial Offices

John Wiley & Sons, Inc., 605 Third Avenue,
New York, NY 10158-0012, USA

Jacaranda Wiley Ltd., G.P.O. Box 859, Brisbane,
Queensland 4001, Australia

John Wiley & Sons (Canada) Ltd., 22 Worcester Road,
Rexdale, Ontario M9W 1L1, Canada

John Wiley & Sons (SEA) Pte Ltd., 37 Jalan Pemimpin 05-04,
Block B, Union Industrial Building, Singapore 2057

Co-Publisher's Offices

Sayce Publishing, 57 Marlborough Road,
St Leonards, Exeter EX2 4LN, Great Britain

A catalogue record for this book is available from the British Library

ISBN 0 471 93417 8

Printed in Great Britain by Ebenezer Baylis & Son Ltd., The Trinity Press, Worcester

Contents

The Contributors   vii

Preface   ix

Part 1   INTRODUCTION   1

1.   The Farming Systems Research Movement and On-Farm Research   3
      R. Tripp

2.   An Overview of the Cases of On-Farm Research   17
      R. Tripp

Part 2   ON-FARM RESEARCH CASE STUDIES   37

3.   Climbing Bean Introduction in Southern Rwanda   39
      W. Graf, J. Voss and P. Nyabyenda

4.   On-Farm Maize Research in the Transition Zone of Ghana   63
      G. Edmeades, A. A. Dankyi, K. Marfo and R. Tripp

5.   Alley Farming in South-Western Nigeria: The Role of Farming Systems   85
      Research in Technology Development
      L. Reynolds, C. di Domenico, A. N. Atta-Krah and J. Cobbina

6.   Intensifying Rice-Based Cropping Systems in the Rainfed Lowlands   109
      of Iloilo, Philippines: Results and Implications
      G. L. Denning

7.   From Diagnosis to Farmer Adoption: MARIF's Maize On-Farm Research   143
      Programme in East Java, Indonesia
      R. Krisdiana, M. Dahlan, Herianto, C. van Santen and L. W. Harrington

8.   Revealing the Rationality of Farmers' Strategies: On-Farm Maize Research   169
      in the Swat Valley, Northern Pakistan
      D. Byerlee, K. Khan and M. Saleem

9.   On-Farm Research in Support of Varietal Diffusion: Bean Production   191
      in Cajamarca, Peru
      W. Janssen, N. Ruiz de Londoño, J. A. Beltran and J. Woolley

10.   On-Farm Research in Caisán, Panama   215
      J. C. Martínez, G. Sain and J. R. Arauz

11.   Post-Harvest Technology Development with Farmers: The Diffused Light   231
      Storage Case
      R. Rhoades, R. Booth, E. Schmidt and O. Cuyubamba

Part 3   FUTURE DIRECTIONS FOR ON-FARM RESEARCH   245

12.   The Limitations of On-Farm Research   247
      R. Tripp

13.   On-Farm Research in Southern Africa: The Prospects for Achieving   257
      Greater Impact
      A. R. C. Low, S. R. Waddington and E. M. Shumba

14.   Adopters and Adapters: The Participation of Farmers in On-Farm Research   273
      J. A. Ashby

15.   Integrating On-Farm Research into National Agricultural Research Systems: 287
      Lessons for Research Policy, Organization and Management
      D. Merrill-Sands, S. D. Biggs, R. J. Bingen, P. T. Ewell, J. L. McAllister
      and S. V. Poats

16.   Integrating On-Farm Research with Disciplinary, Commodity and Policy   317
      Research: The Potential of On-Farm Wheat Research in Pakistan
      D. Byerlee, P. Hobbs and R. Tripp

Acronyms   339

Index   341

The Contributors

José Roman Arauz (agronomist) is an independent consultant in Chiriquí, Panama

Jacqueline Ashby (sociologist) is Director of the Participatory Research in Agriculture (IPRA) project at the International Center for Tropical Agriculture (CIAT), Colombia

Kwesi Atta-Krah (agronomist) coordinates the Alley Farming Network for Tropical Africa (AFNETA) and is based at the International Institute of Tropical Agriculture (IITA), Nigeria

Jorge Alonso Beltran (breeder/agronomist) works with the Cropping Systems section of the Bean Program at the International Center for Tropical Agriculture (CIAT), Colombia

Stephen Biggs (agricultural economist) is a Senior Lecturer in the School of Development Studies at the University of East Anglia, UK

James Bingen (political scientist) is an Associate Professor in the Department of Resource Development at Michigan State University, USA

Robert H. Booth (plant pathologist and post-harvest technologist) is Assistant Director General, International Cooperation at the International Center for Agricultural Research in the Dry Areas (ICARDA), Syria

Derek Byerlee (agricultural economist) is Director of the Economics Program at the International Maize and Wheat Improvement Center (CIMMYT), Mexico

Joe Cobbina (soil scientist) works with the International Livestock Centre for Africa (ILCA) and is based in Ghana

Oscar Cuyubamba (seed specialist) works with the Instituto Nacional de Investigación Agrícola y Agroindustrial (INIAA), Peru

Marsum Dahlan (maize breeder) is Research Coordinator at the Malang Research Institute for Food Crops (MARIF), Indonesia

A. A. Dankyi (economist) works with the Grains and Legumes Development Board (GLDB), Ghana

Peter Ewell (agricultural economist) is Regional Social Scientist at the International Potato Center (CIP) in Kenya

Glenn L. Denning (agronomist) heads the International Programs Management Office at the International Rice Research Institute (IRRI), Philippines

Cathy di Domenico (sociologist) works with the International Livestock Centre for Africa (ILCA) and is based in Nigeria

Gregory Edmeades (crop physiologist) is coordinator of agronomy and physiology for the Maize Program at the International Maize and Wheat Improvement Center (CIMMYT), Mexico

Willi Graf (agronomist) works in the Agricultural Division of the Swiss Development Corporation (SDC)

Larry Harrington (agricultural economist) is based in Thailand as a Regional Economist for the Economics Program of the International Maize and Wheat Improvement Center (CIMMYT)

Herianto (socioeconomist) works with the Malang Research Institute for Food Crops (MARIF), Indonesia

Peter Hobbs (agronomist) is based in Nepal with the Wheat Program of the International Maize and Wheat Improvement Center (CIMMYT)

Willem Janssen (economist) works in the Bean Program at the International Center for Tropical Agriculture (CIAT), Colombia

Kiramat Khan (agronomist) is Scientific Officer at the Cereal Crops Research Institute (CCRI) in Pakistan

Ruly Krisdiana (socioeconomist) is Head of the Socioeconomics Section at the Malang Research Institute for Food Crops (MARIF), Indonesia

Allan Low (agricultural economist) is a Lecturer at the Development and Project Planning Centre, University of Bradford, UK

Kofi Marfo (economist) works with the Crops Research Institute (CRI), Ghana

Juan Carlos Martínez (agricultural economist) is a Research Fellow at the John F. Kennedy School of Government, USA

Jean McAllister (anthropologist) is a graduate student at Columbia University, USA

Deborah Merrill-Sands (anthropologist) is a Senior Research Officer with the International Service for National Agricultural Research (ISNAR), The Netherlands

Pierre Nyabyenda (breeder/agronomist) is Head of the Food Crop Department at the Institut des Sciences Agronomiques du Rwanda (ISAR)

Susan Poats (anthropologist) is a Social Scientist in the Cassava Program at the International Center for Tropical Agriculture (CIAT)

Len Reynolds (animal scientist) is coordinator of the Small Ruminant Meat and Milk Thrust of the International Livestock Centre for Africa (ILCA) and is based in Kenya

Robert Rhoades (anthropologist) is Professor and Head of the Department of Anthropology at the University of Georgia, USA

Nohra Ruiz de Londoño (agronomist) is an Associate Researcher in the Economics Section of the Bean Program at the International Center for Tropical Agriculture (CIAT), Colombia

Gustavo Sain (agricultural economist) is based in Costa Rica as a Regional Economist for the Economics Program of the International Maize and Wheat Improvement Center (CIMMYT)

Mohammed Saleem (botanist) works with the Cereal Crops Research Institute (CCRI) in Pakistan

Ella Schmidt (anthropologist) is a graduate student at the University of California, USA

Enos Shumba (agronomist) is Head of the Agronomy Institute in the Department of Research and Specialist Services (DR&SS), Zimbabwe

Robert Tripp (anthropologist) is Assistant Director of the Economics Program at the International Maize and Wheat Improvement Center (CIMMYT), Mexico

Charles van Santen (agricultural economist) is Program Leader in Human Resources Development for the United Nations Economic and Social Commission for Asia and the Pacific (ESCAP), based in Indonesia

Joachim Voss (anthropologist) is Senior Program Officer with the Environmental Policy Program of the International Development Research Centre (IDRC), Canada

Stephen Waddington (agronomist) works with the Maize Program of the International Maize and Wheat Improvement Center (CIMMYT) and is based in Zimbabwe

Jonathan Woolley (agronomist) is Training Coordinator for the Maize Program of the International Maize and Wheat Improvement Center (CIMMYT), Mexico

Preface

Farming systems research (FSR) has been a dominant feature of international agricultural research for the past decade and a half. Its starting point is an understanding of the rationality of farmers' practices and it entails an interdisciplinary effort in describing current farming systems, designing alternative technologies and testing them under farmers' conditions. Research that embodies these characteristics has been promoted by a wide range of international agricultural research centres, universities and donor agencies, and most national agricultural research systems in developing countries have some experience with FSR.

Despite this experience, there are several serious gaps in the literature. First, FSR has been plagued by a profusion of jargon and hampered by battles among competing methodologies. There have been few attempts to establish common ground or to look for a set of principles that can guide this type of research. Second, although considerable resources have been invested in FSR, virtually nothing has been written about its impact. There is no clear idea of the extent to which FSR has actually been responsible for bringing better technologies to farmers. Lastly, an increasing number of questions are being raised about the future of FSR, but only recently have suggestions begun to be made on how to improve its efficiency. This book is an attempt at a reasoned discussion about FSR, demonstrating some of its accomplishments and analysing its weaknesses.

The book was not originally conceived to respond to such a comprehensive set of objectives. Several years ago, a few people in the Economics Program at the International Maize and Wheat Improvement Center (CIMMYT) discussed the idea of writing an article or a working paper that would put together a few 'success stories' — descriptions of cases where FSR had led to the development of improved technology. It was soon obvious that the cases known to CIMMYT were themselves quite varied, and that an even more interesting range of cases existed in other institutions. It was obvious as well that rarely were we describing a straightforward research process, and cases needed to be described in some detail. The cases exhibited common features that were useful in deriving principles for conducting research, but equal emphasis had to be given to allowing for contrasting approaches.

The challenge of providing a useful range of cases was matched by the difficulties in uncovering well-documented instances of actual technology adoption. Many interesting stories of FSR ended without demonstrating benefits to farmers. The problems of documenting adoption (admittedly only one possible measure of FSR success) helped crystallize the growing criticism of FSR, and were the principal motivation behind a decision to expand the book's objectives to look more critically at FSR. Thus, several people who had thought carefully about the shortcomings of FSR were invited to make suggestions for improving the efficiency of the agricultural research process.

It is appropriate to focus on the research process, rather than on FSR *per se*. This book addresses planned change in agricultural research and looks at ways in which information about farmers' conditions and problems is used to set research priorities. This is the central theme of FSR, but is neither an invention nor the exclusive property of any particular research movement. That technology generation should be based on client needs is obvious, but how research should be organized to achieve this end is not necessarily clear. One way of examining the issue is to look at a set of research activities that has had this focus. This is the purpose behind a book that describes the experiences of a type of FSR that we call on-farm research. The book is not a 'how-to' manual, but rather a collection of detailed accounts of the flow of research in projects aimed at location-specific technology generation. It describes how decisions were taken and how research was managed. It addresses the successes and failures of this type of research and looks towards the future.

The principal actors in the book are researchers and extension agents working for government institutions in developing countries, and personnel from international agricultural research centres that support the development of those institutions. This focus should not be taken to imply that public sector agricultural research and extension is the only, or even the most important, source of innovation and change for developing country farmers, but it is certainly a force whose full potential has yet to be realized. This is one of the reasons why the FSR movement has been concentrated among these institutions. Conclusions about the organization of farmer-oriented research should be useful to a wide audience, but it should be clear that the work described in this book is dedicated to improving the efficiency of public sector agricultural research and extension in developing countries.

The ideas and concerns that led to the development of this book have a long and complex history. However, I wish to highlight an early and important source of inspiration. In the mid-1980s, a group of people in the CIMMYT Economics Program, including Derek Byerlee, Mike Collinson, Edgardo Moscardi and Donald Winkelmann, began working to ensure that the farmers' point of view was represented in the research agenda, and their example helped many of us to see the importance of building responsive national agricultural research services. In addition, the work of many people and institutions, some of whom are represented in this book, has contributed to our ability to conduct effective research in farmers' fields.

In evaluating the wide range of experience represented in this book, I have sought advice and comment from a number of people, including Mike Collinson, L. Van Crowder, Tony Fischer, Larry Harrington, Peter Hildebrand, Richard Jones, Clive Lightfoot, Michael Morris, H.J.W. Mutsaers, Graham Thiele, Greg Traxler, Charlie Wedderburn and Jonathan Woolley. All the contributors have been very patient in responding to questions and providing information and ideas.

In the early stages of assembling the material for the book, a number of chapters profited from the editorial skills of Kelly Cassaday. The general editing, organization and production of the book was undertaken by Kay Sayce and her colleagues, Helen Mills and Sergio Falzoi. Thoughout the process, innumerable drafts of chapters were modified, retyped, mailed to authors and reviewers and pushed towards final form, all under the management of Beatriz Rojón.

However, most of the work represented in this book is that of national programme researchers and extension agents. Their task is immense and their rewards are small. They deserve our admiration and support.

ROBERT TRIPP
ECONOMICS PROGRAM, CIMMYT

Part 1

INTRODUCTION

Planned Change in Farming Systems
Edited by R. Tripp
© 1991 R. Tripp
A Wiley-Sayce Co-Publication

1

The Farming Systems Research Movement and On-Farm Research

R. TRIPP

The urgency of improving the productivity of small-farm agriculture in the developing world is widely recognized. Not only is this essential for providing an adequate food supply to the world's poor, but a dynamic agricultural sector is also a key to more balanced and robust economic development. The challenge of making small-farm agriculture more efficient is difficult, especially because it depends on improving production from a large number of farms operating under a wide range of conditions, constraints and objectives. The task is shared by many people, including farmers, policy makers and academics, but an important part of the burden falls on agricultural researchers and extension agents. Numerous institutions, public and private, national and international, conduct agricultural research and extension in developing countries. Their work is driven by a variety of incentives — intellectual, humanitarian, economic and political. Indeed, the difficult objectives, varied institutions and complex motives that shape agricultural research impose serious constraints on meeting the challenge in a coordinated fashion.

The coordination of agricultural research and extension efforts is facilitated by shared scientific paradigms and a common body of knowledge. In addition, people working towards a common goal often look for ways to organize their concerns, observations and experiences. From time to time, such searches for order and direction in agricultural or rural development produce what might be called a 'movement'. To be successful, a movement must have broad appeal, offering a common vision to theoreticians and a common procedural framework for practitioners. It must provide comfortable niches for the academic, the field worker and the administrator. Such a movement is never monolithic, and usually brings with it schools and factions, devotees and skeptics, experts and hangers-on. It may represent only a reorganization of already existing ideas and knowledge, but it contributes justification, order and a sense of unity to the task of agricultural development.

This book is concerned with one such movement: farming systems research (FSR). The FSR movement, already a decade and a half old, has influenced how agricultural technology is developed throughout the world. The movement is so broad and loosely defined that it may be impossible to grasp in its entirety. This book does not attempt to synthesize the entire FSR movement, much less to pass judgement. But FSR has produced a rich legacy of field experience that deserves scrutiny. It is appropriate to look in detail at some of this research in order to document the accomplishments of FSR and to analyse its strengths and weaknesses.

It is disturbing that the energy and imagination generated by FSR and other such movements are often dissipated by the inevitable changes in fashion and interest which affect the development field. The tragedy is not that these movements fail to achieve everything they set out to accomplish, but that too few of the lessons they produce are articulated or used. FSR is part of a long history of rural development movements. It arrived on the scene after production campaigns to promote 'modern agriculture', the community development movement and integrated rural development. It competes for attention with other movements such as training and visit (T & V) extension, price policy reform and structural adjustment lending. And it is already being challenged by the growing interest in sustainable agriculture (*see* Figure 1.1), high hopes by some donors that the private sector will become a more important force for development, and increasing support for non-government organizations (NGOs).

Figure 1.1   Frequency of topic in titles of AGRIS Abstracts, 1976-87 (3 year moving average)

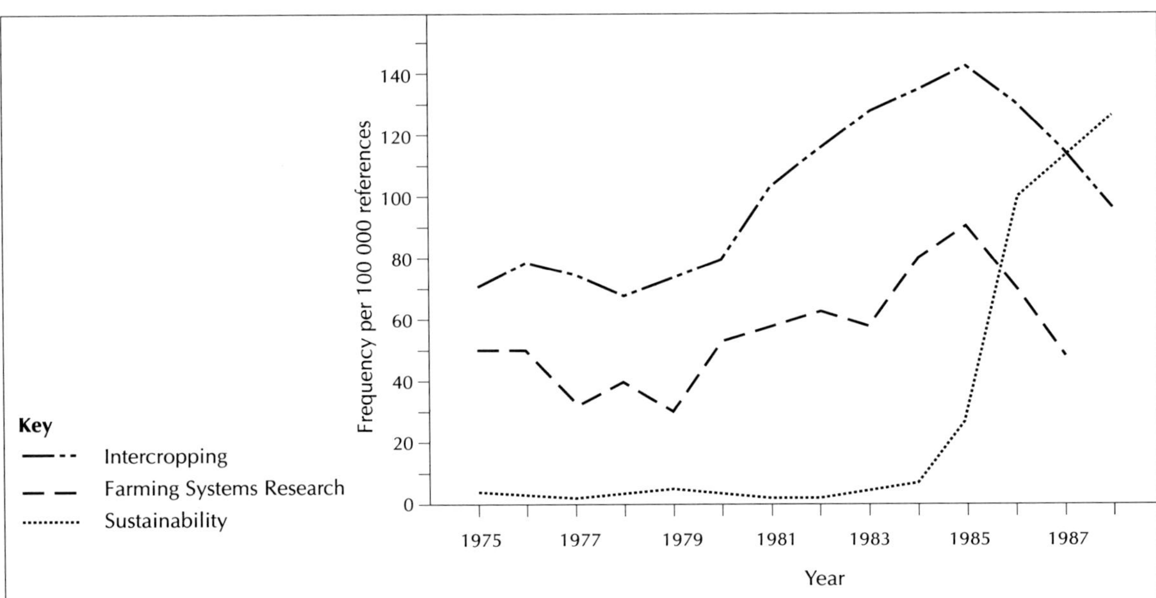

This book is not by any means a postmortem on FSR, however. No matter what terminology may be in vogue a decade from now, the FSR movement has made a number of lasting contributions. One particularly important product of the FSR movement is the idea that planned agricultural change needs to be organized around an understanding of farmers' conditions and priorities. This book emphasizes that it is essential for this idea to survive, independent of changes in development theory or practice. The idea is certainly not original to FSR, but FSR articulates it with particular force and clarity, although at times it has been less than perfectly implemented. FSR has developed a number of successful agricultural technologies, some of which are examined here. There have also been failures and false starts, and this book suggests ways of improving the efficiency of farmer-centered technology development. The goal of the book is therefore to identify instances where FSR has had an impact in farmers' fields, to describe characteristics common to these successful efforts, and to understand why success is not more frequent.

CHANGE IN FARMING PRACTICES

Before examining methods to promote change in farming practices, it is important to recall that all farming practices are subject to adjustment, change and evolution. In considering the process of agricultural development, it is easy to assume that peasant farmers are bound by tradition to follow an unchanging set of practices passed on to them by previous generations. Nothing could be further from the truth. Any responsible study of farming systems must place considerable emphasis on the dynamics of the systems and their susceptibility to change (e.g., Ruthenberg, 1980).

A variety of theories attempt to explain how farming systems evolve (Grigg, 1974). One of the most prominent is Boserup's (1965) contention that increasing population density stimulates innovation in agricultural practices. Population growth increases the frequency of land use, which in turn encourages change in agricultural technology. Pingali et al. (1987) test this theory for a specific technology, showing that population density and market availability are primary factors in determining tillage practices in sub-Saharan Africa.

Changing economic conditions are also crucial determinants of technological change. The development of markets strongly influences change in agricultural practices, as farmers readily respond to commercial opportunities and price fluctuations (Askari and Cummings, 1976). However, institutional factors often greatly affect the efficiency of markets and their influence on technological development. Such factors include political structures, access to resources and information dissemination (Piñeiro et al., 1979; de Janvry and Dethier, 1985).

Apart from these major demographic and economic forces, there are other factors which motivate farmers to change. One of the most important of these is farmers' own curiosity and their propensity to experiment (Biggs and Clay, 1981). At times, farmers experiment by using only the resources immediately at hand, but very often contact with neighbouring groups stimulates farmers to borrow and adapt ideas, cultivars or techniques (Knight, 1974). The capacity of peasant farmers to take advantage of new opportunities and adapt them to their own conditions is illustrated by the rapid and spontaneous growth of the cocoa industry in Ghana in the early part of this century (Hill, 1963).

In addition to developing or adapting new practices themselves in response to new conditions and local initiatives, farmers are also encouraged to take up new technology by government research and extension programmes, foreign donor development schemes, private voluntary projects and a growing number of commercial firms (Pray, 1983). An immense literature examines the results of these efforts (e.g., Feder et al., 1985) and documents the long history of theoretical discussion on the adoption process (Rogers and Shoemaker, 1971).

There is no shortage of examples of farmers responding to the development of new technology. The most spectacular success in recent years was associated with the Green Revolution, particularly in irrigated areas of Asia where new rice and wheat varieties, grown under higher levels of purchased inputs, raised yields and reduced food grain deficits (Hanson et al., 1982; Herdt and Capule, 1983). The massive scale of adoption and the apparent homogeneity of this seed-fertilizer technology may mask the remarkable range of farmer adaptation, however. Some farmers intensified and expanded food grain production, leading to important changes in rural welfare (Leaf, 1984); others took advantage of the increase in food grain productivity to diversify out of cereal crop production (Bradnock, 1984); and in many cases even those farmers in marginal areas were able to utilize elements of the new technology (Hussain, 1987).

A large number of less publicized but significant instances of technological change in small-farm agriculture have also taken place. These include the adoption of new varieties, such as the spread of hybrid maize in eastern and southern Africa (Gerhart, 1975); the introduction of new crops, such as specialized vegetable production in the Guatemalan highlands (von Braun et al., 1989); and the adoption of new management practices, such as chemical weed control (DeWalt et al., 1982; Bassett, 1988).

Despite the undeniable evidence of continual change in farming practices, considerable dissatisfaction remains concerning the ability of researchers and extensionists to effectively provide advice and technology to resource-poor farmers. Many crop varieties released for farmers still find very limited acceptance (Kydd, 1989). Production campaigns and agricultural development projects often fail (Cassen, 1986). Efforts to develop technology for small farmers frequently meet with low adoption (Ahmed and Kinsey, 1984; Goodell, 1984), and sometimes new inputs remain unused or are misused (Litsinger et al., 1980). Marked inefficiency in planned agricultural change is one of the principal reasons for the growth and development of the FSR movement.

THE FARMING SYSTEMS RESEARCH MOVEMENT

FSR has attempted to improve the success ratio of planned change in agricultural technology development. Since its appearance in the mid-1970s as a strategy for conducting agricultural research, FSR has been taken up in various forms by international research organizations, donor agricultural programmes and projects, and national agricultural research systems. It has been embraced at one time or another by many different institutions, been promoted as a new discipline, attracted support from a range of (not always compatible) interests, and become part of the vocabulary of international agricultural development.

Although the FSR movement is quite young, its origins go back to numerous types of academic and applied research. As Brush and Turner (1987) illustrate in an excellent review of the history of the farming systems concept, a range of disciplines, from geography to anthropology to systems theory, provided the intellectual underpinning for FSR. In addition, various examples of applied research in tropical agriculture dating back almost a century express many of the concerns associated with FSR (Simmonds, 1985).

The FSR movement also owes much to the debate that followed the Green Revolution. The initial success led to a number of questions and concerns regarding the possibility of bringing similar change to less favoured environments and the necessity of emphasizing equity and distributional issues in technology development (e.g., Falcon, 1970; Griffin, 1975). At the same time, research efforts in many parts of the world emphasized the complexity of local farming systems and the need to understand farmers' criteria in adopting or rejecting new technology. These included the work of Bradfield (1966) and Harwood (1979) on multiple cropping in the Philippines, Collinson's (1972) farm management studies in East Africa, and Norman's (1977) research in Nigeria. Several projects in Latin America contributed to establishing FSR as a research strategy as well, including the Caqueza integrated rural development project in Colombia (Zandstra et al., 1979), the Puebla project which attempted to bring new technology to maize farmers in central Mexico (Jiménez, 1970), and Rockefeller Foundation support for establishing a national agricultural research institute in Guatemala to focus on generating technology at the farm level (Hildebrand, 1976).

Most of the international agricultural research centres (IARCs) became involved in FSR during the 1970s. A major review of FSR at four of the centres in 1977 was followed by an attempt to provide some definitions and coordination for FSR (CGIAR/TAC, 1978). The IARCs pursued a range of FSR approaches that produced a rich variety of research methods and results but also contributed to growing uncertainty regarding the place of FSR within agricultural research institutions. In the course of the following decade, farming systems efforts in the IARCs were variously rethought, renamed, dismantled or reorganized. Eventually, a workshop on FSR held by the IARCs in 1987 moved towards a more practical consensus on the nature and conduct of this kind of research (ICRISAT, 1987).

During the 1980s, FSR also came to prominence in the portfolios of several donors. The United States Agency for International Development (USAID) played a leading role, funding more than 75 agricultural projects involving FSR (USAID, 1989), sponsoring a comprehensive review of FSR methods (Shaner et al., 1982) and establishing a Farming Systems Support Project at the University of Florida. The International Development Research Centre (IDRC) supported a number of FSR projects and activities, most notably the Asian Rice Farming Systems Network (ARFSN). The World Bank increased its lending for adaptive agricultural research (World Bank, 1985), and the Food and Agriculture Organization (FAO) of the United Nations launched a major effort in FSR (FAO, 1989). Many bilateral donors have also sponsored FSR projects.

At the same time that the IARCs and donors expanded their FSR activities, national agricultural research and extension institutions increased their commitment to a farming systems approach. This was partly a response to pressure from external agencies, but it also grew out of internal initiatives by individual scientists, administrators and the academic community. The result is that most national agricultural research systems now have some experience with FSR (Merrill-Sands et al., 1989).

CONTRIBUTIONS AND PROBLEMS OF FARMING SYSTEMS RESEARCH

The immense range of activities described as FSR has occasioned several excellent reviews that attempt to bring some order to the subject (Simmonds, 1985; Merrill-Sands, 1986), which is now firmly established in the literature of agricultural development (e.g., Caldwell, 1987; Upton, 1987). No attempt is made here to list all the characteristics that at one time or another have been proposed to constitute FSR. It is more important to focus on how FSR has contributed to the way that agricultural research is organized and conceived. Those contributions can be usefully classified as perspectives and methods.

Perspectives

FSR is, more than anything, a perspective on research. It asks that researchers take account of the whole farm and see farm family welfare as dependent on a wide range of variables. Small farms typically produce a number of crops and animals for subsistence and/or sale. Managing this array of enterprises, temporally and spatially, leads to complex farming systems. Concentrating on any one enterprise without understanding its place in the system is unlikely to produce successful results (*see* Table 1.1 *overleaf*). In some cases, the farming systems

Table 1.1 Classification of farming system interactions

Type of interaction	Example
Direct interaction between crops	
Interactions in space	Interactions arising from intercropping
Interactions over time	Conflicts in planting crop in relation to harvest of previous crop
	Carry-over of soil structure and crop residues from previous crop
	Carry-over of soil fertility from previous crop
	Carry-over and build-up of weed seeds and other pest populations from previous crop
Interactions between crops and livestock	Use of crops and crop residues for fodder
	Use of farmyard manure as crop nutrient source
	Use of animals for draught power
Resource competition and complementarity	Conflicts in labour use and cash needs between enterprises, including non-farm enterprises
	Competition for irrigation water between enterprises
Meeting multiple objectives of farm households	Choice of multiple crops, livestock species and production practices to manage risk
	Planting and storing food crops to balance seasonal food needs, and off-farm work to meet seasonal cash needs

Source: Byerlee and Tripp, 1988

perspective has led researchers to address a large number of enterprises and variables in a system simultane-ously. But in many other cases, it has served as a framework for concentrating on particular crop enterprises within a larger farming system (Byerlee et al., 1982).

As a consequence of this perspective, agricultural researchers have broadened their criteria for assessing the acceptability of new technology. Apart from considering conventional measures of yield or profitability, researchers more frequently design and assess a new technology in the light of other criteria, such as the importance of household food requirements, the cooking and storage qualities of new varieties, the role of labour and draught power constraints, the compatibility of new crops and practices with intercropping and rotation patterns, and the importance of risk in farmers' decision making. An understanding of the complexity of farming systems and the trade-offs that farmers must make in taking production decisions has inspired a shift away from searching for optimal technologies and towards identifying acceptable compromises (Collinson, 1981).

The farming systems perspective has also encouraged a growing acceptance of a problem-oriented approach to planning agricultural research (Tripp and Woolley, 1989). Agricultural programmes are now less frequently planned from the top down, while more effort is being made to understand local farming conditions and problems as a basis for planning research. Related to this change in strategy is a move towards more participation by farmers themselves in identifying research priorities; the experience of FSR has encouraged debate and produced alternative proposals for ensuring more meaningful farmer input in the research process (Chambers et al., 1989).

Lastly, the farming systems perspective has fostered a more realistic look at the targeting of agricultural research programmes. It is now widely recognized that the world is not divided into 'progressive' and 'traditional' farmers and that differences in adoption rates can often be explained by the natural and socioeconomic conditions of particular farming systems rather than by psychological traits of individual farmers (Perrin and Winkelmann, 1976). More emphasis is also placed on identifying appropriate client groups for research before the research programme is under way (Shaner, 1984).

Methods

The second major contribution of FSR has been to develop new methods for adaptive agricultural research. Gathering information on farming systems requires data collection techniques that go well beyond standard farm management surveys. The range of FSR procedures is quite broad, although a common sequence is usually proposed; two examples are given in Figures 1.2 and 1.3 (*overleaf*). The sequence usually includes diagnosing farmers' practices and problems, planning an experimental programme, testing technological alternatives in farmers' fields, assessing the results, and developing and extending recommendations.

One of the more innovative diagnostic techniques to emerge has been 'rapid rural appraisal' (Carruthers and Chambers, 1981), in which researchers interview farmers informally in their fields. Researchers are encouraged to seek out representative farmers and to hold an open interchange of ideas and observations. The interview follows certain broad guidelines but uses no questionnaire. Such techniques have found widespread application (Khon Kaen University, 1987) and have become established procedure in a number of national research programmes (Hildebrand, 1981). Complementing the development of these informal techniques has been a move towards more efficient and better-focused methods for formal surveys (Byerlee et al., 1980).

The spread of FSR has also motivated a great increase in the amount of experimentation done in farmers' fields. This work has generated considerable experience in the design and management of on-farm experiments (Zandstra et al., 1981; Hildebrand and Poey, 1985; Mutsaers et al., 1986). Researchers are more familiar with the trial designs that are most appropriate for experiments in farmers' fields, the selection of treatments and treatment levels for these experiments, and the involvement of farmers in managing experiments (Lightfoot and Barker, 1988). Methods for the economic analysis of on-farm experiments are now well established (CIMMYT, 1988).

In addition to the development of specific research methods, the adoption of FSR has contributed to significant changes in the management of agricultural research. One of the most important changes is a move towards more interdisciplinary research. Addressing production problems in the context of a farming system

Figure 1.2   Stages of on-farm research

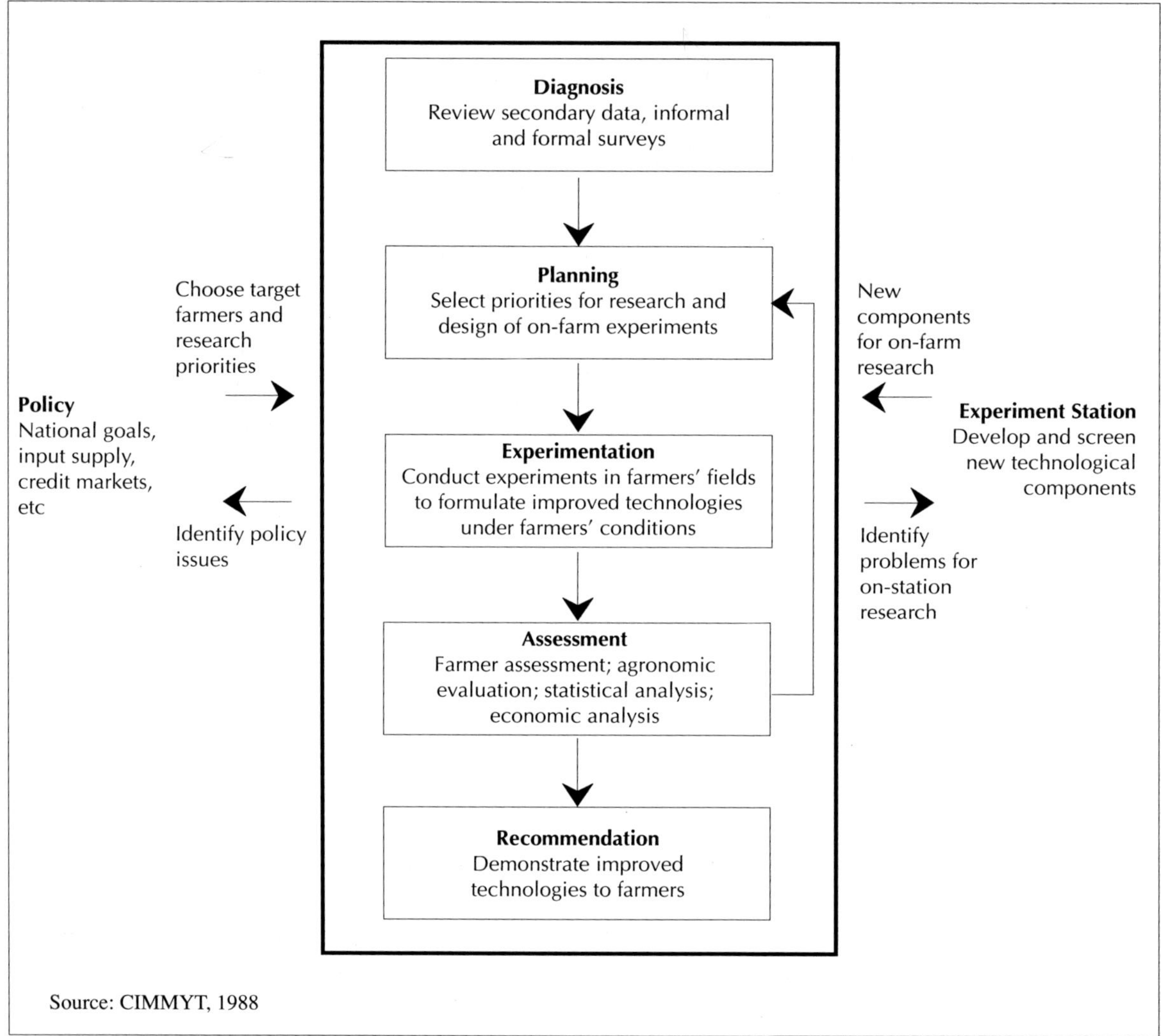

Source: CIMMYT, 1988

necessarily requires the participation of researchers from a range of disciplines. The implementation of interdisciplinary research within national research systems is far from perfect, but one advance that can be credited in part to the FSR movement is a wider recognition of the role of social science in agricultural research (Rhoades, 1984; Byerlee and Tripp, 1988). A second major change in research management that can be attributed to FSR is a more location-specific approach to research. Conducting this type of research requires increased field presence and an assignment of staff to particular farming systems and areas of a country. This has led to the development of a number of institutional arrangements for decentralizing research responsibilities (Gostyla and Whyte, 1980; Kean and Singogo, 1988).

Figure 1.3   Stages of the on-farm cropping systems research methodology

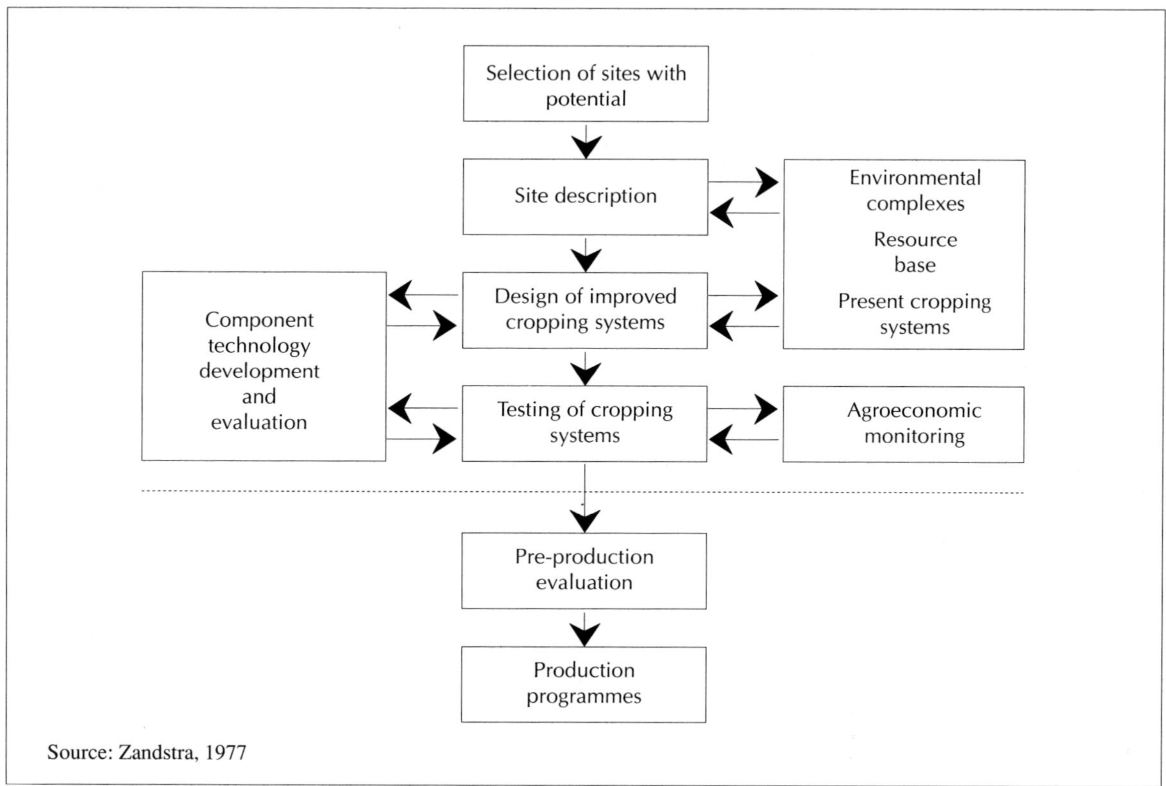

Source: Zandstra, 1977

CRITICISMS OF FARMING SYSTEMS RESEARCH

As is true of most movements, for a time FSR attracted almost unquestioning allegiance and seemingly unlimited donor support. In recent years, however, funding for FSR has become more restricted and the movement has been subject to increasing criticism. Although even its critics are willing to acknowledge many of its contributions, FSR now faces a number of difficult questions about its future role.

Some of the problems stem from the very characteristics that distinguish FSR from more conventional research procedures (Tripp et al., 1990). Its emphasis on holism has separated FSR from commodity research. Its location specificity has isolated it from the rest of the research establishment by sending small teams to work independently. Its attention to resource-poor farmers has sometimes been interpreted to mean that FSR is appropriate only for the most marginal farming populations. And its emphasis on experimentation in farmers' fields has provoked criticism that the results of FSR do not meet rigorous research standards.

Further criticism of FSR derives from the way it has been introduced. The promises made on behalf of FSR have sometimes been so extravagant that a negative reaction is only to be expected (Farrington, 1988; Coulter, 1989). When FSR has captured the major share of research resources, this excessive attention has led to divisiveness within research organizations (Herdt, 1987). Debates on minor issues of terminology or procedure within the FSR movement have not helped its cause either, attracting the attention of critics who may be unfamiliar with the substance of FSR but who find an easy target in such bickering.

Lastly, there is growing concern about the efficiency of FSR as a research strategy. Some question the costs of placing FSR teams in the field for long periods (Upton, 1987; Fielding, 1988). But most important is the demand to see, after more than a decade of effort, some evidence of results in the hands of farmers (Herdt, 1987).

ON-FARM RESEARCH

The debate over the FSR movement and its future encompasses a wide range of issues, including the appropriate scope of agricultural development programmes and the role of donors and international organizations in such efforts. The term FSR has taken on so many different meanings and interpretations and engendered so many kinds of activity that a global assessment is virtually impossible. The purpose of this book is to extract some practical conclusions regarding the conduct of adaptive agricultural research and to look at the ways in which research incorporating a farming systems perspective and utilizing field research techniques developed by FSR produces useful results for farmers.

Such an analysis will profit from the precise use of terminology. One of the less attractive by-products of movements such as FSR is a profusion of vocabulary to describe differences in approach, philosophy or method. Within the IARCs alone, a number of terms have been used. Several IARCs have had 'farming systems research' programmes (CGIAR/TAC, 1978; ICRISAT, 1987); a number have preferred the term 'on-farm research' (Byerlee et al., 1982; Woolley et al., 1988); the International Rice Research Institute (IRRI) has developed a comprehensive methodology known as 'cropping systems research' (Zandstra et al., 1981); the International Potato Center (CIP) has promoted research under the term 'farmer-back-to-farmer' (Rhoades and Booth, 1982); and an analysis by the International Service for National Agricultural Research (ISNAR) of the management of research related to FSR in national programmes has adopted the term 'on-farm client-oriented research' (Merrill-Sands et al., 1989).

The profusion of terms is frustrating, but to some extent represents honest differences of opinion and experience regarding concepts and procedures. Harrington et al. (1989) argue that many of these differences correspond to the variety of conditions under which FSR is conducted. The IRRI cropping systems methodology, for instance, is aimed at intensifying cropping patterns in lowland areas of Asia where rice is the only appropriate crop during the wet season; the method therefore emphasizes the study and improvement of rice-based cropping patterns. On the other hand, much of what has been called 'on-farm research' has been conducted in rainfed areas where a single crop may not be dominant, where the most promising research opportunity is not necessarily to intensify cropping, and where significant improvements often can be made in the management of a particular crop. It is not surprising that research in such diverse areas has generated different sets of methods.

Harrington et al. (1989) place particular importance on the similarities between these methods rather than their differences. They promote a general approach to adaptive agricultural research in which methods and procedures are selected according to their utility rather than because of their place in a particular research scheme; such an approach underlies the analyses in this book. A pragmatic examination of FSR processes must necessarily focus on a subset of the FSR movement; thus the movement itself will be referred to here as FSR, while the type of research discussed will be described as 'on-farm research' (OFR). This distinction is not made to escape from the criticisms aimed at FSR. Although the movement has included a varied and at times inconsistent set of activities, most of the work described here as OFR reflects the same concerns about efficiency, integration and impact that are directed at the FSR movement. Similarly, the term OFR is not chosen to impose a particular set of methods or vocabulary on the discussion that follows. Indeed, the cases of OFR presented here are described in a variety of ways by the authors; similarly, while most authors use the term OFR, others favour variations of this term. The distinction between OFR and FSR is intended to focus the discussion on a class of FSR that fulfills certain criteria.

In the first place, the cases given in this book are all instances of commodity-specific research. One of the main issues in the FSR movement has been the debate over the breadth of research that is to be carried out. Although all variants of FSR demand a systems perspective, some proponents of FSR maintain that research should focus on a limited number of commodities in the context of a larger system, while others propose that research should include a large number of farm enterprises simultaneously. OFR takes the former approach.

Related to the commodity focus of OFR is an emphasis on incremental change in farming systems. OFR has successfully questioned the strategy of promoting technological packages to farmers. The preponderance of evidence shows that, even when farmers adopt packages, they do so in a step-by-step fashion (Byerlee and Hesse, 1986). Thus all the cases in this book show a careful selection of research priorities and illustrate the strategy of introducing gradual change, characteristics which are not necessarily elements of FSR in general. One school of thought holds that the real challenge for FSR is to develop and promote completely new farming systems (Simmonds, 1985; Fresco and Westphal, 1988). The distinction between the two views is not absolute; in certain cases, simple step-by-step changes probably will not have an impact in farmers' fields (e.g., Nagy et al., 1988). The cases discussed in this book, however, focus on the possibility of incremental change in farming practices.

A third characteristic of the cases is a problem-solving approach that begins with an analysis of farmers' conditions and priorities. In theory, most proponents of FSR share this approach. (It is rare to find examples of successful agricultural research that did not take some account of farmers' conditions.) All the cases in this book illustrate how data obtained in the field (using a variety of survey and experimental techniques) are reviewed continuously to set and adjust research priorities. The research methods described in the cases respond to the high level of field presence demanded by OFR and reflect sustained interchanges between farmers and researchers.

A key issue affecting the future of OFR is its cost-effectiveness. Research administrators need a better idea of when the extra expense and effort implied by OFR might be justified. When does careful attention to farmers' circumstances pay dividends in the form of technology adoption? To help answer that question, the cases share one additional characteristic — they all provide good evidence of adoption. Data are presented to show that farmers changed some of their practices as a result of the research that was carried out. This characteristic is emphasized here because nearly all discussions of OFR stop short of looking at adoption. So much attention has been spent describing systems and suggesting how they might be improved that the question of results is often ignored. These cases attempt to redress that imbalance.

Chapter 2 introduces the cases by summarizing their similarities and differences. The chapters in Part 2 present the cases themselves. The concentration on instances where OFR has brought about change should be useful for drawing conclusions regarding the choice of appropriate research procedures. However, the reader should not conclude that OFR projects which have yet to provide new technologies are necessarily failures. And most important, it should be clear from the outset that although all the cases of OFR described here had successful outcomes, this level of success is not yet a common characteristic of OFR. Thus the chapters in Part 3 provide a critical examination of the limitations of OFR and suggest ways of improving its performance.

References

Ahmed, I. and Kinsey, B.H. 1984. *Farm Equipment Innovations in Eastern and Central Southern Africa*. Aldershot, UK: Gower.

Askari, H. and Cummings, J.T. 1976. *Agricultural Supply Response: A Survey of the Econometric Evidence*. New York, USA: Praeger.

Bassett, T.J. 1988. Breaking up the bottlenecks in food crop and cotton cultivation in northern Côte d'Ivoire. *Africa* 58: 147-73.

Biggs, S.D. and Clay, E.J. 1981. Sources of innovation in agricultural technology. *World Development* 94: 321-36.

Boserup, E. 1965. *The Conditions of Agricultural Growth*. Chicago, USA: Aldine.

Bradfield, R. 1966. *Toward More and Better Food for the Filipino People and More Income for Her Farmers*. New York, USA: Agricultural Development Council.

Bradnock, R.W. 1984. Agricultural development in Tamil Nadu: Two decades of land use change at village level. In Bayliss-Smith, T.P. and Wanmali, S. (eds) *Understanding Green Revolutions*. Cambridge, UK: Cambridge University Press.

Braun, J. von, Hotchkiss, D. and Immink, M. 1989. *Nontraditional Export Crops in Guatemala: Effects on Production, Income and Nutrition*. IFPRI Research Report 73. Washington DC, USA: IFPRI.

Brush, S.B. and Turner, B.L. 1987. The nature of farming systems and views of their change. In Turner, B.L. and Brush, S.B. (eds) *Comparative Farming Systems*. New York, USA: Guilford Press.

Byerlee, D., Collinson, M. et al. 1980. *Planning Technologies Appropriate to Farmers: Concepts and Procedures*. El Batan, Mexico: CIMMYT.

Byerlee, D. and Hesse de Polanco, E. 1986. Farmers' stepwise adoption of technological packages: Evidence from the Mexican altiplano. *American Journal of Agricultural Economics* 68(3): 519-27.

Byerlee, D. and Tripp, R. 1988. Strengthening linkages in agricultural research through a farming systems perspective: The role of social scientists. *Experimental Agriculture* 24: 137-51.

Byerlee, D., Harrington, L. and Winkelmann, D. 1982. Farming systems research: Issues in research strategy and technology design. *American Journal of Agricultural Economics* 64: 897-904.

Caldwell, J.S. 1987. An overview of farming systems research and development: Origins, applications, and issues. In Gittinger, J.P., Leslie, J. and Hoisington, C. (eds) *Food Policy*. Baltimore, USA: Johns Hopkins University Press.

Carruthers, I. and Chambers, R. 1981. Rapid appraisal for rural development. *Agricultural Administration* 8: 407-22.

Cassen, R. 1986. *Does Aid Work?* Oxford, UK: Clarendon Press.

CGIAR/TAC. 1978. *Farming Systems Research at the International Agricultural Research Centers*. Rome, Italy: TAC Secretariat, CGIAR.

Chambers, R., Pacey, A. and Thrupp, L.A. 1989. *Farmer First: Farmer Innovation and Agricultural Research*. London, UK: Intermediate Technology Publications.

CIMMYT. 1988. *From Agronomic Data to Farmer Recommendations: An Economics Training Manual*. (revd edn) El Batan, Mexico: CIMMYT.

Collinson, M. 1972. *Farm Management in Peasant Agriculture*. New York, USA: Praeger.

Collinson, M. 1981. A low cost approach to understanding small farmers. *Agricultural Administration* 8: 433-50.

Coulter, J.K. 1989. Research for agricultural development in sub-Saharan Africa. *Tropical Agriculture Association Newsletter* 9: 1-11.

DeWalt, B., DeWalt, K., Adelski, E., Duda, S., Fordham, M. and Thompson, K. 1982. *Cropping Systems in Pespire, Southern Honduras*. INTSORMIL Report No. 1. Lexington, USA: University of Kentucky.

Falcon, W.P. 1970. The Green Revolution: Generations of problems. *American Journal of Agricultural Economics* 52: 698-710.

FAO. 1989. *Farming Systems Development*. Rome, Italy: FAO.

Farrington, J. 1988. Whither farming systems research? *Development Policy Review* 6: 323-32.

Feder, G., Just, R.E. and Zilberman, D. 1985. Adoption of agricultural innovations in developing countries: A survey. *Economic Development and Cultural Change* 33: 255-98.

Fielding, D. 1988. A system that now seems inappropriate. *International Agricultural Development* 84: 11.

Fresco, L.O. and Westphal, E. 1988. A hierarchical classification of farming systems. *Experimental Agriculture* 24: 399-420.

Gerhart, J. 1975. *The Diffusion of Hybrid Maize in Western Kenya*. El Batan, Mexico: CIMMYT.

Goodell, G. 1984. Bugs, bunds, banks and bottlenecks: Organizational contradictions in the new rice technology. *Economic Development and Cultural Change* 33: 23-41.

Gostyla, L. and Whyte, W.F. 1980. *ICTA in Guatemala: The Evolution of a New Model for Agricultural Research and Development*. Ithaca, USA: Rural Development Committee, Cornell University.

Griffin, K. 1975. *The Political Economy of Agrarian Change*. London, UK: Macmillan.

Grigg, D.B. 1974. *The Agricultural Systems of the World*. Cambridge, UK: Cambridge University Press.

Hanson, H., Borlaug, N. and Anderson, R.G. 1982. *Wheat in the Third World*. Boulder, USA: Westview Press.

Harrington, L.W., Read, M.D., Garrity, D., Woolley, J. and Tripp, R. 1989. Approaches to on-farm client-oriented research: Similarities, differences, and future directions. In Sukmana, S., Amir, P. and Mulyadi, D.M. (eds) *Developments in Procedures for Farming Systems Research*. Semarand, Indonesia: AARD.

Harwood, R. 1979. *Small Farm Development*. Boulder, USA: Westview Press.

Herdt, R.W. 1987. Whither farming systems? In *Proc. of Farming Systems Research Symposium*. Fayetteville, USA: University of Arkansas.

Herdt, R.W. and Capule, C. 1983. *Adoption, Spread, and Production Impact of Modern Rice Varieties in Asia*. Los Baños, Philippines: IRRI.

Hildebrand, P. 1976. *Generating Technology for Traditional Farmers: A Multidisciplinary Methodology*. Guatemala City, Guatemala: ICTA.

Hildebrand, P. 1981. Combining disciplines in rapid appraisal: The *sondeo* approach. *Agricultural Administration* 8: 423-32.

Hildebrand, P. and Poey, F. 1985. *On-Farm Agronomic Trials in Farming Systems Research and Extension*. Boulder, USA: Lynne Rienner.

Hill, P. 1963. *The Migrant Cocoa Farmers of Southern Ghana*. Cambridge, UK: Cambridge University Press.

Hussain, T. 1987. Innovation, adoption, and choice of technique in a mountain farming system: A study of the diffusion of new wheat varieties in the Gilgit District of Pakistan. PhD thesis. Chicago, USA: University of Chicago.

ICRISAT. 1987. *Proc. of Workshop on Farming Systems Research*. Patancheru, India: ICRISAT.

Janvry, A. de and Dethier, J-J. 1985. *Technological Innovation in Agriculture: The Political Economy of Its Rate and Bias*. CGIAR Study Paper No. 1. Washington DC, USA: World Bank.

Jiménez, L. 1970. The Puebla project: A regional program for rapidly increasing corn yields among 50 000 small holders. In Myren, D. (ed) *Strategies for Increasing Agricultural Production on Small Holdings*. El Batan, Mexico: CIMMYT.

Kean, S. and Singogo, L. 1988. *Zambia: Organization and Management of the Adaptive Research Planning Team (ARPT), Research Branch, Ministry of Agriculture and Water Development*. OFCOR Case Study No. 1. The Hague, The Netherlands: ISNAR.

Khon Kaen University. 1987. *Proc. of 1985 International Conference on Rapid Rural Appraisal*. Khon Kaen, Thailand: Rural Systems Research and Farming Systems Research Projects, Khon Kaen University.

Knight, C.G. 1974. *Ecology and Change*. New York, USA: Academic Press.

Kydd, J. 1989. Maize research in Malawi: Lessons from failure. *Journal of International Development* 1: 112-44.

Leaf, M.J. 1984. *Song of Hope*. New Brunswick, USA: Rutgers University Press.

Lightfoot, C. and Barker, R. 1988. On-farm trials: A survey of methods. *Agricultural Administration and Extension* 30: 15-23.

Litsinger, J.A., Price, E.C. and Herrera, R.T. 1980. Small farmer pest control practice for rainfed rice, corn, and grain legumes in three Philippine provinces. *Philippine Entomologist* 4: 65-86.

Merrill-Sands, D. 1986. Farming systems research: Clarification of terms and concepts. *Experimental Agriculture* 22: 87-104.

Merrill-Sands, D., Ewell, P., Biggs, S. and McAllister, J. 1989. Issues in institutionalizing on-farm client-oriented research: A review of experiences from nine national agricultural research systems. *Quarterly Journal of International Agriculture* 28(3/4): 279-300.

Mutsaers, H.J.W., Fisher, N.M., Vogel, W.O. and Palada, M.C. 1986. *A Field Guide for On-Farm Research*. Ibadan, Nigeria: IITA.

Nagy, J.G., Sanders, J.H. and Ohm, H.W. 1988. Cereal technology interventions for the West African semi-arid tropics. *Agricultural Economics* 2: 197-208.

Norman, D. 1977. Economic rationality of traditional Hausa dryland farmers in the north of Nigeria. In Stevens, R.D. (ed) *Tradition and Dynamics in Small-Farm Agriculture*. Ames, USA: Iowa State University Press.

Perrin, R. and Winkelmann, D. 1976. Impediments to technical progress on small versus large farms. *American Journal of Agricultural Economics* 58: 888-94.

Piñeiro, M., Trigo, E. and Fiorentino, R. 1979. Technical change in Latin American agriculture. *Food Policy* 4: 169-77.

Pingali, P., Bigot, Y. and Binswanger, H. 1987. *Agricultural Mechanization and the Evolution of Farming Systems in Sub-Saharan Africa*. Baltimore, USA: Johns Hopkins University Press.

Pray, C. 1983. Private agricultural research in Asia. *Food Policy* 8: 131-40.

Rhoades, R. 1984. *Breaking New Ground: Agricultural Anthropology*. Lima, Peru: CIP.

Rhoades, R. and Booth, R. 1982. Farmer-back-to-farmer: A model for generating acceptable agricultural technology. *Agricultural Administration* 11: 127-37.

Rogers, E. and Shoemaker, F. 1971. *Communication of Innovations: A Cross Cultural Approach*. New York, USA: Free Press of Glencoe.

Ruthenberg, H. 1980. *Farming Systems in the Tropics*. Oxford, UK: Clarendon Press.

Shaner, W.W. 1984. Stratification: An approach to cost-effectiveness for farming systems research and development. *Agricultural Systems* 15: 101-23.

Shaner, W.W., Philipp, P.F. and Schmehl, W.R. 1982. *Farming Systems Research and Development: Guidelines for Developing Countries*. Boulder, USA: Westview Press.

Simmonds, N.W. 1985. *Farming Systems Research: A Review*. World Bank Technical Paper No. 43. Washington DC, USA: World Bank.

Tripp, R. and Woolley, J. 1989. *The Planning Stage of On-Farm Research: Identifying Factors for Experimentation*. El Batan, Mexico: CIMMYT/CIAT.

Tripp, R., Anandajayasekeram, P., Byerlee, D. and Harrington, L. 1990. Farming systems research revisited. In Eicher, C.K. and Staatz, J.M. (eds) *Agricultural Development in the Third World*. (2nd edn) Baltimore, USA: Johns Hopkins University Press.

Upton, M. 1987. *African Farm Management*. Cambridge, UK: Cambridge University Press.

USAID. 1989. A review of AID experience: Farming systems research and extension projects, 1975-1987. *AID Evaluation Highlights* No. 4. Washington DC, USA: USAID.

Woolley, J., Beltran, J.A., Vallejo, R.A. and Prager, M. 1988. *Identifying Appropriate Technologies for Farmers: The Case of the Bean and Maize Systems in Ipiales, Colombia, 1982-1986*. CIAT Working Document No. 31. Cali, Colombia: CIAT.

World Bank. 1985. *Agricultural Research and Extension: An Evaluation of the World Bank's Experience*. Washington DC, USA: World Bank.

Zandstra, H. 1977. Cropping systems research for the Asian rice farmer. In *Proc. of Symposium on Cropping Systems Research and Development for the Asian Rice Farmer*. Los Baños, Philippines: IRRI.

Zandstra, H., Swanberg, K., Zulberti, C. and Nestel, B. 1979. *Caqueza: Living Rural Development*. Ottawa, Canada: IDRC.

Zandstra, H., Price, E., Litsinger, J. and Morris, R. 1981. *Methodology for On-Farm Cropping Systems Research*. Los Baños, Philippines: IRRI.

2

An Overview of the Cases of On-Farm Research

R. TRIPP

Each case described in this book represents an instance where on-farm research (OFR) has led to substantial change in farmers' practices. This chapter reviews common elements among the cases to identify conditions that contribute to the effective operation of OFR and highlights differences between the cases to emphasize the range of requirements that OFR can impose on the organization of research. As this review is based on only nine cases, caution should be taken in generalizing from these cases or making sweeping conclusions about the progress or conduct of OFR as a whole. However, considerable effort went into identifying these cases of successful OFR and, although similar cases exist, their numbers are not overwhelmingly large.[1]

The relatively few documented cases where OFR has led to substantial adoption give cause for concern. To some extent this apparently limited success may be a problem of insufficient documentation; there is probably more progress than is immediately evident, although it is difficult to assemble the requisite reports and demonstrate impact. But certainly much of the problem lies with the conduct of OFR itself, which lends urgency to the analysis of those instances where OFR has brought change to farming systems.

In the discussion that follows, the cases are referred to by country. Table 2.1 (*overleaf*) gives information on the location, technologies, institutions and size of the research area involved in each case; the order of the cases in the table corresponds to the order in which they appear in this book.

LOCATION SPECIFICITY

One of the hallmarks of OFR is its location-specific approach, which takes the conditions and priorities of particular groups of farmers as the starting point for planning and executing an adaptive research programme. The nature and extent of the location is chosen according to a variety of factors. For example, in Indonesia and Panama, OFR was undertaken to demonstrate the methods and organization of OFR for national research programme staff, as a way of encouraging the national programme to develop and carry out similar work in other areas of the country. In the case of northern Peru, OFR was part of a training course offered by the International Center for Tropical Agriculture (CIAT) to national researchers and extension agents.

Table 2.1   Summary of the main features of the on-farm research cases

Case	Location	Technology	Institutions	Size of research area
Rwanda	Eastern and South Central zones	Climbing beans for farmers growing bush beans	ISAR; MINAGRI; PAP; CIAT	4 villages/communes, with 20 000 families and 31 500 ha of cropped land
Ghana	Brong-Ahafo Region	Improved maize varieties, row planting, improved plant spacing, fertilizer use	CRI; GLDB; MOA; CIMMYT	15 000 families and 35 000 ha of maize
Nigeria	Oyo State	Alley farming of *Leucaena* and *Gliricidia* spp. in yam/maize and cassava/maize fields; use of tree foliage for feeding small ruminants	Ministry of Agriculture and Natural Resources, Oyo State; ILCA	2 villages with about 225 families and 500 ha of cropped land (as focus of research); 5 other villages joined later
Philippines	Iloilo Province, Western Visayas Region	Dry seeding, early maturing rice varieties to intensify cropping patterns	Philippine Ministry of Agriculture; Kabsaka project; RADIP; IRRI	About 130 000 ha of rainfed lowland rice
Indonesia	Malang District, East Java	Insect control and improved plant stand management for maize planted in rotation with maize or upland rice	MARIF; Netherlands Technical Cooperation Project ATA-272; CIMMYT	30 000 ha of farm land and 40 000 families
Pakistan	Swat District, North West Frontier Province	Improved maize variety	CCRI; PARC; CIMMYT	25 000 ha of cropland and 30 000 families
Northern Peru	Cajamarca Department	For beans intercropped with maize: varietal dissemination, change from broadcasting to row planting; increased plant density, fertilizer use	INIPA; CIAT	13 000 ha of beans and about 12 700 families

Table 2.1   **(continued)**

Case	Location	Technology	Institutions	Size of research area
Panama	Caisán, Chiriquí Province	For maize in a maize/bean rotation: chemical weed control and/or zero tillage, row planting and improved plant spacing, elimination of use of inappropriate fertilizers	IDIAP; CIMMYT	1 600 ha of maize and 300 farm families
Central Peru	Mantaro Valley and two coastal sites	Diffused light storage of seed potatoes	INIPA; CIP	Work done in communities representative of about 100 000 ha of potatoes and about 200 000 families

Elsewhere, OFR was part of a wider research effort, often at the national level. The Ghana case describes OFR conducted in one region of the country under the aegis of a nationwide programme of maize technology development and on-farm testing. The cases of Pakistan and Rwanda emerged from collaborative research between national commodity programmes and the International Maize and Wheat Improvement Center (CIMMYT) and CIAT, respectively, but each case reports work from a specific area.

Several cases are examples of efforts to test and refine research methods that could be used in other situations to adapt new technologies to farmers' conditions. The work of the International Potato Center (CIP) and the Instituto Nacional de Investigación Pecuaria e Agrícola (INIPA) in central Peru, for instance, was conducted under a larger programme to develop appropriate technology for storing potatoes. Research by the International Livestock Centre for Africa (ILCA) on alley farming was conducted at several locations in Nigeria with a view to gaining experience and developing methods that could be applied in many West African settings. The Philippine case describes one of many collaborative efforts between the International Rice Research Institute (IRRI) and national research programmes in cropping systems research aimed at intensifying rice-based systems by adapting high yielding, early maturing rice varieties to farmers' conditions.

The size of the area in which the technologies developed through location-specific research might be applicable varies widely among the cases. In Pakistan, a new maize variety achieved high adoption but only in a well-defined micro-environment. The research programmes in Panama, northern Peru and Indonesia were concerned with developing technologies for particular locations as well, but the cases of Ghana and Rwanda describe how national commodity programmes developed technology that was applicable beyond the locations described. The cases from the Philippines and Nigeria represent efforts to develop technologies applicable across broad environments, although the level of success reported in those cases may not be achieved in other locations, given differences in biophysical (Price, 1982) or socioeconomic (Francis, 1987) conditions. Of all the cases presented in this book, the work in central Peru on potato storage was probably applicable across the widest area; the principles developed in that research were eventually applied to other parts of Peru as well as to several other countries.

Thus location-specific OFR can either help adapt technology to the needs of specific groups of farmers or provide experience to further the development of technologies that may have much wider applicability. The need for efficiency argues that the research area should be as large as possible, but several factors can intervene to limit the size of this area. A key factor is the heterogeneity of a research area; compare, for example, the broad applicability of research recommendations developed from research conducted in the transition zone of Ghana to the more limited applicability of recommendations developed in the mountain valleys of Peru or Pakistan. But the northern Peru case illustrates that it is often possible to identify common research themes even in a diverse environment, and the research in Pakistan produced information useful in several countries where maize is planted for both food and fodder. In the Philippine case the issue of heterogeneity was addressed by carefully stratifying the research sites by soil type, landscape and rainfall.

The type of technology on which OFR eventually focuses is another factor that influences the choice of research area. A particularly complex technology may require more intensive interaction with farmers in a small area. In Nigeria, research on alley farming was conducted with a limited number of farmers in two villages as a possible forerunner to a broader extension project. For the sake of efficiency this kind of focused approach must be accompanied by an effort to derive more general principles that can eventually be applied across a wider area. Organizing location-specific research so that it contributes to technology generation across as wide an area as possible is crucial to the future of OFR. This issue is addressed in more detail in the final chapters of this book.

NEW TECHNOLOGY AND A PROBLEM-SOLVING APPROACH

OFR can be used for developing various types of technology. Two types of technological change resulted from the cases described in this book: technologies that improved the efficiency of crop management, and those that required a significant degree of change in the farming system. Despite their differences, both types originated with researchers studying the current farming system and exploring ways of making gradual changes.

The first type of change may involve providing new varieties, changing planting methods, applying fertilizer, introducing weed control or insect control techniques, or improving storage methods. The cases from Panama, northern Peru, central Peru, Ghana, Indonesia and Pakistan are all examples of this type of change. The second type of change involves a more profound shift in the farming system. The cases demonstrate several such changes: the introduction of new leguminous tree species as part of an alley farming system to maintain soil fertility and provide animal fodder, in Nigeria; the introduction of a different crop establishment method and the use of early maturing varieties to open new possibilities for intensifying cropping patterns, in the Philippines; and the introduction of a new variety that requires significant changes in crop management practices, as in the case of climbing beans in Rwanda.

The research agenda for OFR may be determined in two ways. Often, OFR is initiated with a broad mandate to address important production problems in a particular research area. When the Panamanians selected Caisán as an area in which to test OFR procedures as a means of improving maize production, they had no idea that weed control in maize would emerge as the key research priority. Although all the cases reported here started with a commodity focus, in some instances of OFR the target commodities are selected only after the initial diagnostic research has identified key production problems.

In other instances, OFR was initiated where researchers already had considerable knowledge of priority problems and the availability of novel solutions. In the Philippine case, early maturing rice varieties were being adopted by farmers and cropping systems research was aimed at utilizing the new varieties to intensify cropping patterns. In Nigeria, alley farming of leguminous tree species was known to be a potential solution to the problems of soil infertility and fodder shortages.

But whether the OFR began with a 'clean slate' or with particular issues and technologies that had already been identified, in all the cases researchers had to test their initial hypotheses under farmers' conditions and modify those hypotheses according to what was learned in farmers' fields. In no instance did research proceed without modification; the final results always differed somewhat from researchers' original expectations.

DIAGNOSTIC SURVEYS

Information from various sources helps researchers diagnose problems, identify priorities for on-farm work, and plan the experimental programme (*see* Table 2.2 *overleaf*). Several diagnostic survey techniques are presented in the cases. In many of the cases, an informal survey was one of the first steps in the OFR process. 'Informal' is perhaps a misnomer, particularly if it gives the impression that the surveying is done in a casual or haphazard manner. Considerable experience is required to plan and carry out an informal survey (Rhoades, 1982). Although informal field observations and conversations with farmers offer opportunities for the spontaneous pursuit of a variety of themes, a competently done informal survey needs to be based on a set of guidelines that is discussed and refined at the end of each day (Collinson, 1981). Equally important, the results of the informal survey should be written up and used in research planning.

In some instances an informal survey provides enough information to proceed with on-farm experimentation. The interchange of ideas elicited by the informal surveys in central Peru, for instance, was adequate for initiating on-farm tests of storage techniques. If an informal survey is carefully done it may be unnecessary to follow up with a formal questionnaire (Franzel and Crawford, 1987). However, to provide the additional information that researchers needed, various types of formal surveys were used in most of the cases. In the cases from Indonesia, northern Peru, Pakistan and Panama, results from informal surveys were used in planning questionnaires for a well-defined target area. In Rwanda, informal surveys were used to design production surveys at the national level to gather general information on the range of bean production practices; subsequent surveys focused on the more specific issues of farmers' management and knowledge of climbing beans. In Nigeria, the formal survey focused on practices and farmers' opinions in the target villages only.

A formal survey is useful for several reasons. First, it provides good baseline data against which the progress of adaptive research can be measured, as is shown in the case of northern Peru. Second, it presents the opportunity to quantify the most common practices in an area and to form a more precise idea of what constitutes 'typical' or 'average' farmer practice. This information is particularly useful for selecting variables for experimentation. The surveys in Rwanda clarified the ways in which farmers assessed new bean varieties, which in turn helped determine the strategy for varietal testing. Third, the results of a formal survey provide a better understanding of variations in farming practices, as in the Pakistan case, where the survey showed the exceptional importance of fodder supply to farmers with very small landholdings.

A fourth use of formal surveys is to develop a more precise estimate of farmers' perceptions of production problems. The survey in Panama showed quite clearly that the weed control problem was foremost in farmers' minds. The survey in Nigeria showed that farmers were more interested in improving soil fertility than in reducing fodder shortages. But there are also limits to farmers' ability to recognize production constraints (Tripp, 1989). Indonesian farmers expressed little concern about shootfly damage, although it turned out to be

Table 2.2   Use of surveys and other methods of data gathering

Case	Types of diagnostic survey	Other studies
Rwanda	Initial bean production surveys in several areas; survey of farmers' cultural practices in climbing bean areas and of farmer knowledge in non-climbing bean areas	Follow-up studies and evaluations with farmer groups
Ghana	Informal surveys in conjunction with establishment of on-farm experimental programme; formal diagnostic survey in a similar area of the country	—
Nigeria	Demographic survey of target villages; single-visit questionnaire in target villages; group visits and discussions	Follow-up studies on farmers' use of hedgerow trees, supplementary feeding of sheep and goats, participation of women farmers and importance of land tenure for technology adoption
Philippines	Baseline survey done for cropping systems research; no surveys included in the applied research	Adoption study
Indonesia	Informal survey; formal questionnaire	Field surveys on shootfly incidence; field study of plant population and yield
Pakistan	Informal survey on farming system; maize harvest surveys; formal survey on maize management, utilization and farming system	Multiple-visit surveys on maize plant population management; survey of seasonal fodder consumption
Northern Peru	Informal survey as part of training course, followed by formal survey conducted by course participants and instructors; formal survey repeated in seven other areas by local researchers	Follow-up studies on diffusion of improved variety; market studies
Panama	Informal survey and formal questionnaire before beginning experiments	—
Central Peru	Informal interdisciplinary surveys with farmers in Mantaro Valley; similar surveys in other areas	—

one of the main problems in maize production. When first asked about the feasibility of growing climbing beans, farmers in Rwanda emphasized the problem of obtaining staking materials, but the lack of staking materials did not prove to be a major constraint to the adoption of the new varieties.

Lastly, well-conducted formal surveys can contribute to further exploring the causality of production problems, which helps identify potential solutions (Tripp and Woolley, 1989). In Indonesia, researchers initially believed that high planting densities in maize were related to the practice of using thinned maize plants to feed animals. After the survey revealed that using maize thinnings for fodder was of minor importance, the need for fodder was eliminated as a possible constraint to improving plant stand management. In Pakistan, however, maize thinnings were an important source of fodder, and farmers' management of plant populations had to be understood in those terms.

OTHER DIAGNOSTIC STUDIES

Although, in theory, OFR proceeds in a well-defined sequence that leads from informal and formal surveying to planning, designing and managing a series of on-farm experiments, the practice of OFR is usually more complicated. An initial diagnosis, alone, is usually not sufficient for determining final research directions. Formal and informal surveys may be done at more than one stage of the research process, and the experiments themselves are, of course, an important source of diagnostic information that is used to refine research priorities. But more precise information than can be provided by a survey may be required to diagnose and understand production practices and problems; Table 2.2 provides examples of data collection methods used in the cases. Careful field observations are often useful. In Pakistan, the unusual system of maize plant stand management required a study in which researchers visited a series of fields periodically, counted plant populations and recorded farmers' operations and use of thinned plants. In Indonesia, the discovery of shootfly infestation as a possible production constraint led to periodic field visits to measure and quantify shootfly damage.

Special diagnostic studies are also useful for assessing the acceptability of proposed technologies, offering the opportunity to make mid-course adjustments in the research programme. The Nigeria case provides two good examples. Because land tenure customs might have had a severe effect on the adoption of alley farming, a special study was done to ensure that the technology would be acceptable in south-western Nigeria. Another study was carried out to explore ways of increasing participation by women farmers. At times, even more interactive procedures help reveal farmers' concerns (Norman et al., 1988); in Rwanda, group meetings assisted in gaining farmers' confidence and eliciting their opinions on new bean varieties.

Continuous monitoring of farmers' practices also reveals information that is important to the OFR programme. The testing of new bean varieties in Rwanda was accompanied by consumer acceptability studies to make sure that the new bean types were consistent with local preferences. In Nigeria, farmers' use of the alley crop for fodder, mulch and other purposes was carefully observed through a series of studies.

Although good diagnosis is essential to OFR, encompassing not only farmer surveys but also diagnostic experimentation, field observations and other studies, elaborate diagnostic studies are sometimes done without a clear concept of how the resulting data will be used. The cases described in this book avoided that tendency. The Rwanda case, for instance, argues in favour of a sequence of data collection activities based on well-defined information needs. The northern Peru and Philippines cases advocate monitoring OFR continually rather than depending excessively on an initial diagnosis. These considerations of choice of diagnostic technique are crucial to implementing OFR efficiently.

ON-FARM EXPERIMENTS

On-farm experiments are the heart of any OFR programme. Their design and management should reflect the dynamic nature of OFR, rather than act as a mechanical test of possible technological interventions. The cases represent a wide variety of experiments (*see* Table 2.3 *overleaf*), and provide several examples in which the experiments were used to identify priorities for further work. An exploratory series of factorial on-farm experiments in different agroecological zones of Ghana pinpointed important factors for further experimentation before any informal survey was done. In Rwanda, a series of 'minus-one' experiments examining a number of factors was useful in identifying priority yield constraints. Although diagnostic surveying had been carried out in Panama before experimentation began, factorial experiments still served a valuable diagnostic function by identifying or confirming key factors for further research.

Most of the cases demonstrate how on-farm experimentation sharpens the research focus. Priority themes are identified rapidly and become the focus of experimental work. In addition, the number of treatments per experiment tends to decline over time. Once the on-farm experiments in Ghana produced basic information on fertilizer response, research concentrated on methods and timing of fertilizer application. Counterbalancing the progression towards a narrower focus, however, are the new research themes uncovered in the course of experimentation, which also shape the agenda for further research. The importance of shootfly control became evident to researchers in Indonesia when their first set of experiments was devastated. Sometimes the new themes are unexpected, as in the Panama case where problems with soil insects during the experimentation led to research on insect control. In other instances, the movement from one research theme to another is more natural, as in the Rwanda case where research on climbing beans led to experimentation with various agroforestry options for producing staking materials.

One of the most debated issues in OFR is the management of non-experimental variables. The degree of control exercised by researchers depends on the purpose of the experiment. In the initial maize variety trials in Ghana, for instance, a fairly large number of varieties were planted in farmers' fields under uniform fertilization and weed control provided by researchers. Once the number of varieties narrowed, they were tested under the more varied management schemes of individual farmers. In most of the cases, non-experimental variables in the on-farm experiments were kept at the level of representative farmers. An exception is the case from Panama, where experiments on fertilizer use were conducted using weed control practices that farmers were expected to adopt by the time that information on fertility was available. Hence a hypothesized adoption sequence was used to select non-experimental variables. A similar example is provided by the fertilizer experiments in Indonesia, which were conducted using more optimal plant populations. In northern Peru, farmers broadcast beans (but row planted maize); researchers row planted beans in small plot trials, but included broadcasting in later verification experiments.

In some cases, researchers managed the non-experimental variables; in others, those variables remained in the hands of farmers. Farmers took considerable responsibility for the experiments in Rwanda and Pakistan, where varietal trials were established by giving seed of new varieties to participating farmers along with the responsibility to decide how and when to plant them. Other cases contain examples of farmers evaluating and adjusting experiments designed by researchers. For instance, researchers had to accommodate their measurements and observations to farmers' timing of operations in the alley farming experiments in Nigeria.

The Philippines case exhibits a different strategy towards experimental management. Researchers and extension agents managed both the adaptive research trials and the demonstrations with little input from farmers, but a very strong extension component of the programme emphasized working with farmers to teach them the new technology. The case argues, however, that more timely monitoring would have highlighted the adjustments that farmers were making to particular parts of the package.

The on-farm experiments described in the cases often evolved into demonstrations. Some demonstrations were carefully designed and used as part of an extension strategy, as in Ghana, where numerous 'verification/ demonstrations' which included the farmers' practice and two alternatives were planted in farmers' fields,

Table 2.3   Types of on-farm experiments and degree of farmer participation

Case	Main types of on-farm experiments	Farmer participation in experiments	Analysis of experiments
Rwanda	Diagnostic ('minus-one') trials on production constraints; on-farm variety trials (farmer-managed); demonstration of basic technology for growing climbing beans; in second stage, further variety, fertility and intercropping trials, and agroforestry trials (farmer-managed)	In variety trials, farmers were free to manage varieties as they wished; in production constraint trials, farmers managed planting time and density; in validation trials, farmers received seed and instructions but were free to modify management	Yield and disease assessment of on-farm trials; monitoring of farmers' experience with validation trials: management, area planted, labour use, varietal preferences (agronomic and culinary), economic analysis of various agroforestry/staking options
Ghana	Exploratory factorial experiments; screening maize varieties; density and spatial arrangements; NPK response; timing and method of fertilizer applications; verification/demonstrations	Most experiments managed by researchers; farmer-managed variety experiments	Economic analysis; density and fertilizer response functions; monitoring farmers' feedback on verification/demonstrations
Nigeria	Farmer-managed experiments with alley farming; on-farm experiments on management of alley farms; on-farm animal feeding experiments (on-station trials used to get information for technology design early in research process and later to investigate farmers' adaptations to the technology)	In farmer-managed trials, farmers given seed and basic instructions, then encouraged to establish their own alley farms; in standardized on-farm trials, farmers provided labour and managed foliage use	Agronomic monitoring of managed trials; measurements of yield, labour inputs, soil fertility changes and foliage utilization in on-farm trials; livestock performance; assessment of hedgerow establishment; monitoring of new adopters and drop-outs
Philippines	Adaptive trials (extension-managed) on variety, establishment, fertilizer, weed and insect control; package demonstrated on large plots in farmer's fields	Experiments and demonstrations managed by researchers	Evaluation of demonstrations
Indonesia	Exploratory and verification experiments on plant stand management; plant protection with insecticide included in exploratory, insect control and verfication experiments; experiments on N dose and timing, P dose and variety	Non-experimental variables managed by farmers; farmers managed fertilizer in verification trials	Economic analysis; field days with farmer assessment

Table 2.3   **(continued)**

Case	Main types of on-farm experiments	Farmer participation in experiments	Analysis of experiments
Pakistan	Farmer-managed verification experiment to examine new variety, phosphorus application and early thinning; experiments on phosphorus response; variety experiments (on-station experi- on variety by plant density)	Farmers responsible for all operations in verification experiments; phosphorus experiments superimposed on farmers' fields in later years	Verification experiments monitored by agronomists; farmers' opinions of new variety assessed after harvest; economic analysis of fertilizer experiments
Northern Peru	Variety trials; variety by planting system trials; explora- tory trials to investigate estab- lishment problems; fertilizer and *Rhizobium* inoculation	All non-experimental variables managed by farmers; in verification trials, farmers managed the broadcast treatment	For variety trials, evaluation of yield, resistance, maturity, suitability for intercropping, and economic analysis; economic analysis of management trials
Panama	Exploratory factorial experi- ments on density, weed control, N and P; types of chemical weed control; demonstrations on improved plant density and spatial arrangements, chemical weed control and zero tillage	Non-experimental variables managed by farmers in some cases; farmer-managed demonstrations	Economic analysis of experi- mental results in choice of herbicide, tillage method and planting method
Central Peru	Small demonstration and test stores in highlands; research/ demonstration stores in coastal locations	Experiments and demonstrations managed by researchers and extension agents; farmers encouraged to adapt principles	Stored seed potatoes analysed for viability, weight loss and sprouting, then planted to measure yield differences

mostly under the supervision of extension agents. In other cases, such as the Philippines one, adaptive experiments themselves became the focus of field days or demonstrations. Relatively formal extension demonstrations of potato storage methods in Peru served as a point of departure for farmer experimentation. In Panama, demonstrations of zero tillage were managed by farmers and exhibited considerable individuality.

OFR requires good relations with experiment station research. In some of the cases, research at the experi- ment station complemented on-farm experiments. The experiment station may offer a more controlled environ- ment for the initial stages of experimentation requiring careful management or observation. In Nigeria, much of the early research on alley farming was done on the experiment station, which was also the site of further research on the rationale of some adaptations farmers had made to the technology. In Pakistan, the inter- relationships between planting density, maize variety and soil fertility were examined carefully in experiment station trials. One conclusion of the Philippines case was that research should pursue a strategy that includes the use of experiment stations, semi-permanent field sites, on-farm testing and farmer experimentation.

ANALYSIS OF EXPERIMENTS

The results of the on-farm experiments need to be analysed and assessed using criteria of importance to farmers. Economic analysis may be done using partial budgeting and marginal analysis (CIMMYT, 1988). This technique was employed in a number of the cases: to select the most efficient herbicide in Panama; to assess the profitability of insect control and phosphorus application in Indonesia; and to monitor the acceptability of the fertilizer recommendations being demonstrated in Ghana. Several cases employed more complex economic analyses. In Rwanda, the options for providing staking materials for climbing beans were examined by looking at labour inputs and alternative uses for the stakes. The labour requirements of alley farming were examined in Nigeria through studies on the farm and the research station to measure the extra labour required for activities such as pruning, as well as reductions in the labour needed for weeding.

Agronomic analysis of on-farm experiments took a number of forms in the cases. Monitoring the disease resistance of new bean varieties was particularly important in Rwanda. In Nigeria, the effects of alley farming on the leguminous trees, associated food crops and soil fertility were all carefully measured. Special attention was given in central Peru to the effects of different storage methods on potato seed viability and sprouting.

The research described in the cases makes a strong argument for informal as well as formal assessments of experiments. In the analysis of on-farm trials, farmers' opinions received particular attention; in the Pakistan case, for example, a special survey was designed to assess farmers' opinions of the yield, fodder quality and cooking quality of the new maize variety, and a similar investigation was part of bean varietal research in Rwanda.

FARMING SYSTEM PARAMETERS

OFR treads a narrow line between bringing change to farming systems and working within the limits imposed by those systems. Although sometimes it is unclear *a priori* when a particular factor can be treated as a variable and when it will operate as a parameter for research, certain parameters often emerge fairly early to define the boundaries within which new technologies must be tested (*see* Table 2.4 *overleaf*). These parameters are associated with basic characteristics of the farming system and reflect the complexity, interactions and multiple decision criteria characteristic of small farm agriculture. The importance of these parameters in defining and orienting a research programme is perhaps the major justification for OFR. Indeed, where these parameters are less important, technology development may not depend so much on a farming system perspective.

The management of household food supplies is an important farming system parameter. In many instances, farmers assess new technologies in relation to household food supply requirements, as shown in the research on options for potato storage in central Peru. Farmers were cautious about investing in storage technology for a single purpose and preferred to be able to use stored potatoes for food or seed. In Rwanda, new bean varieties had to be carefully assessed in terms of taste, cooking qualities and edible leaf production, and not simply for yield potential.

Risk management is another parameter that often influences farmers' choice of technology. Fertilizer application methods in Ghana were adjusted to take account of farmers' perceptions of the risk involved in adopting the new fertilizer practice. The susceptibility of new bean varieties to various diseases was a key factor determining their acceptability to Rwandan farmers. Concerns about risk also foster farmer interest in risk-reducing technology, such as insect control in Indonesia.

Table 2.4   Identifying constraints and setting priorities

Case	Farming system parameters	Identification and modification of priorities
Rwanda	Serious household food shortages, particularly before harvest; land shortage; declining soil fertility, particularly in South Central zone; labour constraints for staking climbing beans; most beans grown as intercrops	Initial research interest in fertility and disease control; climbing beans performed well in variety trials and attracted farmer interest; special interest paid to staking materials for climbing beans, leading to agroforestry experiments
Ghana	Labour constraints for maize planting and weed control practices; storage and marketing characteristics of new maize varieties; fertilizer response related to cropping history; risks in stand establishment influence fertilizer practices; some maize intercropped	Best-bet maize varieties identified rapidly; emphasis placed on developing practical recommendations for planting and fertilizer application; weed control research postponed; fertilizer recommendations adjusted according to changing prices
Nigeria	Fertility maintenance in fallowing and rotations; importance of small ruminants; labour availability; intercropping is the most common practice	Soil fertility found to be a more important concern for farmers than fodder production; women as important clients of research
Philippines	Soil type and rainfall pattern determine rotation possibilities in different areas of Iloilo; labour shortage and neighbours' harvest schedule determine turnaround time between crops; seeding method dependent on labour availability, especially for weeding	Early cropping systems experimentation identified wet seeding as the most appropriate technology for most of Iloilo, but subsequent on-farm tests showed dry seeding to be superior in certain environments
Indonesia	Planting date for post-rainy season crop determined by crop and labour availability in previous season; planting date affects insect incidence	Initial interest in low adoption of improved maize varieties, but no clear differences between old and new recommended maize varieties; plant population and nutrition observed to be problems; hypotheses that high planting densities were caused by fodder needs or poor seed quality rejected in favour of importance of shootfly
Pakistan	Maize equally important for both food and fodder; livestock an important source of income; small farm size and long winters lead to fodder shortage; maize planting may be delayed by management of cash crops	Farmers' plant population management and weeding and thinning practices thought to be inadequate, but subsequent analysis showed them to be efficient for managing maize for grain and fodder; densities varied for local and improved maize

Table 2.4   (continued)

Case	Farming system parameters	Identification and modification of priorities
Northern Peru	Beans intercropped with maize; local market presents opportunities for certain bean types; labour shortages because of seasonal migration; cash shortage	Initial interest in disease resistance, more efficient plant type, yield and establishment; the need to row plant some trials led to discovery that row planting a feasible technology
Panama	Weed control method in maize needs to be compatible with following bean crop; early planting of maize important for establishment of following crop; increasing costs of labour	Interviews and observations showed main problem was weed control; plant spacing and density had to be improved concomitantly with weed control; improved weed control could be linked to reduced tillage to address erosion problems
Central Peru	Various food and feed uses for potatoes damaged in storage; preference for flexibility in storage facilities	Initial interest in potato storage changed to focus on importance of seed potato storage, particularly for new varieties; if farmers on coast could store own seed potatoes, costs of production lowered

Other crops in the farming system also set limits and define parameters for research design. The techniques used to control weeds in maize in Panama had to be compatible with farmers' management of the subsequent bean crop. Pest control methods in maize in Indonesia were determined partly by late planting, occasioned by the management of the previous crop in the rotation. Crop rotations provided both opportunities and parameters for the introduction of early maturing varieties in the Philippines.

The management of intercrops was a key factor in several of the cases. All experimentation with improved bean management in northern Peru took place in the presence of a maize intercrop, and researchers had to select innovations compatible with that system. The management of climbing beans in Rwanda was adapted to suit farmers' intercropping practices, and alley farming research in Nigeria was done in the context of local intercropping practices. In Ghana, the degree of adoption of improved maize technology differed markedly in monocropped and intercropped systems.

The presence of animals in a farming system may set limits and define research opportunities. The species and management of small ruminants in Nigeria were two of the most important factors determining how farmers would manage alley farming. The great importance of maize as a fodder crop in Pakistan was a key factor for identifying new maize varieties or changes in management practices.

Farm resources, such as the supply of labour, also present parameters for the development of new technology. The labour involved in staking climbing beans in Rwanda was an important factor in differential adoption patterns between labour-scarce and labour-abundant areas of the country. In the development and demonstration of new planting and fertilizer application techniques in maize in Ghana, careful attention was paid to the labour demands involved. In the Philippines, the labour demands of the wet-seeded rice technology limited its acceptability to those households with adequate labour.

RESEARCH PRIORITIES

This review of the cases has indicated that the identification of research priorities in OFR is a dynamic process. Tentative priorities are identified early in the research but are often modified later. For example, the importance of weed control in Panama and seed potato storage in central Peru emerged very early as research issues, but in some of the other cases key priorities were identified after research was in progress. Interdisciplinary research in Nigeria focused on a technology that addressed soil fertility, the farmers' main priority, but simultaneously provided an opportunity for the production of supplementary fodder. In Rwanda, initial interest in variety selection and disease control in bush beans was replaced by a focus on climbing beans, after on-farm trials had revealed the potential of climbing bean varieties.

In most of the cases, some factors originally thought to be important were discarded or assigned lower priority in the research. Experiments in Indonesia suggested that newly released maize varieties, when grown under representative conditions, produced yields similar to those of the older improved varieties already used by many farmers. In Ghana, although weed control was recognized as an important factor in maize production, several years of experimentation yielded no significant improvements that could be recommended to farmers. Efforts to improve farmers' plant stand management and fertilizer use in Pakistan were also unsuccessful.

SEQUENCE OF ADOPTION

Most of the cases confirm that technological change is best approached in a sequential manner (*see* Table 2.5). Sometimes a strategy of sequential technology adoption emerges clearly from the OFR. In Panama, for instance, adopting improved weed control practices was a prerequisite for changes in planting methods and later in tillage practices. In Indonesia, insect control was necessary before improvements in plant stand management could be contemplated. Although recommendations were demonstrated to Ghanaian farmers in the form of a small package of practices, farmers were encouraged to try individual elements of the package. The experience in the Philippines showed that farmers would adopt some elements of a package and modify others.

Success in introducing particular innovations often prepares the ground for further opportunities. In the Nigeria case, researchers first made sure that farmers could manage the combination of food crops and multipurpose trees in an alley farm and recognized the benefits of forage, before promoting the establishment of more intensively managed fodder gardens. The introduction of early maturing rice varieties in the Philippines led the way to experimentation with alternative crop establishment methods and new types of crop rotations. Researchers in Rwanda adopted a conscious strategy of introducing climbing beans in such a way that farmers could begin by planting small areas and gradually expand.

In northern Peru, bean farmers responded to new market opportunities and the characteristics of a new bean variety by changing some of their production practices independently of the OFR. This serves as a reminder that OFR not only generates its own adoption sequence but also that it is in the midst of other changes in the farming system, resulting from farmer initiative, changing markets and other factors. OFR must always be able to deal with the dynamic nature of technological change. A farming system is not static; rather, it is a 'moving target' (Maxwell, 1986), subject to continual transformation by researchers, farmers and the environment.

The dynamic nature of the farming systems addressed by OFR can be illustrated by considering the example of fertilizer practices in several of the cases. In Pakistan, OFR was carried out with farmers who were already moving towards a reasonable level of nitrogen application, and hence no additional research on this theme was necessary. In Panama, farmers were finding the official fertilizer recommendation unacceptable, and the OFR

Table 2.5   Technology adoption sequence and modification

Case	Adoption sequence	Farmer modifications to technology
Rwanda	Farmers experimented with small quantities of new bean varieties; climbing beans were adopted gradually, especially by smaller farmers; more beans were grown in the second season	Farmers chose varieties not only on basis of grain yield but also for grain cooking qualities and production and palatability of leaves; farmers experimented with various intercrops with climbing beans, particularly sweet potatoes and bananas; farmers did not accept row planting of climbing beans
Ghana	Farmers tended to adopt variety, row planting and fertilizer in a logical sequence	Sighting poles found to be more practical than string for line planting; effectiveness of farmer practice of applying basal fertilizer after germination confirmed in experiments
Nigeria	Declining soil fertility and low fertilizer prices led to widespread demand for chemical fertilizer, but rising prices and poor infra-structure limited availability and use; alley farming was adopted as a method to improve soil fertility and to address rising costs of fertilizer; increasing use of foliage from alley farms for fodder; increasing interest in more intensive fodder production	Management of hedgerows (spacing, structure, pruning) modified by farmers; farmers experi-mented with alley farms in a wide variety of crops; farmers used hedgerow trees for purposes beyond those originally envisioned
Philippines	Early maturing, high yielding rice varieties were adopted first; labour shortages and availability of herbicides had led to shift from transplanting to wet seeding of rice; shift to dry seeding increased rice yields, yield stability and cropping intensity; farmers usually adopted parts of technology package	Farmers used less fertilizer than recommended; only farmers with light soils used pre-emergent herbicides
Indonesia	Control of insects was necessary before farmers adopted improved planting practices, lower seed rates and more efficient fertilizer rates	—
Pakistan	New maize variety adopted when it fulfilled farmers' grain and fodder requirements	—
Northern Peru	New variety, introduced at the time OFR programme began, stimulated several changes in management practices	Shallower planting of new variety; development of market potential for new variety; adoption of fungicides as commercial opportunities increased; change in planting distances

Table 2.5   (continued)

Case	Adoption sequence	Farmer modifications to technology
Panama	Short-term priority to improve weed control and plant density; once farmers were familiar with improved weed control this was linked to reduced tillage; longer-term objective to test and monitor more efficient fertilizer levels	Light tractor harrowing substituted for manual chopping of crop residues in zero tillage by farmers with labour shortage; homemade shield devised by farmers to help direct application of contact herbicide
Central Peru	Adoption of new potato varieties caused some problems for seed storage, which led to adoption of diffused light storage methods; on the coast, adoption of these methods allowed farmers to be less dependent on purchased seed, which in turn allowed earlier planting; future challenges for storage research included aphid-transmitted viruses and potato tuber moth (in warmer climates)	Farmers adopted principle of diffused light storage and used local materials to construct a variety of seed stores or to improve other storage techniques

supported their perceptions. In northern Peru, some farmers experimented with fertilizer without interacting with research or extension, while in Ghana the OFR served farmers who had been exposed to several unsuccessful fertilizer campaigns in the past.

Finally, progress in OFR often leads to second-generation challenges. The successful introduction of improved potato seed storage methods in Peru led to increased attention to aphid-transmitted viruses and the potato tuber moth in stored seed. The successful introduction of chemical insect control in Indonesia motivated researchers to consider more environmentally sound pest control methods. In Panama, intensification of maize production led to research on long-term strategies for managing soil fertility.

FARMERS' MODIFICATIONS TO TECHNOLOGY

The importance of farmers' participation in OFR is illustrated in the cases by numerous instances of farmer-initiated adjustments and changes to the technologies being tested. Sometimes farmers took the ideas being tested and modified them to suit their own conditions, as in central Peru, where the principle of diffused light storage was adapted to a wide range of household storage arrangements. When farmers in Nigeria found the planting distances and pruning schedules for fodder crops in alley farming experiments to be inconvenient, they modified them to suit their own labour availability. Farmers in the Philippines made significant modifications in the use of inputs for new crop rotations being tested.

In other cases, on-farm experimentation stimulated new lines of enquiry by farmers. Farmers in northern Peru began to experiment with changes in bean planting distances. The introduction of climbing beans to areas of Rwanda where only bush beans had been grown led to several innovations by farmers in staking and planting methods. The success of alley farming with maize in Nigeria led farmers to test the idea with other traditional intercropping mixtures as well. In addition, farmers found several new uses for the leguminous tree intercrop.

EXTENSION

The cases of OFR described in this book make it abundantly clear that a strong extension effort is necessary in any adaptive research programme (*see* Table 2.6 *overleaf*). In none of the cases did the new technology spread of its own accord, and in several cases, including Pakistan and Panama, lack of extension limited the diffusion of new technology. The rigidity of the extension system in Indonesia was a problem for establishing effective communication between researchers and extension staff.

In some instances researchers made an explicit effort to work with the extension service to disseminate new technology. The case of central Peru is a good example; various training programmes on seed potato storage were organized for extension. The OFR in Ghana led to a nationwide programme of demonstrations designed to be managed by the extension service. The strategy included programmes for training extension agents and the production of bulletins summarizing the research results for extension agents and farmers. Much of the early work in adapting the new rice seeding methods in the Philippines was done by extension agents through a special programme to disseminate the new technology through field days, farmer classes and other means; later, extension was carried out through an integrated rural development project.

Extension of research results was handled in other cases by identifying effective extension or community development programmes and channeling the OFR results through them. The importance of women farmers in Nigeria led to a special study and the development of extension strategies to involve more women farmers in the OFR. In Rwanda, rather than concentrating exclusively on the government extension service, several community development projects that had worked well with farmers were chosen as partners for the OFR effort.

Lastly, in certain cases the researchers themselves organized extension activities. These activities proved to be effective but, as demonstrated by the Pakistan and Panama cases, limited resources and personnel prevented them from having an impact over a wider area.

INTERACTIONS WITH THE POLICY ENVIRONMENT

Interactions between OFR and the policy environment are found in several of the cases (*see* Table 2.6). The policy environment is often taken as a parameter in OFR, and research proceeds on the assumption that technological alternatives must be developed within the limits of current policy. Current input and output prices are generally used in an economic analysis of on-farm experiments, and the choice of treatments is often limited by unavailability of particular inputs.

Policies that ensure good input supply are certainly an important factor in the success of OFR. Good markets for chemical inputs made the development of recommendations for insect control in Indonesia and weed control in Panama much more effective. Poor fertilizer supply and distribution in Ghana, on the other hand, hampered the development and diffusion of fertilizer recommendations.

Seed was a critical input in much of the OFR described in the cases, many of which suffered from significant limitations in seed supply and distribution. In Rwanda and Pakistan, the lack of assured supplies of seed led researchers to encourage farmers to multiply and distribute seed; multiplying planting materials of fodder crops was also problematic in Nigeria. The government seed company in Ghana could not supply sufficient quantities of seed of new maize varieties.

Occasionally, OFR can encourage improvements in policy at the local level. In Panama, OFR demonstrated the inadequacy of government requirements for the credit package for maize production and encouraged the development of a more reasonable credit programme.

Table 2.6   Role of extension and the policy environment

Case	Extension participation	Interactions with policy environment
Rwanda	Close collaboration with extension project (PAP) and with local government extension agents; training; and development of extension materials	—
Ghana	GLDB staff responsible for managing experiments and demonstrations; increasing participation of extension agents from the MOA; extension bulletins produced and updated	Seed supply increased, but still insufficient; fertilizer supply and distribution unreliable; fertilizer subsidies being removed, causing changes in fertilizer recommendations
Nigeria	Little adoption of technology until a community focus with extension involvement was adopted; extension agent helped organize farmer-managed trials, gave demonstrations and arranged visits to farmers' fields and experiments; female researchers promoted technology with women farmers	Local government ruling that animals must be confined increased the importance of fodder production; removal of government subsidy on fertilizer led to reduced usage by small farmers
Philippines	Technology disseminated through Kabsaka project, which was aimed at demonstrating and promoting dry seeding, and RADIP project, which included the technology in an integrated rural development effort; studies showed adoption of the new technology influenced by degree of contact with extension agents	—
Indonesia	Extension agents participated in many of the trials and helped organize field days; further extension participation was limited by the centralized system of research-extension communication	Markets for fertilizers and pesticides functioned fairly well; subsidies on fertilizer slowly being removed
Pakistan	Researchers organized a seed multiplication and distribution system in two villages; the new variety spread to some neighbouring villages; lack of extension participation limited further diffusion	Introduction and diffusion of new maize varieties limited by inadequate seed supply system
Northern Peru	Joint collaboration by INIPA extension agents and researchers	Seed production system limited diffusion of variety

Table 2.6   **(continued)**

Case	Extension participation	Interactions with policy environment
Panama	Farmer field days organized by IDIAP agronomist in conjunction with farmer groups; little direct involvement of extension service	Credit bank willing to change fertilizer requirements for loan package based on results of experiments; farmers organized in groups to buy inputs; some herbicides not readily available in area; type of fertilizer available depends on supplies to nearby potato growing region
Central Peru	Comprehensive training in diffused light storage techniques given to extension agents resulted in spread of the technology to much of the country	—

CONCLUSION

This overview of the cases highlights two important and, at first sight, contradictory, prerequisites for conducting OFR successfully. Good OFR demands both an exceptional degree of flexibility in research design and the application of rigorous agricultural science. Adaptive research that emphasizes one of these characteristics to the exclusion of the other is not likely to produce useful results.

The importance of flexibility is evident in all the cases, both in the selection of formal research methods and in structuring interactions between farmers and researchers. A careful selection of data-gathering techniques, including surveys, experiments and other studies, was directed at a limited set of priorities that evolved as the research progressed. Research methods were chosen to fulfill well-defined goals, not merely to generate information. The diversity of problems and methods in OFR may require the participation of specialists from disciplines representing a range of biological and social sciences. As the cases indicate, the balance may shift from one group of disciplines to another, depending on the issues under examination, but the key to successful collaboration is maintaining a systems perspective that respects the farmers' point of view.

Ample scope must exist for farmers and researchers to interact. Formal methods for conducting OFR can contribute to this interaction, but unstructured opportunities for farmers to participate in, and take control of, the research are also essential. Farmers often change or modify the definition of research problems. In addition, they may offer unexpected criteria for judging the acceptability of new technologies and often contribute novel ways of testing technologies and adjusting them to suit their conditions. Unless researchers are open to, and respectful of, farmers' contributions, OFR becomes rigid, unimaginative and unproductive. Because so much flexibility is demanded, both in the choice of research methods and in researcher-farmer interaction, no simple formula can be given for carrying out OFR. The choice of strategies and methods depends very much on the individual situation. But experience, such as that represented in this book, is increasingly available to guide the conduct of OFR, and respect for farmers as partners in technology generation is an attitude that can be learned.

Counterbalancing the need for flexibility in OFR is the requirement for high-quality agricultural science. The technical requirements for developing solutions to the production problems of resource-poor farmers are so demanding that a haphazard offering of innovations and ideas will not suffice. Where OFR is conducted in specific locations by small research teams, care must be taken that the investment of research resources and skills is adequate. Where OFR is part of a broader commodity or disciplinary research strategy, on the other hand, it will be successful only when the technical requirements posed by location-specific conditions are taken into account. In both instances, good technical and analytical skills in agricultural science are essential.

The need for flexibility and rigour means that OFR is neither simple nor quick. In all the cases reported here, research was conducted over a 3-5 years before new technologies were delivered to farmers. The kind of research described in the cases requires well-trained people with good access to resources, requirements which have important implications for the efficiency of OFR. Successful OFR is usually cost-effective, as in the Panama and northern Peru cases. However, there are still too few documented examples of OFR bringing improved technology to farmers, which implies that a challenge remains for the methods and perspective of OFR to make a more effective contribution to agricultural research.

Notes

1.   The major exception is rice-based cropping systems research in Asia, which is under-represented in this book; Chapter 6 presents one case from the Philippines. Other successes in the intensification of cropping patterns in lowland areas following the introduction of new rice varieties are reported in Magor (1984), Mathema (1986) and Siwi et al. (1986).

References

CIMMYT. 1988. *From Agronomic Data to Farmer Recommendations: An Economics Training Manual.* (revd edn) El Batan, Mexico: CIMMYT.

Collinson, M. 1981. A low cost approach to understanding small farmers. *Agricultural Administration* 8: 433-50.

Francis, P. 1987. Land tenure systems and agricultural innovation: The case of alley farming in Nigeria. *Land Use Policy* 4: 305-19.

Franzel, S. and Crawford, E. 1987. Comparing formal and informal survey techniques for farming systems research: A case study from Kenya. *Agricultural Administration and Extension* 27: 13-33.

Magor, N. 1984. *Potential in Rainfed Transplanted Rice Production in North-East Bangladesh.* Dhaka, Bangladesh: Bangladesh Rice Research Institute.

Mathema, S.B. 1986. Adoption and effects of technologies generated by the Cropping Systems Program in Nepal. PhD thesis. Los Baños, Philippines: UPLB.

Maxwell, S. 1986. Farming systems research: Hitting a moving target. *World Development* 14: 65-77.

Norman, D., Baker, D., Heinrich, C. and Worman, F. 1988. Technology development and farmer groups: Experiences from Botswana. *Experimental Agriculture* 24: 321-31.

Price, E.C. 1982. Adoption and impact of new cropping systems in Iloilo and Pangasinan, Philippines. In *Report of a Workshop on Cropping Systems Research in Asia.* Los Baños, Philippines: IRRI.

Rhoades, R. 1982. *The Art of the Informal Agricultural Survey.* Social Science Dept Training Document 1982-2. Lima, Peru: CIP.

Siwi, B.H., Ismail, I.G., Basa, I., Syarifuddin K, A., Sultoni Arifin, M., Djauhari, A., Syam, M., Mundy, P. and McIntosh, J.L. 1986. *The Impact of Cropping Systems Research in Indonesia.* Indonesia: CRIFC.

Tripp, R. 1989. *Farmer Participation in Agricultural Research: New Directions or Old Problems?* Discussion Paper 256. Sussex, UK: IDS.

Tripp, R. and Woolley, J. 1989. *The Planning Stage of On-Farm Research: Identifying Factors for Experimentation.* El Batan, Mexico: CIMMYT/CIAT.

Part 2

ON-FARM RESEARCH CASE STUDIES

3

Climbing Bean Introduction in Southern Rwanda

W. GRAF, J. VOSS and P. NYABYENDA

Beans are the most important annual crop in the agricultural production systems of Rwanda. In 1984, beans occupied 302 356 ha during the short rainy season (September to December) and 181 000 ha during the long rainy season (February to June), either as a sole crop or associated with other crops. This represents one third of the total cultivated area in the country (MINAGRI, 1985). Average yields in 1984 were about 750 kg/ha. The most important bean producing areas are the East and North Central agroecological zones (*see* Figure 3.1 *overleaf*), with the East zone being the only area in Rwanda which regularly produces a surplus (Clay and Dejaegher, 1987).

The average Rwandan consumes almost 50 kg of beans per year, and 60% of the protein in the average diet is derived from dry beans (MINIPLAN, 1988). Since 1960, beans have retained their important position in Rwandan farming systems despite the fact that between 1960 and 1984 productivity per unit of land did not increase (Delepierre, 1985). They are consumed mainly as dry beans, cooked for several hours in water. Fresh bean, snap bean and leaf consumption are also important, but this importance varies from region to region according to wealth and the availability of dry beans (ISAR, 1985). Leaf consumption is particularly high in areas with a deficit in dry bean production.

A national survey on trade and consumption revealed that only about 5% of Rwanda's bean production is commercialized (MINIPLAN, 1988). Specific studies suggest that many poor farming families consume all their bean seed before the following planting season, forcing them to buy seed from richer neighbours at high prices (MINAGRI, 1987; CIAT, 1988; Sahli, pers. comm.). Thus, there is a capital flow from the poorer to the richer farmers, which the poorer farmers try to compensate for by selling their labour to wealthier neighbours or by migrating temporarily to areas in which there is a high demand for agricultural labour.

In the past 20 years, efforts to meet the food demands of the growing population have included expansion into marginal areas (Delepierre, 1985) and unofficial importations from Uganda and Zaïre (World Bank, 1983). Most of the expansion has occurred in the lower parts of the East zone, where there is erratic rainfall and a high risk of crop failure because of drought, and on the slopes of the Zaïre/Nile Divide, an area characterized by acid soils with poor cation exchange capacity. The expansion has kept food production at a level of self-sufficiency

Figure 3.1   Agroecological zones of Rwanda

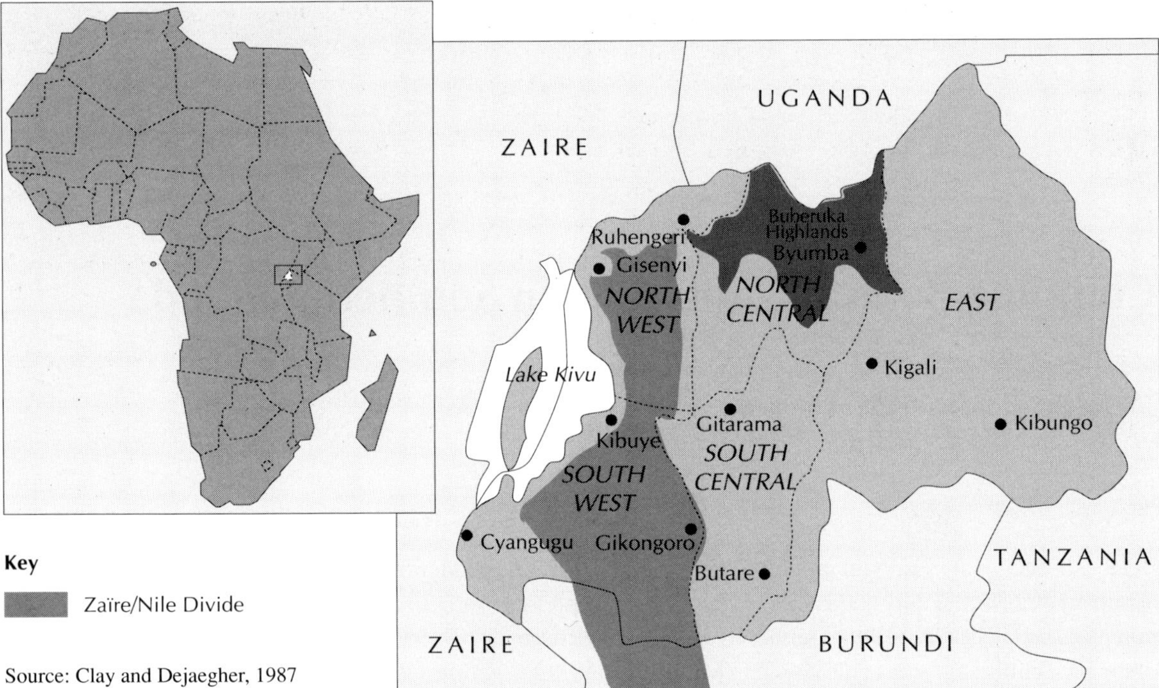

Key

Zaïre/Nile Divide

Source: Clay and Dejaegher, 1987

in good years, but Rwanda is a country where one bad season can create major food supply problems for the rural population. This is compounded by social and regional inequalities (MINIPLAN, 1988).

In addition to expanding into marginal areas, farmers have started growing other crops. On the slopes of the Zaïre/Nile Divide, for instance, they have shifted from beans to peas and soybeans, which are better adapted to acid, infertile soils. Tuber crops have become more important as well, with potatoes being grown as a cash crop and cassava and sweet potatoes being grown for home consumption. Farmers have also responded to a Ministry of Agriculture (MINAGRI) campaign to plant avocado trees. In the East zone, bananas are more important than in other areas. Their stable yield, even in years with prolonged dry spells, and their importance as a source of household income through the sale of beer, are the main reasons for their popularity. This is relevant to bean production as beans are well suited to intercropping with bananas (Graf and Ndorehayo, 1989). The cropping patterns in the East and South Central zones are shown in Figure 3.2.

In general, bean production in Rwanda is greater in the short rainy season (*see* Table 3.1 *overleaf*). The heavy rainfall during the long rainy season leads to high disease incidence in the central areas, while in the eastern areas the rains often stop abruptly at pod filling and the onset of the long dry season results in low yields (Graf and Trutmann, 1989). In these areas it is mainly bush beans of determinate and indeterminate types that are grown. In the northern parts of Rwanda, however, climbing beans account for almost the entire bean production and the average yield is high. They are preferred because the heavy rainfall makes bush beans difficult to dry at harvest and increases their susceptibility to some of the most devastating bean diseases, including anthracnose (*Colletotrichum lindemuthianum*) and ascochyta blight (*Phoma exigua* ssp.).

The difference in performance between climbing beans and bush beans led researchers at the Institut des Sciences Agronomiques du Rwanda (ISAR) to consider introducing climbing beans into other areas of the country in order to intensify bean production. However, early attempts failed to persuade farmers in the area around Rubona, ISAR's main research station in the South Central zone, to grow climbers, although high yields were obtained on the research station. In 1984, the International Center for Tropical Agriculture (CIAT), with funding from the Swiss Development Corporation (SDC), placed an interdisciplinary team of five scientists at Rubona to set up a regional bean research network involving Rwanda, Burundi and Zaïre. The CIAT team identified on-farm research (OFR) as being a particularly weak element in these countries' national research systems and placed considerable emphasis on stimulating OFR in interdisciplinary teams throughout the region. An important objective of this CIAT effort was to strengthen collaboration with locally based extension agents in the development of adaptive research programmes for promising bean technologies (Voss and Graf, 1991).

In this chapter, we describe how the research conducted by ISAR and CIAT led to the successful transfer of climbing bean technology from northern Rwanda to other areas of the country, with modifications being made according to the needs of farmers in the target areas.

Figure 3.2   Relative importance of major food crops in the East and South Central zones of Rwanda, 1984

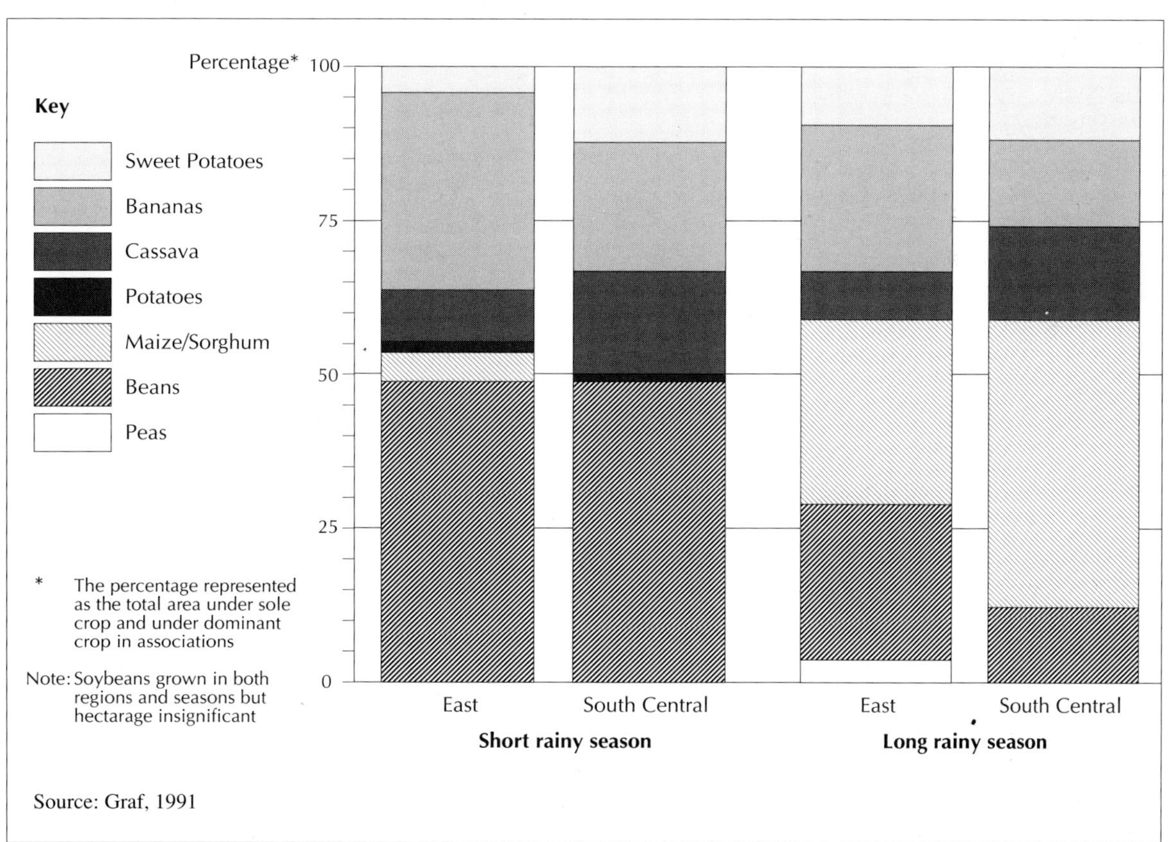

Table 3.1   **Seasonal differences in bean production in the agroecological zones of Rwanda, 1984 (in ha)**

| | Agroecological zone | | | | |
	North Central	South Central	South West	North West	East
Short rainy season (Sept. to Dec.)	39 508	41 038	64 233	54 542	103 034
Long rainy season (Feb. to Jun.)	36 157	24 931	27 211	30 321	62 813

Source: MINAGRI, 1985

DIAGNOSIS

Diagnostic strategies for a national commodity programme involving OFR differ from those for a farming systems project with a location-specific mandate. For a farming systems project, the definition of the zone of intervention is a precondition for the project's establishment (Norman, 1980). A commodity programme, on the other hand, must select out of a vast region the particular areas in which the technologies developed by the programme are likely to have an impact. One of the main goals of the diagnostic process is to define these areas (*see* Table 3.2).

Starting in 1984, CIAT, in collaboration with the ISAR Legume Programme and MINAGRI extension projects, conducted diagnostic studies in all the main bean producing areas in the country, focusing primarily on production problems. As it would have been impossible to carry out in-depth studies of farming systems in a large number of areas at the same time, preference was given to rapidly obtaining adequate knowledge about problems and potentials in many areas, rather than attempting to know everything about one particular area. This was a practical application of the principle of sequential data collection. Additional information was collected only when it seemed likely that this information would benefit the research process (Byerlee et al., 1982).

Table 3.2   **Goals and diagnostic tools in the ISAR Legume Programme**

Goal	Tools
Understand general bean production patterns	Surveys, interdisciplinary monitoring tours
Identify bean production problems	Surveys, diagnostic trials, interdisciplinary tours
Quantify bean production problems	On-farm diagnostic trials
Identify potential impact areas	On-farm variety trials (exploratory)

Farmers' experience with new bean varieties

Beans are generally grown in complex varietal mixtures (ISAR, 1984) and intercropped with a wide range of other crops. The diagnostic studies showed that farmers' management of varietal mixtures was far more sophisticated than previously thought. Almost all farmers selected different mixtures for different soil and intercropping conditions, based on detailed knowledge of the agronomic and culinary properties of the component varieties. In a sample of 42 households, the number of varieties in these mixtures varied between six and 30, with an average of 20 (Voss, 1990).

The farmers were very open to experimentation with new varieties. Many of them regularly tested varieties acquired from neighbours or purchased in the market (*see* Table 3.3), usually by planting a single variety in a small plot near the house so that its performance could be closely observed. If it performed satisfactorily, it was then often tested on several different soils and in various crop associations. In the general survey, 78% of the 150 farmers interviewed preferred testing in pure stand to testing in mixture. The studies also showed that farmers selected specific mixtures of bean varieties according to soil fertility (*see* Table 3.4). For example, a mixture sown on a poor soil would contain more small grains than a mixture sown on a richer soil (Voss, 1990).

Table 3.3   **Percentage of farmers testing varieties on their own initiative, Ruhengeri, North West zone, 1984 (N = 150)**

Frequency of testing	% farmers
Often test new varieties	40
Sometimes test new varieties	52
Never test new varieties	8

Source: Voss, 1990

Table 3.4   **Percentage of farmers sowing different mixtures according to growing conditions, Ruhengeri, North West zone 1984 (N = 150)**

Growing conditions determining mixtures	% farmers
Fertile soils	65
Infertile soils	61
Association with bananas	45
Others	14

Source: Voss, 1990

Production constraints

The main problem facing agriculture in Rwanda is the ever increasing pressure on land. The annual population growth rate is 3.7%, and the current population density is 250-800 people/km^2. Some 92% of the population lives in the rural areas and produces most of its own food, as well as depending on farming activities for most of its income (MINIPLAN, 1988). The average size of a farm in Rwanda is less than 1 ha, with 43% of farms having less then 0.5ha of crop land (MINAGRI, 1985). Identifying technological innovations which will increase the productivity per unit of land in a sustainable manner is thus an urgent national priority (MINAGRI, 1986).

The diagnostic studies indicated that diseases and soil fertility were the most important agronomic factors limiting yields in the main bean producing areas. Significant interactions occur between these two factors as fertilizer use seldom results in yield increases under heavy disease pressure. An extensive series of diagnostic trials using 'minus-one' designs showed that the yield potential for bush beans would probably be about just under 2000 kg/ha when the highest levels of inputs were applied (*see* Figure 3.3).

These agronomic findings were consistent with the survey findings on the farmers' perception of limiting factors. After high yield-potential, farmers said that the most important agronomic characteristic of the new bean varieties was 'resistance to rain' (disease resistance), followed by earliness and tolerance of poor soils (ISAR, 1987).

Figure 3.3   Yield losses in bush beans resulting from diseases, pests and suboptimal plant nutrition, South Central zone, Rwanda, 1986-87* (Mugusa village; bordering East zone)

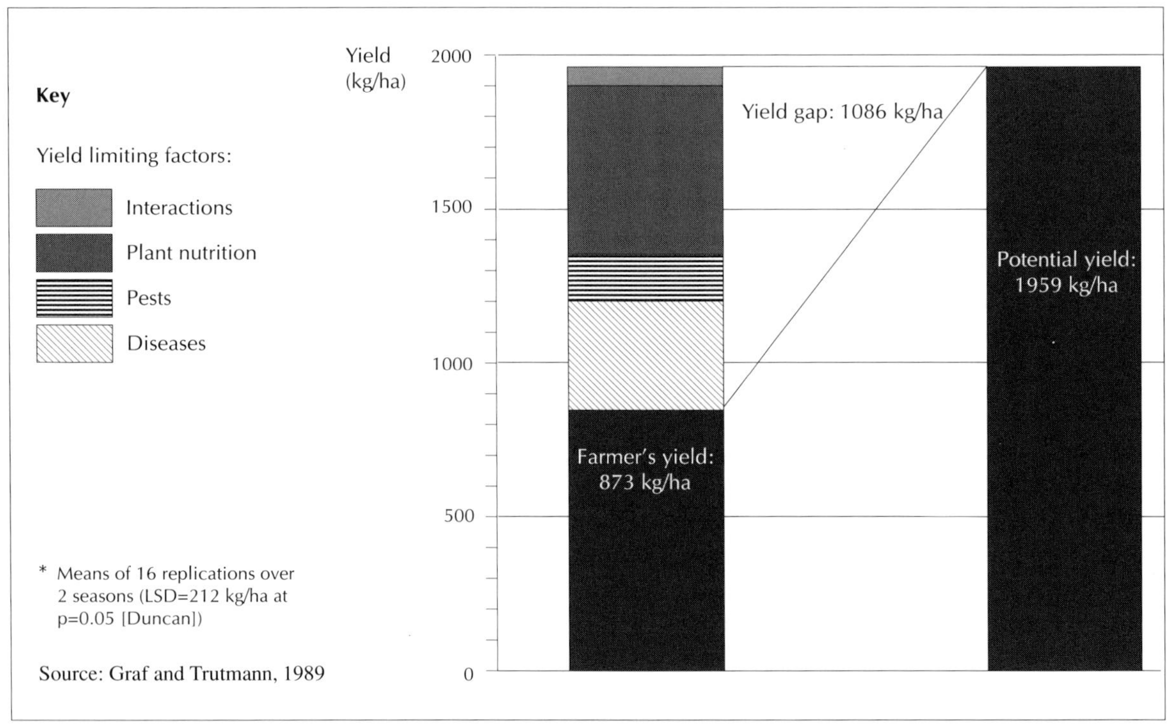

On-farm variety trials

An extensive network of on-farm variety trials, conducted in close collaboration with extension, was a key tool for collecting information about the dynamics of the bean production system (ISAR, 1985-88; Voss and Graf, 1991). The variety trials were strictly farmer-managed; farmers were randomly selected and did not apply specific cultural practices appropriate for climbers, except for staking. The results indicated that climbing beans were a promising technology for much of the South Central zone and the higher altitudes of the East zone. In the bush bean trials, the new varieties rarely outyielded the local mixtures in the relevant zones (ISAR, 1985). However, when a climbing bean treatment was included, its potential on fertile soils became apparent. It was concluded that with better knowledge of managing climbers, farmers would be able to double bean yields per unit of manured land by switching to climbing varieties. This assumption was supported by on-station data which suggested a 100% yield advantage of climbers over bush beans (Nyabyenda, 1986).

The diagnostic trials were jointly managed by researchers and farmers. The researchers carried out the spraying and fertilizer application, but they let farmers decide on planting time and density (Graf and Trutmann, 1989). The results confirmed the potential for climbing beans in many areas of Rwanda, as well as in Burundi and Zaïre. They showed that, for bush beans, fertilizer application often did not increase yields, that the feasibility of regular spraying of fungicides was highly questionable and that the probability of being able to promote a high input package combining both technologies was very low. At the same time, the researchers had to take into account the need for a substantial increase in productivity in order to meet future food demands.

An outbreak of root rots in bush beans in 1987 and 1988, aggravated by poor agronomic practices and heavy rainfall at the beginning of the main planting season, reduced bush bean yield close to zero in both years. Malnutrition and emigration to eastern Rwanda and Tanzania were the main consequences. Climbing beans grown in trials were not affected by root rots, probably because of their different root structure.

PLANNING

Following the diagnosis, the researchers had a choice between various research options known to offer potential solutions to production problems in Rwanda's main bean producing areas. These options were exposed to a screening process that roughly followed the criteria suggested by Tripp and Woolley (1989). For climbing beans the following factors were considered:

- *Disease control.* In general, climbing beans are less affected by disease because of the disease escape mechanism of climbing, which allows better aeration and limits the development of foliar pathogens.

- *Soil fertility considerations.* The high soil fertility required for climbing beans to significantly outyield bush beans would encourage farmers to concentrate their efforts (manure and labour) on a small area to achieve maximum productivity. This was congruent with the need for system intensification and would make more land available for tuber production while maintaining adequate protein levels in the diet. Evidence from other areas supported the assumption that resource-poor farmers could increase their total output by following such an intensification strategy.

- *Chance of success.* In view of earlier failures, the prospect of successfully introducing climbing beans into non-climbing bean areas seemed less than favourable. However, it was generally felt that the earlier efforts had not benefited from a critical mass of resources, that collaborating farmers may have been poorly selected, and that farms had become significantly smaller in the intervening period; also, the technology was known and it was successful in northern Rwanda.

- *Ease of carrying out research programme.* It would be fairly easy to explore the potential of climbing beans with a number of carefully selected farmers, and later work with them on technologies suited to their farming conditions.

- *Compatibility with farming systems.* Little was known about the effect of climbing bean introduction on traditional cropping systems. It was feared that climbing beans might reduce the extent of intercropping because of their competitiveness. However, it was clear from surveys in other areas that climbing beans performed well in primary association with bananas; also, it is under bananas where the most fertile soil is usually found.

- *Inputs and institutional support needed.* In the first stage of diffusion, seed availability would be a limiting factor, and institutional efforts would be needed to quickly overcome this. At a later stage, fertilizer might be needed to sustain productivity at a high level. A considerable effort involving adaptive testing, demonstration and extension would also be required.

- *Profitability and social desirability.* In the first phase of adoption no purchased inputs would be needed and thus profitability would probably be high, depending on the amount of manpower available in the household as returns to labour would often be no higher than with bush beans. The initial diagnostic survey, however, had shown that most farmers considered land constraints to be more important than labour constraints and thus it was felt that climbing beans would be particularly attractive for smaller farmers.

- *Stability of farm productivity.* In areas with erratic rainfall patterns, climbers might represent a higher risk because of their generally longer growth cycle. However, the deeper root system would counteract this to some extent, as would the possibility of cultivating them during the long rainy season. If farmers did not intercrop climbers to the same extent as they intercropped bush beans, climbing bean cultivation might have adverse effects on stability.

- *Other options.* A similar analysis was done on the research options for bush beans. Seed treatment against bean fly and seed-borne diseases was actively researched and promoted because of its low input character and the high chance of success (CIAT, 1988). NPK fertilization and foliar applications of fungicides were dropped as options because of the high level and cost of inputs needed for success. To some extent, these more conventional approaches were also being tested and promoted by national and international organizations. Breeding for disease resistance in bush beans continued to be a standard feature of the crop improvement programme. A small on-station research programme was initiated to explore the feasibility of various techniques of green manuring and their integration into bean cropping systems.

From this analysis, climbing beans emerged as having the greatest potential for meeting the overall goal of increasing bean production per unit of land with a technology that required few purchased inputs and favoured smaller farmers because of the high labour and management requirements. However, researchers felt that additional information was needed in order to plan the experimental phase; in particular, they decided that surveys had to be conducted to assess the cultural practices of farmers in climbing bean areas and the knowledge of farmers in non-climbing beans areas about climbing beans.

The survey in northern Rwanda showed that farmers used highly sophisticated techniques to grow climbing beans. Special attention was paid to staking techniques as it was assumed that stake availability, as well as the manpower necessary to cut, transport and place stakes, could be a prime reason for farmers to reject the technology. The strategy farmers use in the north is to grow hedges of elephant grass (*Pennisetum purpureum*) on anti-erosion lines. These hedges are usually planted adjacent to the fields of climbing beans, which reduces the time needed to transport stakes (CIAT, 1987).

The survey conducted in the central plateau region indicated that only about 5% of the farmers had ever tried to grow climbing beans and that their knowledge about these beans was poor. A more targeted survey conducted among farmers who were already growing some climbing beans indicated considerable potential as all these farmers felt that they could double yields on fertile soils. They said that the major constraints they faced were lack of seed and staking material. As mentioned earlier, a treatment involving climbing beans was added to a limited number of on-farm bush bean trials in the area around ISAR's main research station. Farmers were questioned about the potential acceptability of the technology and the main problems likely to emerge if they grew climbing beans on a larger scale. Not surprisingly, they cited stake availability as the main problem; however, it was not clear whether the problem was truly one of availabilty or, in fact, of the extra labour required. Labour requirements are considerably higher for climbers than for bush beans, as indicated in Figure 3.4 (Brewster, 1988).

Figure 3.4   Labour requirement for bush bean versus climbing bean cultivation, Buberuka Highlands, Rwanda

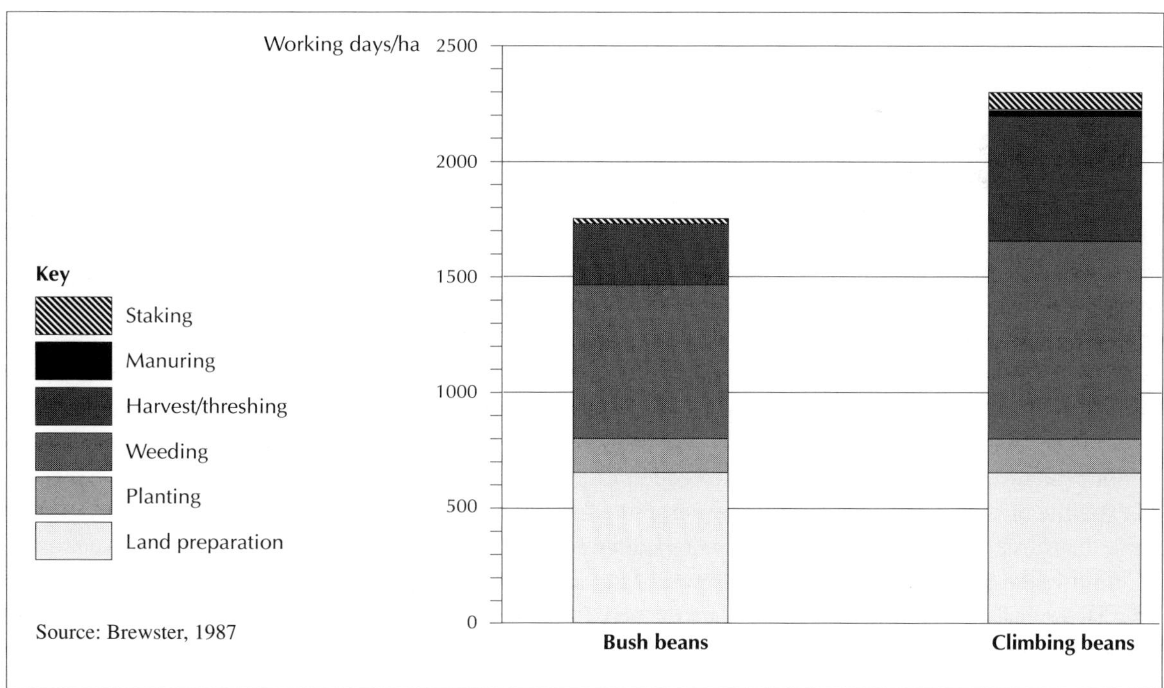

It should be pointed out here that the close researcher-farmer collaboration necessary for an effective OFR programme is particularly difficult to achieve in Rwanda. Because of the country's feudal history, it is traditional among farmers to be prudent when answering questions posed by authority figures, such as extension agents. As the government still has the right to demand a certain amount of communal labour (*umuganda*), farmers may have been reluctant to say that they were concerned about the extra work required for fear that they would simply be ordered to do the work. Furthermore, in the Rwandan tradition questions are not answered directly until it is clear why the questions are being asked. To avoid giving a correct response without knowing the implications, the person questioned will often not come to the point or may even give wrong information (Erny, 1983). For all these reasons, it can be misleading to rely on questionnaires in individual interviews.

EXPERIMENTATION

The information from the initial diagnosis and the additional information on cultural practices in northern Rwanda and on the knowledge base in non-climbing bean areas were incorporated into the design of the experimental phase.

Selection of target areas and collaborating farmers

Two target areas were chosen, one in the East zone and the other in the South Central zone. The first area encompassed Ntyazo and Ntongwe 'paysannats' (settlements); the second encompassed Karama and Rukondo villages. Both areas were served by the Project Agropastoral de Nyabisindu (PAP), a German-funded extension project. PAP was charged with the selection of 79 farmers from the four sites (*see* Table 3.5). In the early stages of the research, however, Ntongwe and Rukondo were dropped as research sites in order to save on costs.

The research was carried out by staff from ISAR's Legume and Farming Systems Programmes and the CIAT Regional Bean Programme. Close collaboration with PAP and MINAGRI extension staff was an important feature of the research process. The main factors influencing the selection of the target areas were easy access from the main research station, presence of collaborative extension project/communal services, and suitability of selected areas for the research topic. The first two factors are logistical, reflecting the reality of OFR being conducted by station-based commodity programmes. Logistical factors are generally more important for site selection than agroecological and agroeconomic ones. In an analysis of common problems in farming systems programmes, Lightfoot (1988) showed that logistical factors were a major reason for errors in trial establishment.

In agroeconomic, socioeconomic and climatic terms, there are significant differences between the East and South Central zones. Figure 3.2 shows the relative importance of main crops in each zone. With regard to farm sizes, there are more farms above 1 ha in the East zone than in the South Central zone. The East zone has more erratic rainfall and higher temperatures and evapotranspiration than the South Central zone; of the five seasons of experimentation in Ntyazo, during one season the rains were insufficient to allow climbing beans to yield more than bush beans; during another, the rains started so late that farmers decided not to plant climbing beans.

Soils in the South Central zone tend to be acid and infertile, particularly in the western part near the Zaïre/ Nile Divide where Karama and Rukondo are located. In Ntyazo and Ntongwe, which were settled by farmers about 20 years ago, soils are more fertile. The paysannats are equipped with a minimum infrastructure provided by the government. Initially, the land in these paysannats was evenly distributed (2 ha per family) and the

Table 3.5   Research sites included in the initial study of farmer acceptability of climbing bean technology

Village/paysannat	No. of farmers	Zone
Ntongwe	19	East
Ntyazo	20	East
Karama	22	South Central
Rukondo	18	South Central

division of land after the death of the family head was forbidden; thus, sons who did not get land from their parents were faced with serious social and economic problems. In recent years, however, the policy has changed and in many cases the land is divided. All families in the paysannats are required to grow coffee, with the government providing technical assistance and inputs. The villages in the South Central zone do not benefit from such support and the farming systems have developed more naturally although coffee is a semi-obligatory crop in this zone. Farmers who have coffee trees on their farm are not allowed to eliminate the crop and are obliged to follow certain cultural practices.

There are also differences in the way the extension services operate in the two zones. It was evident to the researchers that the general attitude of extension agents towards farmers was more authoritarian in Ntongwe and Ntyazo than in Karama and Rukondo, largely because their role as 'coffee police' in the paysannats. As no particular emphasis was given to selecting a representative sample of farmers to carry out experiments, the farmers chosen by the extension agents in Karama had substantially bigger farms than the average for the region; in Ntyazo the farm sizes were below average but off-farm activities were more important than for farmers with farms of average size (*see* Table 3.6)

Table 3.6   Average farm size, age of household head, number of persons per household and percentage of household revenue from off-farm activities[1] for the East zone (Ntyazo) and South Central zone (Karama) and among collaborating farmers

	Average for the zone		Average among farmer collaborators	
	South Central	East	South Central [2]	East[3]
Farm size (ha)	0.90	2.00	1.57	1.00
Age of household head	50.3	45.2	47.8	49.8
Persons per household	4.8	5.3	5.0	6.3
Off-farm revenue (%)	41.1	20.5	17.2	33.0

Notes: 1. Including trade and official functions.
       2. N = 22.
       3. N = 18.

Introduction of climbing beans

The first phase consisted of introducing a basic technology similar to that used by farmers in northern Rwanda. Local varieties, known to be acceptable and reasonably well adapted to the target areas, and a standard set of cultural practices were used. Varietal trials began in 1985, and on-farm climbing bean trials started in the following year. The objectives were to increase the knowledge base of a large number of farmers about climbing bean cultivation, to evaluate the potential for extension and to involve farmers in the design of second-generation technologies that would meet farmer needs and thus be more likely to be adopted on a large scale.

Each of the 79 collaborating farmers received 500 g of seed of the climbing bean cultivars Gisenyi 2- (large grain, white with black stripes) and C10 (large purple grain), along with a pamphlet explaining the cultural practices required to successfully grow climbing beans. In addition, local extension agents received training at the research station on how to grow climbing beans and later trained the farmers in the presence of research personnel.

The farmers were closely observed during the first few seasons. In individual interviews, the stake problem emerged as the main constraint to climbing bean adoption, especially in Ntyazo. Nevertheless, a pattern of climbing bean adoption soon evolved, with important differences emerging between the two agroecological zones. In Ntyazo, farmers brought more land under climbing beans than was the case in Karama, but in Karama there was more farmer-to-farmer distribution of small quantities of seed (ISAR, 1987) (*see* Table 3.7). When PAP offered climbing bean seed for sale in the two zones, in the South Central zone over 500 farmers purchased up to 2 kg of seed at a price higher than the market price for consumable beans, but in the East zone there was little demand for seed. In meetings with farmers it became obvious that the main problem in the East zone was not stake availability but rather the work involved in staking. Surveys indicated that a higher proportion of farmers needed external labour to complete agricultural tasks in Ntyazo than in Karama. This was mainly because of the greater involvement in off-farm activities and the high frequency of malaria, which considerably reduced labour availability.

Researchers found it difficult to communicate with collaborating farmers, especially in the Ntyazo area, and decided to try to select more representative farmers for further experiments. They also decided to put more emphasis on group meetings as opposed to individual follow-up surveys, partly to reduce costs as individual

Table 3.7   **Area sown with climbing beans by collaborating farmers in the second season after introduction in Ntyazo, Ntongwe, Karama and Rukondo**

Village/paysannat	Area with climbing beans (ares)	Area with bush beans (ares)	% of climbing beans in total bean area
Karama	1.57	26.8	5.5
Rukondo	2.49	27.9	8.2
Ntyazo	6.50	59.3	9.9
Ntongwe	3.50	38.1	8.4

Source: ISAR, 1987

surveys were very time consuming and partly to overcome communication problems. In group meetings in farmers' fields it is easier to get a better idea of farmers' reactions, especially by listening to conversations among farmers and by guiding the meeting from a discussion on broad issues to one focusing on more specific issues. The researchers invited not only farmers from the initial group of 79 to the meetings but also the neighbours of climbing bean farmers and farmers who adopted climbing beans later.

Varietal selection

The ISAR plant breeding programme identified several new climbing varieties which held promise for the target areas. Although the absolute yields of the varieties tested in on-farm trials in 1987 and 1988 did not show any one of them to be clearly superior (*see* Table 3.8), G 2333, introduced through CIAT from Mexico and later named Umubano, turned out to be the most preferred cultivar in Karama. Whenever farmers obtained it they abandoned the other varieties and grew it in pure stand, saying that they would like to mix it with other grain types if these types performed equally well.

Table 3.8   **Results of on-farm variety climbing bean trials in Ntyazo and Karama in the short rainy season, 1988-89 (kg/ha)**

Varieties	Karama		Ntyazo		Days to maturity in Ntyazo
	1988	**1989**	**1988**	**1989**	
G 2333	2936A[1]	2214AB	1594A	3391	87
G 858	2225A	1736BC	852B	—	108
G 685	2724A	2393A	1450AB	3369	104
G 13671	—	1504C	—	3878	83
Gisenyi 2-	2813A	1897ABC	1983A	2674	89
C 10	2346A	—	1463AB	—	89

Note: 1.   Means marked with same letter do not differ at $p = 0.05$ (Duncan).
Source: ISAR, 1989

The reason for Umubano's popularity became clear in group meetings. In the Karama area, farming families are obliged to eat leaves in order to get through the period of food shortage after the long dry season between sowing and harvesting; at the meetings, the farmers said they considered the leaves of Umubano to be superior to those of other varieties. Two seasons after Umubano had been introduced, it was already accepted by a large number of farmers in Karama. Confidence among farmers in the new technologies grew and the discussions at group meetings became more open.

The staking problem

While the on-farm work was in progress, on-station research was initiated to address the stake availability problem. This research focused on agroforestry techniques rather than on growing *Pennisetum* hedges, partly because of the competition between *Pennisetum* and adjacent crops but also because of the potential benefits of agroforestry species in terms of biological nitrogen fixation.

As farmers in the Karama area began to recognize the advantages of climbing beans and became aware of the outstanding results, particularly with Umubano, the issue of stake availability became less important. Those who adopted climbing beans now considered soil fertility and manure availability to be the main factors limiting crop expansion and said they would always be able find enough stakes. However, the effect of stake gathering was already visible in areas where groups of farmers grew climbing beans; *Grevillea robusta* trees, for example, were being stripped of most of their branches. Thus, continued research on agroforestry systems was justified, although farmers were showing only moderate enthusiasm about growing *Calliandra*, *Leucaena* or *Sesbania* in hedges to produce stakes and did not accept the idea of starting trials on the use of living maize as stakes.

PAP had long been active in promoting hedges of agroforestry species, mainly for fodder production and the prevention of soil erosion, but had met with only limited success. So as not to repeat the PAP experience, it was decided to introduce agroforestry 'tree by tree' without promoting any particular layout in the field and by discussing management practices with groups of farmers during key periods. Each farmer received 5-10 *Calliandra calothyrsus* trees and some *Sesbania magrantha* seeds, and was free to decide how to integrate them into his/her system. In general, the *Calliandra* trees were distributed along contour lines but never in compact lines of trees. They were often planted in protected corners of the farm to prevent damage by cattle and goats. A particular concern with *Calliandra* is that until the young plants are at least 18 months old and have been coppiced twice they are easily destroyed by livestock; when farms are close to each other, although a farmer growing *Calliandra* may control his/her goats, neighbouring farmers may be less careful.

With *Sesbania,* however, there is a low risk of loss and this, coupled with the low initial investment because of the plant's high self-seeding rate, made it more attractive than *Calliandra*. Although the rate of reproduction with *Sesbania* is such that it can become a weed, farmers did not consider this a problem as it was easily removed using normal weeding practices. A comparison of the characteristics of *Calliandra*, *Sesbania* and *Leucaena* is given in Table 3.9.

The experience regarding stake availabilty in the Ntyazo area differed from that in Karama. In Ntyazo, where there was strong farmer participation in experimentation process, the farmers acknowledged the high yield potential of climbers but continued to complain about the stake problem, although the area exports *Pennisetum* stakes for fence building and these stakes are also used for mulching coffee. From the discussions in group meetings, it appeared that they were not convinced that climbing beans were worth the extra labour required for their cultivation. This was extremely important as they would have to hire labour for stake management tasks and there was not yet the same degree of population pressure on the land as in the Karama area.

In the light of these observations, a trial was designed involving the simultaneous association of maize with climbing beans. The bean variety used was G 13671, which had passed the ISAR varietal testing scheme as a semi-climbing bean with encouraging yield results under moderate rainfall conditions. It was initially rejected by farmers because it was difficult to weed and the grains were destroyed by rot if the rains continued after the plants had reached maturity. Staking G 13671, however, made it much more acceptable because this kept the seed pods off the ground, and thus it was included in variety trials with wooden stakes as well as in association trials with maize.

Table 3.9   Comparison of agroforestry species used for stake production in Ntyazo and Karama

| Species | Stakes | Production | | Management | | Acceptability |
		Start of production	Biomass	Establishment	Rejuvenation	
Calliandra calothyrsus	+++	12 months	+++	+	+++	++
Leucaena leucocephala	++	18 months	++	+	+++	+
Sesbania magrantha	++	7 months	+	+++	+++	+++

Note: +++ very high/very easy.
        ++ moderately high/moderately easy.
        + low/difficult.

In the first season the results from the on-farm trials were not very encouraging. This was because the maize had been planted too late, which meant that, with an interval of 4 weeks between maize and bean planting, the sowing date for beans had also been delayed. The on-station results indicated that a 2-week interval held more promise. Interestingly, several farmers preferred the maize/bean association despite the poor trial results, stating that they would try the system out with shorter planting intervals. They also preferred the short-cycled climbing bean variety G 13671 over the elite climber Umubano, although Umubano had a more acceptable grain type. The difference in varietal preference between Ntyazo and Karama derived mainly from the differences in rainfall patterns.

Economic analysis

After several seasons of experimentation with the different options for climbing bean cultivation, it seemed necessary to analyse the existing information from an economic perspective. Such an analysis at this stage can provide useful information to help set priorities for further research, although it may be too early to judge the performance of new technologies in farmers' fields.

Agroforestry options

As some of the agroforestry options involved large investments by farmers as well as by local authorities (nurseries, distribution of seedlings), it was important to estimate their potential benefits before launching extension efforts. In Karama, the village was subsidizing stake production by allowing farmers to cut branches from the trees bordering communal roads. The current market prices for *Pennisetum* stakes were quite high. It was calculated that a field of *Calliandra* with a planting density of 10 000 plants/ha and a production of 16 stakes/yr from each plant would result in a far higher return (320 000 FRw) than a climbing bean crop with a bean production of 3000 kg/ha (105 000 FRw). Although there was a market for stakes in some areas of Rwanda (Sperling, pers. comm.), it was thought unlikely that farmers would specialize in stake production. Farmers generally produce such inputs themselves as soon as prices are high enough to cover production costs on farm. In addition, the priority given by farmers to food security would not allow for such specialization.

Fertilization

In Karama, the on-farm trials involved the application of both organic and inorganic fertilizer in the early stages of bean growth. In terms of yield the results were encouraging but farmers rejected the proposed technology because of its labour requirements. A partial budget analysis for the tested treatments showed their assessment to be correct (*see* Table 3.10).

Table 3.10   Partial budget analysis of on-farm fertilizer placement trials, Karama (means of 20 repetitions over 2 seasons)

Treatment	Yield (kg/ha)	Var. costs (FRw/ha)[1]	Net benefit (FRw/ha)
Additional 15 t farm manure[2]	2341	15 400	66 535
Additional 15 t farm manure + 110 kg DAP[2]	2611	22 110	69 275
Farmers' practice[3]	1995	—	69 825

Notes: 1. US$ 1 = 80 FRw.
        2. Applied at stage of three trifoliates.
        3. Ranges between 20 t and 40 t compost/manure/ha.

A new series of trials with mineral phosphorus fertilization applied at planting was proposed, based on encouraging results in exploratory trials (Rwandanga, 1989) but complex institutional and economic problems may make any mineral fertilization technology unfeasible in the immediate future. However, farmers did modify their fertilization strategies by applying compost and manure over a small area under climbing beans rather than spreading it over a larger area under bush beans.

Strategies to minimize research costs

ISAR's Legume Programme has followed a strategy of systematically collaborating with extension agencies in order to carry out OFR. The objectives of this strategy are to:

- take advantage of the knowledge of extension staff in the research process, leading to better research efficiency;

- shorten the time lag in the transfer of technologies from research to extension;

- reduce research costs by taking advantage of existing extension services in terms of personnel and infrastructure.

Although extension agents in Rwanda play the role of 'coffee police', which inhibits the establishment of a good relationship with farmers (World Bank, 1977), by carefully selecting agents who have managed to establish a good rapport with farmers this problem may be minor compared to the advantages of working with extension.

In the climbing bean case, the costs of OFR and extension activities were further reduced by the following measures:

- basing undergraduate students (from schools of agronomy) locally to follow the research process, thereby reducing transport costs for research personnel;

- switching to a system of group meetings rather than individual interviews and surveys;

- choosing cost-efficient methodologies, such as yield estimations by farmers;

- after the first follow-up survey, dropping Ntongwe and Rukondo as research sites in order to save on the high costs of transport and perdiems.

McIntire (1989) questions the superiority of extensive methods of data collection over intensive ones. He argues that intensive data collection reduces the rate of research error and that the benefits of this outweigh the additional costs of data collection. Even if these points are valid, the magnitude of the initial investment needed to collect information has a strong influence on the decisions taken by directors of national institutes, and the only way to prevent OFR programmes from being terminated in periods of budget difficulties is to reduce costs on such items as transport and perdiems.

The value of increased precision in a particular evaluation should be carefully weighed against the costs of achieving this precision. In the climbing bean case, having the farmers estimate yields was a sensible approach as their decisions about adoption would ultimately be based on their own evaluations. Similarly, making locally based extension agents responsible for counting the number of farmers growing climbing beans and estimating the area sown to the crop may have meant that figures were less precise than would have been the case if research personnel had carried out the task, but it was considered that an increase in precision would not have justified the increased costs.

Although collaboration with PAP and MINAGRI extension services was weakened by staff changes, mainly within PAP and the Ministry rather than in ISAR, the common goal of transferring a useful technology to farmers kept the work going. An example of this is the strategy adopted by the mayor of Karama, who grew climbing beans on the communal fields and dried the harvest in front of the communal office in order to allow farmers to 'steal' some seeds for cultivation.

Figure 3.5 (*overleaf*) shows the distribution of costs between research and extension during the research process, including the contribution by the Institut Africain de Développement Economique et Social (INADES). Research benefited most from the collaboration in the experimentation and extension phases.

Figure 3.5   Relative contributions of research and extension services to costs of climbing bean study (1986-90)

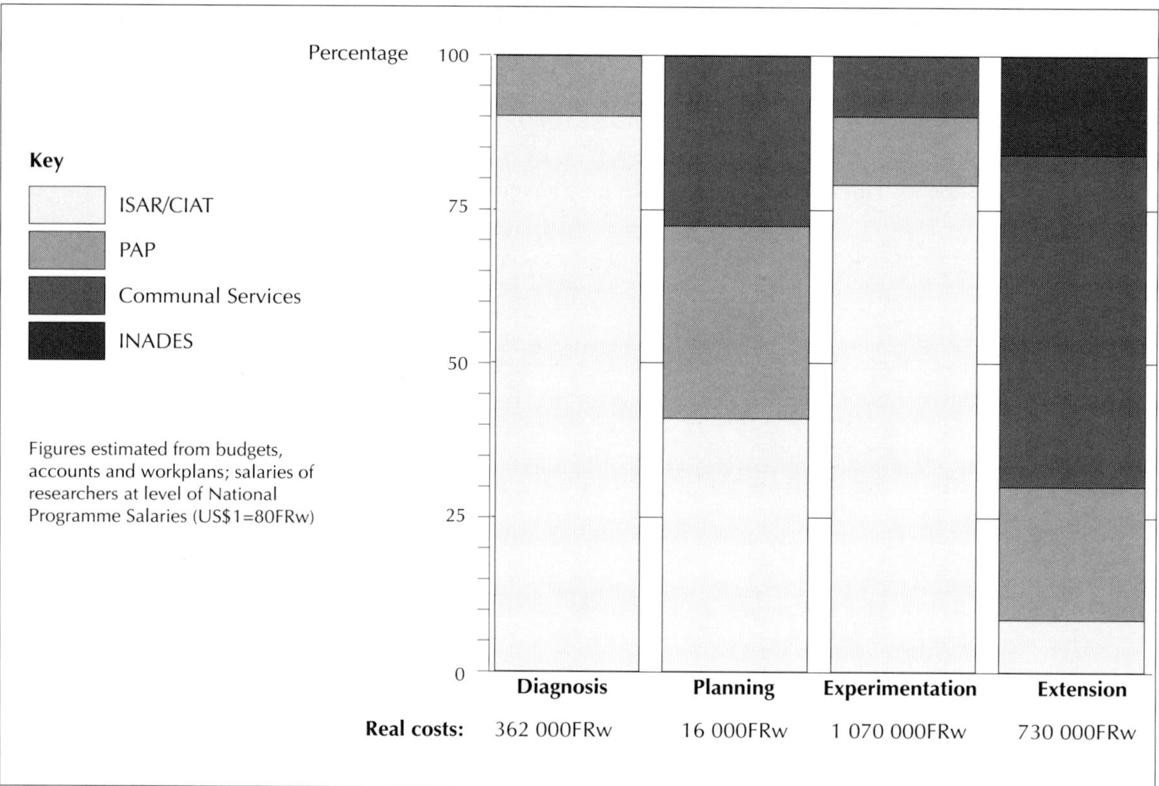

DIFFUSION AND ADOPTION

It is a common belief among research and extension staff in Rwanda that farmers are conservative and not very receptive to innovations (Gasarasi, 1985). That this is a serious misconception has been proved by the dynamic nature of Rwanda's farming systems. Cowpeas were important 20 years ago, but have now been completely replaced by beans. Similarly, fingermillet has been replaced by wheat and potatoes. Crops such as cassava and sweet potatoes, which are now staple crops, were introduced less than 100 years ago and avocado trees were introduced less than 50 years ago. This is an indication not only of the innovative spirit of Rwandan peasant farmers but also of their ability to evaluate complex innovations and to fit them into their farming systems. Recognition of their analytical and practical skills guided the research process in the climbing bean study.

A discussion on technology adoption only a few years after the on-farm experimentation was initiated may seem premature but our observations indicate that climbing beans are entering the exponential phase of diffusion. We believe that the conclusions we draw here, based on existing adoption patterns, will provide useful guidelines for future extension strategies.

The diffusion of the technology in the Rwanda case was effected through two channels. The first was the extension effort, which was based on the knowledge acquired during the research process and was organized to demonstrate the proven technology to the farmers. The second was through the farmers collaborating in the OFR process who adopted the proposed technology and extended it to their friends, relatives and neighbours. The following discussion focuses mainly on the latter, prefaced by a brief outline of the formal extension materials used to transfer the technology.

Extension materials

In 1989 MINAGRI began to actively promote climbing beans in non-climbing bean areas and the Minister himself mentioned the technology in public speeches at various occasions. In the same year an extension booklet was produced for use by extension agents (INADES, 1988). It included interviews with two farmers, one from a conventional climbing bean area and the other from Karama.

The woman chosen for interview in Karama had obtained the first climbing bean seed from a neighbour who was among the first group of 79 selected farmers. Two seasons later, she had planted two fields, covering 7 ares and representing her total bean area, to climbing beans. She began experimenting with various associations which she normally used with bush beans, such as growing sweet potatoes on mounds in the bean field. She also took advantage of the manure she had applied for the climbing beans to plant more maize than she would have planted in a bush bean field. After a further two seasons, she had identified the most productive agronomic practices to suit her conditions and stated that climbing beans outyielded bush beans grown under the same conditions by 100%. This was identical to the researchers' findings.

Adoption by collaborating farmers

In 1990, as the rains in the Ntyazo area did not begin till November, farmers planted only small areas of bush beans and decided against climbing beans. The following data therefore derive mainly from the Karama area, where rains were sufficient throughout the experimentation period. In order to monitor the adoption of climbing beans and to analyse the adoption pattern, 22 of the initial 40 farms in the South Central zone were surveyed.

It became obvious that the farmers chosen as collaborators in this zone (whose farms tended to be above average size, as shown in Table 3.6) were not the most receptive group of farmers for climbing beans. Among the 22 collaborators, six had abandoned the crop and two had not extended the area under climbing beans at all. The analysis revealed that there was greater adoption of the climbing bean technology by the farmers with smaller farm sizes and a smaller area under food crops (*see* Table 3.11). Other findings showed that climbing beans were adopted more quickly in areas inhabited by poorer farmers.

Table 3.11   Differences in degree of adoption of climbing beans in different sectors of Karama

Sector	Farm size (ares)	Area under food crops (ares)	% climbing beans of total area under beans
Nyanzoga	86.5	55.5	30.3
Ngoma	218.7	126.0	20.5

It may seem questionable that extension agents were allowed to select better-off farmers for the initial study, but most of these farmers played an important role in demonstrating the potential of climbers to their neighbours and it seems that the unrepresentative choice of farmers in the early stages did not hamper diffusion. In addition, these farmers were more able to absorb the risks associated with testing a new technology. Greater emphasis is now being placed on resource-poor farmers, among whom there is a higher demand for climbing beans.

The impact on nutrition

As noted earlier, food shortages in Rwanda, especially among poor farming families, has made leaf edibility an important criterion for varietal selection. The yield evaluations described earlier also made it clear that farmers appreciated the possibility of harvesting large amounts of leaves from climbing bean plots. On 90% of the plots sown with Umubano, farmers had picked the leaves, compared to only 10% of the bush bean plots. The crucial role that climbing bean leaves can play in periods of food shortage is reflected in the following statement made by a woman farmer at a group meeting: 'Without the leaves of Umubano, I do not know how we would have got over the last 2 months'. Bean leaves are an excellent source of protein and vitamin A.

Cultural practices

At the beginning of the research process there was concern about the potential effect of climbing beans on farmers' intercropping practices. It was feared that farmers would grow the competitive climbers as a monocrop and that, as a result, production stability at farm level would suffer. This fear has proved unfounded. To date, 71% of climbing bean fields are intercropped and farmers are experimenting with various new combinations. However, farmers in Karama have not accepted the idea of using maize as living stakes, although beans were associated with maize in 15% of collaborators' fields in the area (*see* Table 3.12). In the immediate future the dominant association in the Karama area is likely to be the one with sweet potatoes. On small mounds (one per

Table 3.12　**Crops associated with climbing beans by farmers growing climbers since 1986, Karama, short rainy season 1990 (N = 27)[1]**

Associated crops	No. of fields	% of fields
Sweet potatoes	10	37
Maize	4	15
Banana	13	48
Others	6	22
Climbing beans monocrop	8	29

Note:　1. Up to three associated crops per field possible.

7 m^2, each mound measuring about 1 m in diameter), farmers plant sweet potatoes with the objective of producing tubers for consumption and vines for planting in the following season.

A major impact on the stability of bean production may be that as many beans can be cultivated during the long rainy season as during the short rainy season. This would enable farmers to balance production problems in one season with increased cultivation in the other and guarantee a constant supply of beans throughout the year. In the group meetings some farmers talked enthusiastically about their initial experiences with climbers during the long rainy season.

The agroforestry techniques are still being evaluated by farmers. In group meetings and individual interviews farmers put less emphasis on stake availability as a key limitation for climbing bean extension. However, ensuring that climbing bean cultivation does not result in environmental degradation through trees being stripped of their branches remains an important priority. In the meetings, farmers were more interested in low investment agroforestry techniques, as in the case of *Sesbania* cultivation, than in techniques which required considerable effort in the establishment phase, as in the case of *Calliandra* and *Leucaena*.

It was recommended that farmers plant climbing beans as soon as the rains started, before sowing their bush beans. This recommendation is now being followed by all climbing bean growers. The predominant planting practice for bush beans is to plant at random at densities between 250 000 and 500 000 seeds/ha, depending on soil fertility (CIAT, 1985). For climbing beans it was recommended that planting in rows, preferably with the rows oriented east/west to improve light penetration, would be beneficial and that densities should be as low as 160 000. The first recommendation was rejected by all farmers who had grown climbing beans for more than one season. As to the second recommendation, farmers in group meetings stated that they would plant at a lower density than had been the case initially and would reduce this still further to allow for more intercropping.

Varietal preferences

The strong preference for Umubano was reflected in a randomly selected sample of farmers for yield evaluations. Out of 26 farmers, 46% grew Umubano, 15% grew Urunyumba 3 (released by ISAR prior to the start of the project) and 39% grew Gisenyi 2-, the main variety distributed by the extension service. In group meetings most farmers indicated that they would grow only Umubano in the following season, giving as reasons for this decision its high yield potential, its resistance to anthracnose, its acceptable grain type with good taste and, in particular, its tasty leaves. They also liked its relative earliness and the moderate vigour compared to Gisenyi 2-, but some of the richer farmers said they would continue growing Gisenyi 2- as well, mainly for its taste and the prestigious large grain type.

Adoption by non-collaborating farmers

It should be emphasized at this stage that research and extension staff advocated that farmers should initially plant only small areas of climbing beans, to familiarize themselves with the crop and its requirements. Research and extension would then assist them to develop second-generation technology. Again, the guiding principle behind this was to utilize the farmers' ability to develop and test new systems.

In the Karama area, after four seasons 520 farmers out of a total of 6600 farming households were reported to be growing climbing beans; in the three sectors where most activity was concentrated, more than 10% of the farming households were growing them. With the growing demand for seed, the farmers who first adopted Umubano are now being rewarded by being able get more beans for consumption in exchange for seed; the rate of exchange is 1 kg of climbing bean seed (mostly Umubano) for 2 kg of beans for consumption. Obviously, once other farmers start producing their own seed, this situation will change. In group meetings, climbing bean growers claimed to have requests for seed from 10 to 18 other farmers, mostly for Umubano.

In order to evaluate yields achieved by non-collaborating farmers, a simple method suggested by Poate (1988) was used. Farmers were asked to estimate the yield of a particular field using their own units of measurement, such as baskets; the full basket was weighed and the field on which the crop had been grown was measured (*see* Table 3.13). In 1991, a survey on farmers' knowledge and use of climbing beans was conducted among 121 farmers in the Karama area. It revealed that 100% of the farmers knew what climbing beans were

Table 3.13   Climbing bean yield compared to bush bean yield, Karama, short rainy season 1990

Variety	Mean yield (kg/ha)	Standard deviation	N
Climbers			
Gisenyi 2-	1321B[1]	754	10
Umubano	3096A	1091	12
Bush beans			
Local mixture	453C	390	15

Note: 1. Figures marked with the same letter do not differ at p = 0.05 (Duncan).

Table 3.14   Results of survey on knowledge and use of climbing bean technology

	% farmers	No. farmers (N = 121)
Know climbing beans	100	121
Plant climbing beans	47	57
Varieties used (N = 57):		
Umubano	59	34
Gisenyi 2-	47	27
Urunyumba	23	13
Others	14	8

Source: Graf, 1991

and 47% of them had planted climbing beans; the main varieties used were Umubano (59%) and Gisenyi 2-(47%), either in pure stands or in mixtures (*see* Table 3.14). Rapid diffusion of climbing beans had occurred in neighbouring localities and several other zones of Rwanda which had access to climbing bean technology through the ISAR/CIAT bean programme (Graf, 1991).

CONCLUSION

This study illustrates that many solutions to farmers' problems may be found by working with the farmers and that researchers can play an important role in identifying and describing these solutions and making them known to the target group. The technological change described in the study occurred at a time when farmers themselves were actively looking for solutions to declining bean production.

The dynamics of a farming system requires researchers to base their planning largely on assumptions about how the system will evolve and to ensure that the innovations they propose will suit future systems. This implies that, during the diagnostic phase, potentials and trends should carry at least the same weight as constraints. Research based solely on actual constraints is in danger of being outdated by the time it produces results.

In the Rwanda case, as the basic technology was already used by farmers under circumstances similar to those of the target group, technology development could be carried out at farm level. Complemetary on-station research was responsible for greatly increasing the returns to this effort by identifying the most productive and acceptable varieties. In testing agronomic innovations, it was more efficient to work directly with farmers than to go through station trials.

Because Rwandan agriculture is characterized by subsistence farming, most of the research benefits in this case will remain with the producers and will not go to merchants or, as is often the case, to urban consumers in the form of reduced prices. A striking feature of this effort to expand the area under climbing beans is that the overwhelming majority of farmers who have adopted the new technology are those with minimal resources and hence the greatest need to increase their production to meet household food demands.

References

Brewster, M. 1988. *Bean Production Systems and the Relationship with Recommendation Domains in the Highlands of Rwanda*. Executive Summary of MSc thesis. Fayetteville, USA: University of Arkansas.

Byerlee, D., Harrington, L. and Winkelmann, D. 1982. Farming systems research: Issues in research strategy and technology design. *American Journal of Agricultural Economics* 64(5): 897-904.

CIAT. 1984-89. *Annual Reports*. Cali, Colombia: CIAT.

Clay, D.C. and Dejaegher, Y.M.J. 1987. Agroecological zones: The development of a regional classification scheme for Rwanda. *Tropicultura* 5(4): 153-58.

Delepierre, G. 1985. Evolution de la production vivrière et les besoins d'intensification. In *Proc. 1er Seminaire National sur la Fertilisation des Sols au Rwanda, Kigali, June 1985*. Kigali, Rwanda: MINAGRI/FAO.

Erny, P. 1983. L'esprit de l'éducation au Rwanda ou le 'caractère national' décrit par un groupe d'étudiants. In *Genève-Afrique* XXI(1) 1983: 27-54.

Gasarasi, L. 1985. Introduction des innovations en matière de fertilisants en milieu rural In *Proc. 1er Seminaire National sur la Fertilisation des Sols au Rwanda, Kigali, June 1985*. Kigali, Rwanda: MINAGRI/FAO.

Graf, W. 1991. Innovation in small farmers' agriculture and the role of research in the case of Rwanda. PhD thesis. Zurich, Switzerland: Institute of Agricultural Economics, Federal Institute of Technology.

Graf, W. and Trutmann, P. 1989. Results and methodology of diagnostic trials in Rwanda. In *Proc. of First Regional Workshop on Bean Improvement in Eastern Africa, Kampala, June 1986*. CIAT Africa Program Working Paper No. 1. Cali, Colombia: CIAT.

Graf, W. and Ndorehayo, V. 1989. Agroforestry research in bean based cropping systems: A review of research methodologies and results. In *Proc. of Workshop on Soil Fertility Research for Bean Cropping Systems in Africa, Addis Ababa, September 1988*. CIAT African Workshop Series No. 3. Cali, Colombia: CIAT.

Graf, W., Dessert, K. and Nyabyenda, P. 1987. Résultats et méthodologies des essais d'adaptabilité en milieu rural au Rwanda. *Bull. Agri. du Rwanda* July 1987: 212-16.

INADES. 1988. *Ibishyimbo Bishingirirwa*. Série des Publications de INADES Formation. Kigali, Rwanda: INADES.

ISAR. 1984-89. *Rapport Annuels*. Butare, Rwanda: ISAR.

Lightfoot, C. 1988. On-farm trials: A survey of methods. *Agricultural Administration and Extension* 30(1988): 15-23.

McIntire J. 1989. Survey costs and rural economics research. In Matlon, P. , Cantrell, R., King, D., Benoit-Cattin, M. (eds) *Coming Full Circle. Farmers' Participation in the Development of Technology*. Ottowa, Canada: IDRC.

MINAGRI. 1985. *Resultats de l'Enquête Nationale Agricole 1984*. Kigali, Rwanda: Ministère de l'Agriculture des Eaux et des Forêts.

MINAGRI. 1986. *4me Plan Quinquennal de Développement 1987-1991*. Kigali, Rwanda: Ministère de l'Agriculture et de l'Elevage.

MINAGRI. 1987. *Resultat d'une Enquête sur le Niveau de Commercialisation du Haricot à l'Echelon du Producteur*. Kigali, Rwanda: Ministère de l'Agriculture des Eaux et des Forêts.

MININTER. 1987. *Monographie de la Commune Karama*. Kigali, Rwanda: Ministère des Affaires Interieures.

MINIPLAN. 1988. *Enquête National sur le Budget et la Consommation (Milieu Rural)*. Vol. 2-4. Kigali, Rwanda: Ministère du Plan.

Norman, D.W. 1980. *The Farming Systems Approach: Relevance for the Small Farmer*. Michigan State University Rural Development Papers No. 5. East Lansing, USA: Department of Agricultural Economics, Michigan State University.

Nyabyenda, P. 1986. Le développement variétal du haricot au Rwanda. In *Proc. Seminaire Régional sur la Production et l'Amélioration du Haricot dans les Pays des Grands Lacs, Bujumbura, May 1985*. Gitega, Burundi: IRAZ.

Poate, D. 1988. A review of methods for measuring crop production from smallholder producers. In *Experimental Agriculture* (1988): 1-14.

Rwandanga, D. 1989. *Effet de Différentes Doses de Phosphore sur le Rendement d'une Variété de Haricot Volubile (Gisenyi 2-) en Commune de Ngoma, Préfécture de Butare*. Mémoire. Butare, Rwanda: University Nationale du Rwanda.

Tripp, R. and Woolley, J. 1989. *The Planning Stage of On-Farm Research: Identifying Factors for Experimentation*. El Batan, Mexico: CIMMYT/CIAT.

Voss, J. and Graf, W. 1991. On-farm research in the Great Lakes Region of Africa. In Schoonhoven, A. and Voysest, O. (eds) *Common Beans: Research for Crop Improvement*. Cali, Colombia: CIAT.

Voss, J. (in press). Farmer management of varietal bean mixtures in Central Africa: Implications for a technology development strategy. In Lewinger-Moock, J. (ed) *Sustainability, Diversity and Farmer Knowledge*. New York, USA: Cornell University Press.

World Bank. 1977. *Rwanda: Etude du Secteur Agricole*. Internal Report No. 1-377-Rw. Washington DC, USA: World Bank.

World Bank. 1983. *Rwanda: Agricultural Strategy Review*. Washington DC, USA: World Bank.

Planned Change in Farming Systems
Edited by R. Tripp
© 1991 R. Tripp
A Wiley-Sayce Co-Publication

4

On-Farm Maize Research in the Transition Zone of Ghana

G. Edmeades, A.A. Dankyi, K. Marfo and R. Tripp

This chapter describes an on-farm research (OFR) programme for maize which is part of an agricultural development project in Ghana. The aim of the project is to increase the production of basic grains by strengthening the capacity of national research and extension. Known as the Ghana Grains Development Project (GGDP), it focuses on maize, the most important cereal in Ghana, with cowpea as an additional target crop.

The GGDP was initiated in 1979 and is sponsored by the Government of Ghana and the Government of Canada through the Canadian International Development Agency (CIDA). The executing agency for the Government of Ghana is the Crops Research Institute (CRI), with the Ministry of Agriculture (MOA) and the Grains and Legumes Development Board (GLDB) as cooperating institutions. The International Maize and Wheat Improvement Center (CIMMYT) is the executing agency for the project and has had at least one agronomist from its Maize Program stationed in Ghana since the project started. The International Institute of Tropical Agriculture (IITA) is represented on the project's management committee and since 1985 has had a cowpea breeder/agronomist stationed in Ghana.

The CRI is responsible for crop improvement and agronomic research for the country's most important food crops. Most CRI staff are based at at the institute's headquarters at Kwadaso, near Kumasi in southern Ghana, or at its research station at Nyankpala, in northern Ghana. The GLDB plays a key role in the project through its network of agents based in towns and villages in all the major maize growing regions of the country. The GLDB staff manage adaptive on-farm experiments, organize field days and demonstrations, and market the seed of improved varieties produced by the Board. The MOA has greatly increased its participation in the project in recent years as its extension agents have taken on greater responsibility for promoting newly developed technologies.

An outstanding feature of the project is its national focus. To develop and demonstrate practical recommendations for Ghana's major maize growing areas as quickly as possible by forging stronger research-

Figure 4.1   Agroecological zones of West Africa and the study area

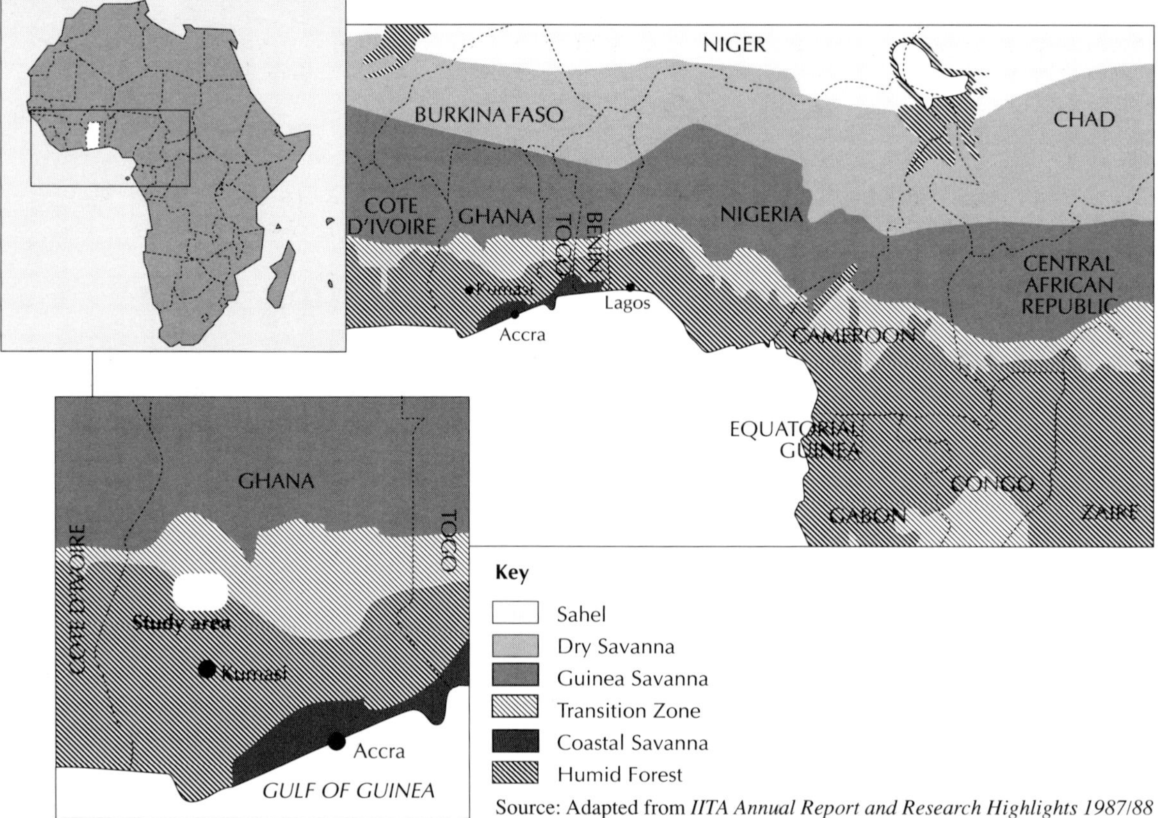

Source: Adapted from *IITA Annual Report and Research Highlights 1987/88*

extension links, an extensive network of researchers and extension agents was trained and organized to manage on-farm experiments and, as recommendations emerged, to conduct field days and demonstrations. This work is monitored and analysed by a group of agronomists, maize breeders and economists from CRI and the GLDB. In any given year there may be 100-150 replicated maize experiments in farmers' fields. These experiments comprise 6-10 basic trial types in two to three replications at each of 6-20 locations throughout Ghana. Trials may be exploratory in nature, or they may be designed to develop response surfaces to inputs or to verify key findings. Annual success rates (the proportion of trials that yield usable results) have ranged from 70% to 86% nationwide. Unreplicated demonstrations, the next step in technology development, are usually conducted at 100-300 sites each year.

The research was originally organized according to agroecological zone, and although there has been a recent trend towards more decentralization and a focus on particular study areas in the country, the project's work is still best thought of in terms of these zones. Maize is grown throughout Ghana, including the coastal savanna, the forest regions of the south, the transition zone between forest and Guinea savanna, and many parts of the northern Guinea savanna. This chapter focuses on maize OFR in the transition zone and on the adoption of new technology in one part of that zone (*see* Figure 4.1).

THE TRANSITION ZONE

The transition zone in Ghana occupies an area of about 45 000 km^2 in a belt stretching from east to west across the centre of the country. It contains few large towns, and although the population is growing rapidly, the total is still thought to be under 1 million people. The total area under maize is about 200 000 ha, including both major and minor cropping seasons. The rainfall and soil characteristics of the transition zone are shown in Table 4.1.

Table 4.1   Mean rainfall totals for major and minor seasons reported from 8 sites over a 20-40 year period, and average soil properties from 36 sites in the transition zone of Ghana

Rainfall		Soils	
Major rainy season (Mar.-Aug.) (mm)	708	Organic matter (%)	1.96
Minor rainy season (Sept.-Dec.) (mm)	435	Cation exchange capacity (meq/100 g)	7.5
		pH	6.4
		Extractable zinc (μg/g)	2.09

Source: Edmeades, 1990

Because of its generally favourable climate, deep and friable soils and relatively less dense forest cover compared to the southern regions, the transition zone has a high potential for maize growing. Much of it is characterized by considerable population growth, as people from more densely settled areas in the north and the south seek land to farm. The proximity to the urban centres in southern Ghana makes it an attractive area for commercial food crop production (Southworth et al., 1979).

At the beginning of the GGDP, an outline of maize production practices in the transition zone was provided by an informal and formal survey conducted in the Kintampo and Mampong-Sekodumasi areas (Bruce, 1980, 1985; Bruce et al., 1980). This initial work was complemented later by further surveys. The survey findings showed that although there were some large commercial maize farms in the transition zone, the vast majority of maize holdings were under 4 ha. Land preparation for the major season was carried out in February or March. Much of it was done by hand, cutting and burning crop residues and weeds, although an increasing number of farmers were taking advantage of tractor hire services. Planting took place with the first rains, usually in April. Most farmers were using local maize varieties at the beginning of the project, but there was growing experience with improved varieties. Planting was done by making holes with a cutlass, planting stick or hoe and the majority of farmers planted in a random arrangement, with the holes about 1 m apart and three or four seeds per hole. Although intercropping was the usual practice with maize in the forest zone, it was less common in the transition zone. When maize was intercropped, it was most often found with cassava. Few farmers used fertilizer on their maize, although most were familiar with it. Maize was weeded once or twice, by hand. The maize harvest for the major season was carried out in August, and farmers sometimes planted again during the minor rains. Crop rotation patterns varied, but fields were sometimes cropped with maize for several years before fallowing.

The initial surveys and field visits indicated that there were a few important priorities for the GGDP to begin work on. Exploratory experimentation carried out in 1979-80 confirmed the importance of variety, weed control, plant population, planting arrangement and fertilization. The almost threefold yield increase resulting from an improved package of practices tested in farmers' fields in 1979 and 1980 was attributed to fertilizer use (54%), improved variety (21%), appropriate plant density (18%) and improved weed control (7%) (Akposoe and Edmeades, 1981). Although all these elements are fundamental to maize production (and formed part of packages that had been recommended by extension programmes in Ghana in the past), several years of careful OFR was needed in order to develop practical recommendations for farmers. The experimentation on monocropped maize in the transition zone is summarized here. A more detailed picture of the type of experimentation carried out in this zone from 1979 to 1984 is given at the end of the chapter, in Table 4.18.

VARIETY

The local maize varieties grown by farmers in Ghana have a long and complex history. Maize was first introduced to the region in the early 16th century by the Portuguese. During the colonial period a number of additional materials were introduced from other parts of Africa and from Latin America. A maize improvement programme was established at Kwadaso in 1956, later becoming part of the CRI. Initial releases were of three improved varieties called Ghana Synthetics. Some varieties, such as Diacol 153, were introduced by foreign aid projects in the 1960s, but were never popular because they were flint-grained instead of dent-grained. In 1968, La Posta, of Tuxpeño origin, was introduced from CIMMYT and quickly established itself as the highest yielding improved variety, later becoming known as CRI La Posta. Other composites were also developed from introduced and local varieties. The best known of these was Composite 4, which was quite widely grown, and Golden Crystal, a yellow-grained variety which was grown in some parts of the transition zone. By 1979, when the GGDP began, the varieties recommended for use in the transition zone were CRI La Posta, Composite 4 and Golden Crystal.

The initial task of the project was to test the wide range of improved materials that were available. In the project's first 3 years, 28 maize varieties were tested in farmers' fields. The number was rapidly reduced to about 10 that were the most promising. The results of 7 years of on-farm testing of these varieties in the transition zone are shown in Table 4.2.

The first three varieties in Table 4.2 are all related to La Posta. They have generally outyielded all other varieties in tests in the transition zone as well as throughout the country, and have shown a good level of stability over sites and years. Ejura (1) 7843, derived from the CIMMYT population La Posta from progeny selected in Ejura; has consistently performed well and was released in 1983 under the name Dobidi. Although results from only one minor season in the transition zone are shown in Table 4.2, the earlier maturing Tuxpeño PB C16 has performed well in the minor season in the forest zone. It became popular for intercropping with cassava, and was released under the name Aburotia. Both Dobidi and Aburotia have since been replaced by streak virus resistant versions of the same general germplasm.

The early experiments with varieties were generally managed by researchers, in farmers' fields, especially when large numbers of varieties were being tested. As the work progressed, farmer-managed variety experiments were conducted whereby farmers were given improved varieties to plant and manage in whatever manner they thought appropriate. The only requirements were that they also planted their own variety in the same field, under the same management, and that they allowed researchers to conduct observations on the plots and to record yields. An example of the yields obtained from these farmer-managed experiments is provided in Table 4.3.

Table 4.2 Grain yield (t/ha at 15% moisture) of selected maize varieties tested in farmers' fields in the transition zone of Ghana during the major rainy seasons from 1979 to 1984 and the minor rainy season of 1982[1]

Variety	1979	1980	1981	1982 Major	1982 Minor	1983	1984
CRI La Posta	5.15	4.38	3.92	4.63	2.21	3.10	4.08
Poza Rica 7843	—	4.95	—	—	—	—	—
Ejura (1) 7843	—	—	4.33	4.73	—	2.39	4.73
Composite 4	4.45	4.23	3.76	4.53	2.24	2.83	4.10
La Maquina 7422	—	4.68	—	—	—	—	—
Poza Rica 8022	—	—	—	—	—	—	4.56
Poza Rica 7822	—	—	4.32	4.59	—	—	—
Tuxpeño PB C16	3.87	—	—	4.04	1.90	—	—
Composite W	4.33	—	4.22	4.87	—	—	—
Suwan 1	—	4.63	3.73	—	—	—	—
Golden Crystal	4.74	4.29	3.70	4.62	1.87	2.88	—
Local	2.63	3.52	2.66	3.54	1.47	1.95	2.02
Mean for location	4.08	4.37	3.91	4.49	1.96	2.69	3.93
LSD (0.05)	0.54	0.47	0.47	0.36	0.46	0.44	0.86
No. of sites	8	9	7	9	5	7	4
No. of entries	8	12	12	12	8	8	8

Note: 1. A dash (—) indicates that the variety was not included in the test in a particular year.

Sources: Akpose and Edmeades, 1980, 1981, 1982; Twumasi-Afriyie and Edmeades, 1983, 1984; Twumasi-Afriyie and Pratt, 1985

Table 4.3 Grain yield (t/ha at 15% moisture) and plant density at harvest of two improved and one local maize variety when grown in farmer-managed variety trials at 28 sites in the forest and transition zones of Ghana, 1985

Variety	Grain yield	Harvest density (plants/ha)
Dobidi (improved)	2.49	27 000
Aburotia (improved)	2.26	27 500
Local	1.81	25 800
LSD (0.05)	0.25	n.s.

Note: n.s. = not significant.

Source: Ampong-Nyarko and Arias, 1986

As the improved varieties were released and farmers gained more experience with them, researchers monitored their acceptability. One of the farmers' main concerns with most improved maize varieties was poor husk cover, which led to bird and insect damage in the field and to significant storage problems as maize is often stored in the husk in cribs. Although the new varieties were clearly recognized as high yielding, the problem of storage remained a handicap. Maize breeding work at CRI has become more oriented towards selecting for improved husk cover, and field research has been initiated on methods for improving farmers' maize storage practices (Dankyi and Donkor, 1986).

The yield advantages of improved varieties over local varieties are clear from Tables 4.2 and 4.3. As shown in Table 4.4, the improved varieties are also more responsive to fertilizer, outyielding the local varieties by 0.76 t/ha under zero fertilizer and by 1.20 t/ha under the higher fertilizer level. Thus the choice of variety is an important factor in improving the efficiency of fertilizer use in Ghana.

Table 4.4　　**Grain yield (t/ha at 15% moisture) of local and recommended maize varieties grown under two fertilizer regimes at 6 sites in Ghana, 1980**

Variety	Fertilizer level	
	Zero	90:60:60 kg[1]
Local	1.71	2.77
Recommended	2.47	3.97

Note:　　1. $N:P_2O_5:K_2O$/ha
Source:　Akposoe and Edmeades, 1981

Breeder seed of the improved varieties was produced by CRI and made available to the state-owned Ghana Seed Company and the GLDB. These two entities then produced foundation seed and commercial seed, selling in a typical year about 300 t and 80 t, respectively, to farmers.

DENSITY AND SPATIAL ARRANGEMENT

One of the most notable findings of the initial surveys and field visits was the low plant populations, which seemed to be the result of at least two factors. First, farmers were used to planting their maize at random, with a large number ōf seeds per hole but often at least 1 m between holes. This practice may not have been too far off the mark for the local varieties and low fertility levels that farmers were accustomed to, but it was not adequate to take full advantage of improved varieties and fertilizer practices. The second factor was the problem of poor stand establishment, caused mainly by birds but also by rodents and soil insects.

The first step was to produce data on adequate plant densities and spacing. An experiment carried out at 10 sites in the transition zone in 1980 examined the interaction between variety and plant density. Three improved varieties, differing in stature and maturity, were planted at three densities (25 000, 50 000 and 75 000 plants/ha). The results, presented in Table 4.5, were used to calculate an optimum plant density for grain yield, which

Table 4.5   **Grain yield (t/ha at 15% moisture) of three improved maize varieties differing in stature and maturity, as affected by plant density when grown at 10 sites in the transition zone of Ghana, 1980**

Variety	Density (plants/ha)		
	25 000	**50 000**	**75 000**
CRI La Posta	3.85	4.35	3.69
Across 7529	3.77	4.26	3.82
Tuxpeño PB C16	3.55	4.34	4.08

Source: Akposoe and Edmeades, 1981

was 40 000 plants/ha for the tall, late maturing CRI La Posta, 42 000 plants/ha for the shorter Across 7529, and 49 000 plants/ha for the very short, earlier maturing Tuxpeño PB C16.

Lodging increased with higher densities and greater interplant competition. This competition resulted in diminished husk cover, leading to increased bird damage. The yield-density relationship showed that a variation in plant density of 30% above or below the computed optimum resulted in yield loss of 5% or less. It was thus considered prudent to recommend densities 10% lower than the computed optimum, to reduce lodging and bird damage. Lower densities would also mean less labour investment for farmers, which was an important consideration for changing planting practices. Taking into account the average stand loss of 25% experienced by farmers, a planting density of about 55 000 plants/ha for the improved varieties was recommended.

Farmers had been used to planting as many as five seeds per hill, and researchers examined the effect of numbers of seeds per hill at different plant densities in several on-station trials. Yields fell only slightly when surviving plants per hill increased from one to two, but the decline became more rapid when the number exceeded two per hill, especially at low plant densities.

Once this information on response to density was established, researchers did not devote more time to additional experimentation on density. Instead, they turned their attention to developing a practical way to improve farmers' planting practices. The major obstacle was the practice of random planting. This was a convenient way of planting, especially on uneven terrain that had been prepared by hand. However, until farmers had a better way of calibrating their plant populations it was unlikely that they would be able to take full advantage of improved varieties, fertilization or weeding practices. Two methods for line planting were demonstrated. The first involved the use of ropes, and although a number of farmers were able to use this method, it involved extra expense and was difficult on fields with many tree stumps. An alternative method was developed using a series of three poles that farmers could sight along in a straight line. With a little practice, farmers found that sighting poles could be used very efficiently. Another problem was measuring the distances between rows and between holes. A simple method was devised taking advantage of the cutlasses that most farmers used in the field. The cutlasses are about 60 cm long, so a recommendation of 90 cm between rows translated into one and a half cutlass lengths, with two-thirds of a cutlass length between holes. This was a simple method for extension agents to explain to farmers.

A more difficult problem was that of stand establishment. Experimental work in 1980 and 1981 examined the effects of overplanting and thinning, seed treatment, planting method (firming or not firming the soil after planting) and seeding depth. The results from the 1980 experiments, given in Table 4.6, clearly showed benefits in stand establishment and yield from deeper planting and firm covering over the seed. Results were less clear in 1981, but it seemed that the seed treatment, Furadan ST, greatly reduced the tendency for birds to attack emerging seedlings.

Table 4.6   Effects of various planting practices on percentage of seedling emergence, plant density at harvest and grain yield (t/ha at 15% moisture) in trials at 5 sites in the transition and forest areas of Ghana, 1980

Treatment	% seedling emergence after 2 weeks	Harvest density (plants/m^2)	Grain yield (t/ha)
Seed number per hill:			
2	70.6	3.37	4.36
4 (thinned to 2)	68.1	4.33[1]	4.72
Seed treatment:			
none	68.8	3.82	4.48
Aldrex T	69.1	3.88	4.60
Planting method:			
cutlass	64.7	3.69	4.48
planting stick	73.3[2]	4.00[2]	4.60
Depth:			
4 cm	60.8	3.55	4.11
8 cm	77.2[1]	4.15[1]	4.97[1]

Notes: 1. Means significantly different at 1% probability level.
         2. Means significantly different at 7% probability level.

Although experimental work has explored other ways in which farmers would be assured of more reliable plant stands, the recommendation most favoured by researchers is that treated seed should be planted at a depth of at least 7 cm and the soil over the seed firmed with the ball of the foot. The availability of appropriate seed treatments, however, has limited the adoption of this practice.

FERTILIZATION

Prior to 1979, several attempts had been made to promote fertilizer use in maize in Ghana. In the early 1970s, for example, a fertilizer project was initiated jointly by the MOA, the United Nations Development Programme (UNDP) and the Food and Agriculture Organization (FAO) of the United Nations. However, fertilizer use in maize fields remained low. Because yield response data for nitrogen (N) and phosphorus (P) were not available for all parts of the transition zone, three N x P factorial experiments were planted in 1980, and six more in 1981 (*see* Table 4.7); nitrogen responses dominated phosphorus responses, and there was no significant N x P interaction for grain yield. It is clear that the sites used in 1981 had a lower natural level of fertility than those used in 1980. The ratio of kg grain produced : kg nitrogen applied was 15 in 1980 and 33 in 1981. When the sites used in 1981 were divided into those which had been cropped for 3 or more years since fallow and those which had been cropped for less than 3 years, the grain : nutrient ratio was 31.3 and 18.3, respectively, showing the increased responsiveness of the more extensively cropped older sites. On the basis of these data, recommendations were developed for reduced or zero fertilizer application on recently fallowed fields and for higher rates on fields which had had a long history of continuous cropping.

Table 4.7   **Grain yield (t/ha at 15% moisture) as affected by four rates of nitrogen and four rates of phosphorus (P_2O_5) when applied to the maize variety CRI La Posta in the transition zone of Ghana, 1980 and 1981**

Nutrient	Grain yield	
(kg/ha)	**1980 (3 sites)**	**1981 (6 sites)**
Nitrogen[1]		
0	4.01	3.07
40	4.61	4.38
80	4.68	4.82
120	4.82	4.83
Phosphorus		
0	4.38	4.05
30	4.46	4.33
60	4.56	4.42
90	4.72	4.31

Note: LSD (0.05) between means of different levels of N or between different levels of P was 0.42 in 1980 and 0.40 in 1981.

Once these experiments had established reasonable levels of nitrogen and phosphorus, attention turned to methods and timing of fertilizer application, which researchers considered to be crucial issues in assuring farmer adoption of fertilizer. Earlier recommendations had included an application of starter fertilizer at planting time followed by sidedressing, but those farmers who used fertilizer tended to apply starter fertilizer late or to make only one fertilizer application. One of the first steps was to examine factors related to the application of starter fertilizer. Table 4.8 (*overleaf*) summarizes the results of experiments over 2 years on timing, placement and type of starter fertilizer.

The results presented in Table 4.8 and those obtained from five additional trials in 1984 showed no significant differences in yields between applying starter fertilizer at planting time or delaying it until 3 weeks after planting. Similarly, there was no difference in yield between a surface application and burying the fertilizer. These results endorsed the farmer practice of combining the starter fertilizer application with the first weeding a few weeks after planting. Combining fertilizer application with weeding requires less labour, and the developing plant canopy and trash from weeding prevent the fertilizer applied near the base of the plant from

Table 4.8   Grain yield (t/ha at 15% moisture) as affected by timing, placement and type of starter fertilizer when applied to the maize variety CRI La Posta in the transition zone of Ghana, 1982-83

Treatment	1982	1983	Mean
Time of application:			
at planting	4.30	3.68	4.07
3 weeks after planting	4.11	3.87	4.02
Placement:			
buried	4.07	3.92	4.02
surface	4.34	3.62	4.08
Form of fertilizer:			
15-15-15 compound	4.37	4.14	4.29
triple superphosphate	4.42	3.58	4.11
ammonium sulphate	3.82[1]	3.60	3.74[1]
Number of sites	7	4	11

Note: 1. Difference between means within factor significant at $p = 0.05$.
Source: Twumasi-Afriyie and Edmeades, 1983, 1984

washing away. In addition, farmers were concerned about the risks of drought or predators at planting and preferred not to apply fertilizer until they were assured of an adequate plant stand.

Recommendations for nitrogen sidedressing were developed partly from trials conducted at six locations in the transition zone in 1980. These trials showed clearly that there was little difference between nitrogen applications at 0, 3 and 6 weeks after planting, but that nitrogen applied at silking was largely wasted. Tassel emergence appeared to be the latest growth stage at which nitrogen should be applied.

WEED CONTROL

Data gathered during the initial surveys and field visits indicated that weeds were a serious problem in maize fields, partly because farmers generally planted a greater area than they could weed effectively. Experimental work was first aimed at developing recommendations for the number and timing of handweedings, and at examining the effectiveness of some of the herbicides that were available in Ghana. Herbicides were rarely completely effective, although the most promising treatment, atrazine + pendimethalin, gave yields equivalent to the handweeded control. Herbicides were examined as a means of reducing labour requirements for weed control to one light handweeding during the crop cycle. Only rarely would they eliminate the need for handweeding. Furthermore, the costs and availability of herbicides seriously affected the possibility of making useful recommendations.

Even experimentation on handweeding was problematic. Variable cropping histories and weed populations made it impossible to produce a single recommendation which was applicable throughout the country's transition zone. It was noted that, on sites which were intensively cropped, weed problems were usually worse than on newly fallowed land, and here two handweedings were found to be necessary, while elsewhere one was sufficient. An extra handweeding would represent a significant investment for farmers and thus they would consider the possible benefits very carefully. The results from weed control experiments were so varied that the weeding recommendations were eventually dropped from the demonstrations given to farmers, in the recognition that all farmers needed to handweed at least once or to use herbicide in combination with handweeding.

This is not to say that weed control in maize cannot be improved. Indeed, it remains as one of the main factors limiting maize yields in Ghana. Judgements regarding weeding are determined by the amount of labour available to the farmer, the condition of the maize field and the length of time that has elapsed since the field was last fallowed. Farmers are aware of the importance of weed competition, but they must balance this against the costs of labour or herbicides. Alternative methods of weed control will be necessary before much progress can be made.

DEMONSTRATIONS

By the end of the 1980 season, GGDP staff felt that they had enough information to begin formulating recommendations for maize in different agroecological zones. These recommendations were tested and demonstrated in what became known as verification/demonstrations (v/d's). A v/d consists of three unreplicated plots, one representing the farmer's practice (Option C) and the other two representing medium technology (Option B) and high technology (Option A) alternatives. The treatments demonstrated in Options A and B of the v/d's planted in the transition zone in 1982 are shown in Table 4.9 (*overleaf*). The plots were each 200 m² and were laid out side by side in a farmer's field.

The extension agent invited neighbouring farmers to the demonstration three times during the growing season, at planting, mid-season and harvest. These field days served to acquaint farmers with the recommended practices and to allow them to judge the results. At harvest, the yields of the three plots were calculated and discussed with farmers. Prior to the field day, the extension agent calculated the costs associated with each technological alternative, and when he had measured the yields at the field day he quickly completed his partial budget and was able to discuss net benefits and costs for each plot with farmers attending the field day. During the early years the v/d's were confined to areas where the GLDB staff were located. Later on, as the MOA became more involved in project activities, coverage widened. The number of successful v/d's carried out in the transition zone increased from 21 in 1981 to 71 in 1982, 77 in 1983 and 93 in 1984, before declining to 75 in 1985.

It is important to point out the dynamic nature of the v/d's. The project had to present recommendations to farmers as quickly as possible, and yet had to be concerned about the adequacy of the data on which they were based. The v/d's tried to address this dilemma in several ways. First, they represented an interaction between the project and the farmers. In the management of field days, extension agents were expected to listen to farmers' reactions and opinions and to incorporate these into their evaluations. For example, as farmers voiced concerns about the management of ropes for line planting, more emphasis was put on demonstrating the use of sighting poles, which farmers found more convenient. Farmer concern about the inadequate husk cover of

Table 4.9   **Comparison of practices in the high (A) and intermediate (B) input options of maize verification/demonstrations in Ghana, 1981-83**

Practice	Option A	Option B
Variety	La Posta	La Posta
Sowing method	Rows, sighting pole, cutlass, firming with foot	Rows, sighting pole, cutlass, firming with foot
Inter-row distance	90 cm	90 cm
Distance between hills	40 cm	40 cm
Sowing depth	7.5 cm	7.5 cm
Seed treatment	Fernasan D	Fernasan D
Seeds per hill	2	2
Thinning	None	None
Starter fertilizer	38:38:38	19:19:19
Starter applied	At sowing[1]	At sowing[1]
Sidedressing of ammonium sulphate at 4 weeks	53:0:0	26:0:0
Handweeding	Twice at 3 and 6 weeks	Once at 4 weeks

Note: 1. Changed in 1984 to 2 weeks after planting.

the new varieties came through clearly at harvest field days, and was a significant factor in reorienting the CRI maize breeding programme.

Second, the v/d's were continually adjusted according to the latest information emanating from the on-farm experimental programme. Until 1984, for example, the v/d's contained two weeding options. Because of a growing realization that a weeding recommendation must be based on site-specific factors, however, this element was dropped from subsequent v/d's. Similarly, the original recommendation for applying starter fertilizer at planting was dropped in favour of an application after emergence, which was more congruent with farmers' practices, caused no apparent yield loss and reduced economic risk.

The results of the v/d's were also submitted to a thorough economic analysis each year, in order to confirm the viability of the technologies being demonstrated and to help adjust recommended fertilizer levels as fertilizer and maize prices changed. The analysis was done using the partial budgeting methods described by Perrin et al. (1979). The acceptable marginal rate of return was set at 100% in 1982 and 1983, reflecting the high rate of inflation and high unofficial interest rates during that period. The results of these analyses are presented in Table 4.10.

Table 4.10   **Maize yields (t/ha at 15% moisture) and economic indices from verification/demonstrations conducted in the transition zone of Ghana, 1982-84**

	1982	1983	1984
Number of sites	71	77	93
Yield, Option A	3.83	2.93	3.73
Yield, Option B	3.20	2.50	3.05
Yield, Option C	1.88	1.58	1.95
Var. costs, Option A[1]	1886	6892	5435
Var. costs, Option B	1592	5876	4158
Var. costs, Option C	694	3314	2467
Net benefit, Option A	6401	18475	3670
Net benefit, Option B	5391	15901	3297
Net benefit, Option C	3470	10673	2829
Marginal rate of return (%)			
Option C -> B	214	204	28
Option B -> A	344	253	29

Note: 1. Currency is Ghanaian cedi.

Source: Twumasi-Afriyie and Edmeades, 1983, 1984

As shown in Table 4.10, there were significant yield advantages with only modest increases in fertilizer applications. Rainfall totals were a little below long-term averages for the major season in 1982, and this was followed by a dry minor rainy season and the worst drought in over 40 years during the major rainy season of 1983. Rains returned to normal in 1984. Under the prices prevailing in 1982-83 the marginal rate of return from adoption of Option B over farmers' practices was well above the cut-off marginal rate of return of 100%. At the same time, fertilizer was extensively subsidized during those years, the cost ratios of maize grain : nitrogen being 0.56 in 1982 and 0.30 in 1983, instead of the more normal ratios of 3-9. In 1984 costs rose abruptly as subsidies for fertilizer were removed (fertilizer prices increased tenfold), maize prices fell because of oversupply and marginal rates of return declined drastically. An additional factor affecting profitability was the fact that many farmers in Ghana engage in share cropping, known locally as *abusua*, whereby one third of the produce may go to the landowner but all costs are borne by the farmer (Robertson, 1987).

The effects of share cropping and increased fertilizer prices were examined for each v/d; the average results of these analyses are presented in Table 4.11 (*overleaf*). It is clear that at grain : nitrogen cost ratios of about 6, Option B is a viable technological alternative for farmers not involved in share cropping. When the ratio rises to 12 or when sharecropping is practised, neither of the recommended technologies holds much attraction for farmers in the transition zone.

The risks associated with each technology were assessed by calculating the proportion of cases in which net benefits fell below zero. In the higher yielding year of 1982, under higher fertilizer costs the risks of obtaining negative net benefits were generally greatest for Option A and least for Option B. In 1983, when drought reduced yields, Option A was the most risky and Option C the least risky.

Table 4.11   Economic indices from verification/demonstrations conducted in the transition zone of Ghana, in the major season 1982-83 under grain : nitrogen cost ratios of 6 and 12, with and without sharing produce on a 2:1 basis

Grain/nitrogen cost ratio	Share cropping	Index	1982	1983
		Number of sites	71	77
6	No	MRR[1] C -> B	132	110
		% net benefits < 0, A	11.3	3.9
		% net benefits < 0, B	5.6	2.6
		% net benefits < 0, C	7.0	2.6
6	Yes	MRR C -> B	55	50
		% net benefits < 0, A	22.5	23.4
		% net benefits < 0, B	16.9	13.0
		% net benefits < 0, C	18.3	13.0
12	No	MRR C -> B	80	58
		% net benefits < 0, A	22.5	24.7
		% net benefits < 0, B	9.8	10.4
		% net benefits < 0, C	15.5	9.1
12	Yes	MRR C -> B	20	11
		% net benefits < 0, A	43.7	44.2
		% net benefits < 0, B	29.6	27.3
		% net benefits < 0, C	25.4	18.2

Note: 1. MRR = marginal rate of return.

ADOPTION

An adoption survey was conducted in May 1986 in the Wenchi, Techiman, Nkoranza and Kintampo agricultural districts of the Brong-Ahafo Region in the transition zone. It was carried out in eight villages which had active extension programmes and where v/d's had been carried out in previous years. In each village, five farmers were randomly selected from the list of people who had attended the v/d, and an additional five farmers were selected from the general population. The objective behind the choice of locations and farmers for the survey was to assess progress in a relatively favoured part of the transition zone, among farmers who were likely to be familiar with the project recommendations. As described below, additional surveys were later carried out in less favoured areas of the country.

Table 4.12 presents a summary of variety adoption, row planting and fertilizer use. In 1986, 81% of the farmers were using an improved variety on at least part of their maize fields; 68% were row planting; and 47% applied fertilizer to at least some of their maize. The adoption of these practices was strongly influenced by cropping pattern, however, with adoption rates being much higher in monocropped maize fields, which had been the initial focus of the project's work. Farmers were asked when they had first started to use each of these

Table 4.12   Adoption of recommended practices[1]

Practice	Ever used (%)	Used in 1986 (%)	Used on largest maize field in 1986 (%)
Improved variety	88.6	81.0	58.2[2]
Row planting	82.3	68.4	57.0
Fertilizer	83.5	46.8[3]	42.9[3]

Note: 1. N = 79.
         2. Farmers who used an improved variety on at least half of their largest maize field.
         3. N = 77.
Source: Tripp et al., 1987

Table 4.13   Farmers' practices in 1986 and attendance at verification/demonstrations

		Farmers following recommended practices (%)		
Attendance[1]	Number	Improved variety	Row planting[2]	Fertilizer[2]
Never	19	78.9	52.6	47.7
1985	41	85.4	51.2	31.7
1984 or earlier	14	78.6	85.7	50.0

Notes: 1. Those who attended a demonstration in 1986 not included in the analysis.
          2. Practice on largest maize field.
Source: Tripp et al., 1987

recommendations; 70% of farmers using improved varieties first began using them in 1983 or later, 63% of those using row planting began in 1983 or later, and 46% of those using fertilizer began in 1983 or later.

One of the purposes of the survey was to examine the effectiveness of the v/d's as an extension tool. Table 4.13 shows that there is little difference in practices between farmers who had attended a v/d and those who had nõt, except for row planting. This lack of correlation between attendance and adoption should not be surprising, however. The v/d's were only one extension activity among many carried out by the GGDP and other agencies, and when farmers were asked how they had learned about the recommendations, extension activities were cited as the most common channel.

The importance of extension is further illustrated by the results of a brief follow-up survey carried out in the same area of the Brong-Ahafo Region in 1989, in villages with no active extension programme. When asked about their practices in 1986 (when the earlier survey had been conducted), farmers recalled a significantly lower rate of adoption of all three recommendations. The use of recommended practices had risen since that time, but the rate of adoption among villages with no extension programme in 1989 was still somewhat lower than that of the villages with extension programmes in 1986.

We now turn briefly to a more detailed examination of the findings of the 1986 adoption survey.

Variety

The new varieties had been widely accepted, with about 50% of the maize area in the study villages planted to improved maize varieties. In addition, over 75% of the farmers using improved varieties had experience in buying commercial seed. The majority of farmers purchased seed the first time they used a new variety, and most of the rest obtained seed from a neighbour or from a demonstration plot.

Prospects for further spread of the new varieties were limited, however, by concerns about storage and cooking quality. Table 4.14 shows that farmers rated the improved varieties as superior with respect to yield (with or without fertilizer), seed quality and resistance to lodging. On the other hand, local maize was rated superior for both storage and cooking. Farmers pointed out that the improved maize was more easily infested by weevils. Although the farmers in this area do not depend to a great extent on maize as a staple food, concerns about cooking quality may be reflected in market prices. Private traders sometimes express a preference for local maize by buying it first or by paying more for it.

Table 4.14   Farmers' opinions on local and improved maize varieties

Characteristic	Number of farmers expressing opinion	Local is better (%)	Improved is better (%)	Same (%)
Yield without fertilizer	66	22.7	74.2	3.0
Yield with fertilizer	63	4.8	93.7	1.6
Lodging resistance	60	28.3	70.0	1.7
Germination	61	3.3	60.7	36.1
Storage quality	72	77.8	13.9	8.3
Cooking quality	55	72.7	23.6	3.6

Source: Tripp et al., 1987

Density and spatial arrangement

The aim of the recommendation for row planting was to improve the spatial arrangement and population of maize fields. The survey showed that those farmers who planted in rows had a higher number of hills per hectare and a lower number of seeds per hill than those who used the practice of random planting.

Field characteristics have a bearing on the decision to row plant. Survey data indicated that older fields and those prepared by tractor were more likely to be row planted. Both these factors are indicative of fields that have fewer obstacles. Row planting does take more time than random planting, if only because more holes per hectare are made, but farmers seemed to manage this method with little difficulty. The survey found no relationship between the type of labour (family or hired) used for planting and row planting. Farmers who said they had difficulty finding labour at planting time were just as likely to row plant as those who said they had no difficulty.

Fertilization

Fertilizer supplies, particularly of ammonium sulphate, had been inadequate in the area. When farmers were able to obtain both fertilizers, most of them made two separate applications (starter fertilizer followed by top dressing), although some farmers mixed the two and applied them at the same time. Timing of application varied, but in general many of the adopters applied the fertilizer later than recommended.

Hardly any intercropped maize received fertilizer, and it is not known if the response of the intercrop to fertilizer would be profitable for farmers. One factor that complicated the comparison of farmers' fertilizer use on inter- and monocropped fields was the fact that intercropped fields tended to be more recently cleared and were thus more fertile. Cropping history was a significant factor affecting a farmer's decision to use fertilizer. As shown in Table 4.15, farmers were far more likely to use fertilizer on an older field than on a newly cleared one, a finding that was consistent with project recommendations.

Table 4.15   Fertilizer use by cropping history of field

Years continuously cropped	Number of fields	Fertilizer applied (%)
0	26	11.5
1-2	17	23.5
3-5	18	55.6
6+	15	86.7

Source: Tripp et al., 1987

The survey found that planting method was also related to fertilizer use. Row planting was more likely to be associated with fertilizer use. One explanation for this might be the relative ease of fertilizer application when maize is planted in rows; another might be the increase in fertilizer efficiency with adequate plant populations.

Adoption sequence

Although 67% of the farmers surveyed had used improved maize varieties, row planting and fertilizer, they had not necessarily adopted all these recommendations at the same time. The data in Table 4.16 (*overleaf*) indicate a step-by-step approach to adoption (Byerlee and Hesse de Polanco, 1986). About 50% of the farmers began by adopting only one of the recommendations, usually fertilization or the improved variety. There is good

Table 4.16   Adoption sequence for farmers who used all three recommendations

Recommendation adopted first	No. of farmers	% of farmers
Fertilizer only	13	24.5
Variety only	10	18.9
Row planting only	4	7.5
Row planting and fertilizer	9	17.0
Variety and fertilizer	4	7.5
Variety and row planting	0	0.0
Variety, row planting and fertilizer	13	24.5
Total	53	99.9

Source: Tripp et al., 1987

evidence that either one of these changes would provide a profitable return to farmers, even if they did nothing else. This is less true for a switch to row planting, and only a few farmers began their adoption in this way.

Of the farmers who adopted two of the recommendations in the same year, most began with row planting and fertilization, which would enable them to profit from the significant interaction of improved plant population with better fertilization. A lower number of farmers began with a combination of the improved variety and fertilization, where an interaction might also be expected. None of the farmers adopted as their first step a combination of improved variety and row planting, where an interaction is probably the least likely.

Adoption studies in other areas

Since the Brong-Ahafo survey, two other adoption surveys have been carried out, one in the transition zone in Ejura District, Ashanti (Dakurah and Arias, 1987) and the other in the forest zone of the Central Region. Table 4.17 compares adoption rates among the three areas surveyed. Ejura is similar to the area surveyed in Brong-Ahafo in that commercial maize growing among smallholders is very common and a considerable amount of extension activity has been carried out there. The lower rates of adoption of GGDP recommendations are therefore somewhat surprising, although there are two likely explanations. First, the survey covered villages served by extension as well as those without extension programmes. Second, the Ejura area has suffered in recent years from Ghana's deteriorating road situation, and it has become more difficult to get inputs into the area and to get maize out to market, which must affect the incentives for adopting new technology.

The results from the Central Region illustrate the importance of different maize cropping systems for technology generation. In the villages studied, maize is almost always intercropped with cassava. Until recently, the GGDP demonstrations have been on monocropped maize. The issue of fertilizer recommendations for such an intercrop is still being researched, and farmers are reluctant to try fertilization. In addition, row planting is more difficult in such an intercrop. Rainfall is also somewhat less reliable in the Central Region. A final point is that maize is a principal staple for farmers in this region, as opposed to Brong-Ahafo where most of the maize is marketed. This means that farmers' taste preferences and storage requirements are different, and the improved varieties currently available are less acceptable.

Table 4.17   A comparison of adoption rates of maize technology in three areas of Ghana

Location	Use improved maize varieties (%)	Use fertilizer on maize (%)	Row plant (%)
Brong-Ahafo Region	81	47	68
Ejura District, Ashanti Region	35	30	52
Central Region	17	3	4

Sources: Dakurah and Arias, 1987; Opoku-Apau et al., 1987; Tripp et al., 1987

CONCLUSION

The GGDP is an example of the application of OFR to a national commodity research effort. A programme of research and extension activities has been managed by a network of staff members from several institutions. The main product has been a set of widely applicable recommendations for the major maize growing areas of Ghana. The research effort was aimed at developing enough information to formulate sound recommendations, while the extension effort emphasized translating those recommendations into language and techniques that farmers found acceptable.

The project has been successful in diffusing improved maize varieties and providing useful information on planting and fertilization methods. These have formed the basis of several types of extension programmes in Ghana and the evidence shows that large numbers of maize farmers have taken up these new practices. Part of the success of the project's strategy has been its concentration on a few basic technological improvements (variety, fertilization and plant population) that are widely applicable to the country's maize farmers.

With this success behind it, maize research and extension in Ghana now faces some difficult challenges. Further increases in maize productivity will probably come from research on issues such as the management of intercrops, control of specific problem weeds and soil fertility maintenance. These issues will demand a combination of more location-specific research and more advanced agronomic research than has characterized the project to date. Such strategies will pull the research service in two directions. Researchers will need to spend more time interacting with farmers and extension agents to acquire local information, and they will also need to invest more time in basic research where there is currently little 'on-the-shelf' knowledge. This course of action is potentially very expensive and research priorities will thus have to be selected with great care.

Lastly, it is necessary to consolidate the gains made by the project to date. The project has led to a reorientation of the extension service and has provided a good example of how extension and research can interact. This momentum needs to be maintained and translated into other commodity research efforts. The project has also provided an example of an efficient system for varietal development and testing. The seed production and distribution system in Ghana, however, has not yet proved to be up to the demands placed upon it and more effort needs to be invested in developing this system. Research and extension now have experience in making recommendations for fertilizer application and adjusting them as prices change. As fertilizer subsidies are removed in Ghana and the fertilizer distribution system is privatized, there is a need to strengthen the links between research, extension and the private sector to promote the most effective use of inputs. The OFR capacity now in place will play an important role in ensuring a more efficient use of the resources of farmers, extension agents, researchers and the input supply system.

Table 4.18   Agronomic maize trials, excluding variety trials, conducted on-farm in Ghana in the transition zone from 1979 to 1983 under the direction of the GGDP

Trial type	Year	Sites	Major factors[1]	Change in recommendation
Factors of production	1979	11	Variety: La Posta/Tuxpeño Weeding: 1/3 Density: 25 000/50 000 Fertilizer : 0/100kgN	Contribution of each factor: emphasize variety, fertilizer and density
	1980	6	Variety: Local/LP or C4 Weeding: 1/2 Density: 25 000/50 000 Fertilizer: 0/90:60:60	Improved package of practices economically attractive
Fertility N and P response	1980	3	N: 0/40/80/120 kg/ha P: 0/30/60/90 kg P_2O_5/ha	Optimal rates N 80-90 kg/ha
	1981	6	As above	Optimal N rate greater on land cropped for >3 years. Optimal P rate 40-60 kg P_2O_5/ha
Times, rates of N sidedressing	1980	6	N: 30/60/90 kg/ha Times: 0/3/6/9 weeks after sowing	Sidedressing must not be applied later than tassel emergence
Times, types placement of starter fertilizer	1982	6	Times: sowing, 3 weeks later Types: 15:15:15/ TSP/ ammonium sulphate. All	Starter fertilizer can be applied by either placement method at 0 or 3 weeks after planting.
	1983	4	balanced at 4 weeks to total 90:60:60 Placement: separate hole/ soil surface	5% advantage in using K- or P-bearing starter fertilizer
	1984	5	Times: sowing, 2/4/6 weeks Placement: on surface/ buried	Results as for 1982-83. Slight advantage in waiting for up to 6 weeks
Response to zinc	1982	11	Zinc: 0/4.2/8.4/12.6 kg Zn/ha Fertilizer: 0/90:60:60	Application of 4-8 kg Zn/ha with starter fertilizer increases yields by 7%
Micro-nutrient and nutrient	1983	9	Copper: 0/13/26 kg $CuSO_4.5H_2O$/ha Zinc: 0/12 kg Zn/ha Phosphorus: 0/90 kg P_2O_5/ha Satellite: 23 kg Mn/ha 0.10 kg Mo/ha as seed dressing	Zinc gives 11% increase increase in transition zone. P increases yields by 16% in areas not receiving regular fertilizer to Mn, Mo or Cu
Variety x density	1980	10	Variety: 3 late varieties Density: 25 000/50 000/ 75 000 plants/ha	Optimum density for late varieties from 40 000-50 000 plants/ha

Table 4.18   (continued)

Trial type	Year	Sites	Major factors[1]	Change in recommendation
Stand establishment	1980	4	Seed number./hill: 2/4 Seed treatment: 0/Aldrex T Method: cutlass, no firming/ stick, firming Depth: 4/8cm	Plant seed with treatment, 8 cm deep and firm over top with foot
	1981	7	As above: seed treatment: 0/Furadan	Furadan gives stand and yield advantage
Weed control	1980	4	8 chemical treatments Handweed control Unweeded	Primagram + handweeding gives best control
	1981	9	12 chemical treatments 2 handweed controls Unweeded	Atrazine + 1 handweeding gives highest yield. Atrazine with pendimethalin best chemical combination, especially for fields with *Rottboelia exaltata*
Weed control x nitrogen	1980	7	N: 0/40/80/120 kg N/ha Weed control: unweeded; 1 hand-weeding at 6 weeks; Primagram 3l/ha; Primagram + handweeding	If fertilizer is to be used, good weed control is essential. N rate optimal between 80-120 kg/ha
Land preparation Zero tillage	1982	5	Slash/no slash Gramoxone: 0.5 l/kg/ha ai Bellater: 0/4 l/ha Handweed: 0/ at 4 weeks Satellite: hoeing + 2 handweedings; scraping + 2 handweedings	Zero tillage with adequate weed control gives same yield as hoeing. Best control with 5 l Paraquat, 4 l Bellater + 1 hand-weeding at 4 weeks, but does not control grassy weeds
	1983	4	Gramoxone + 4 chemical treatments +/- handweeding Gramoxone at 2 times Hoeing + 2 handweedings Scraping + 2 handweedings	Gramoxone 4 l/ha + Primagram 4 l/ha + handweeding at 4 weeks gives yields equal to conventional tillage + 2 handweedings. Pendimethalin + atrazine best residual if *Rottboelia* present
Insecticide (minor season)	1982	4	8 insecticide treatments 1 seed treatment Untreated	Furadan seed or granule treatment controls insects and maize streak
	1983	4	6 insecticide treatments 1 seed treatment Streak-resistant variety with/ without Furadan seed treatment	Streak-resistant variety may double minor season yields. Furadan seed treatment, granules control maize streak, boost yield

Note: 1. Slash (/) within treatments indicates contrast between levels of a factor.

Sources: Akposoe and Edmeades 1980, 1981, 1982; Twumasi-Afriyie and Edmeades, 1983, 1984

References

Akposoe, M.K. and Edmeades, G.O. 1980. *Ghana Grains Development Project First Annual Report 1979.* Mimeograph. Ghana: CIDA.

Akposoe, M.K. and Edmeades, G.O. 1981. *Ghana Grains Development Project Second Annual Report 1980. Part 2, Research Results.* Mimeograph. Ghana: CIDA.

Akposoe, M.K. and Edmeades, G.O. 1982. *Ghana Grains Development Project Third Annual Report 1981. Part 2, Research Results.* Mimeograph. Ghana: CIDA.

Ampong-Nyarko, K. and Arias, F.R. 1986. *Ghana Grains Development Project Seventh Annual Report 1985. Part 2, Research Results.* Mimeograph. Ghana: CIDA.

Bruce, K. 1980. *Maize within the Farming Systems of the Wenchi-Kintampo Agricultural District: Results of an Exploratory Survey.* Mimeograph. Kumasi, Ghana: GLDB.

Bruce, K. 1985. Maize production in Ghana: Agronomic and economic analysis. MS thesis. Fort Collins, Colorado, USA: Department of Agricultural and Natural Resource Economics, Colorado State University.

Bruce, K., Byerlee, D. and Edmeades, G.O. 1980. *Maize in the Mampong-Sekodumasi Area of Ghana: Results of an Exploratory Survey.* CIMMYT Economics Working Paper. El Batan, Mexico: CIMMYT.

Byerlee, D. and Hesse de Polanco, E. 1986. Farmers' stepwise adoption of technological packages: Evidence from the Mexican Altiplano. *American Journal of Agricultural Economics* 68: 519-27.

Dakurah, A.H. and Arias, F.R. 1987. *Cropping Systems in the Ejura District of Ashanti Region, Ghana: Results and Implications of a Formal Study.* Kumasi, Ghana: GGDP.

Dankyi, A.A. and Donkor, S.K.. 1986. *An Informal Survey on Maize Storage in the Mampong/Sekodumasi Area.* GGDP Working paper. Ghana: GGDP.

Edmeades, G.O. 1990. Significant accomplishments of the Ghana Grains Development Project during Phase I, 1979-1983. Paper presented at the Tenth Annual Maize and Cowpea Workshop, 20-23 March, 1990, Kumasi, Ghana.

Opoku-Apau, A., Anchirinah, V.M. and Read, M. 1987. *The Maize-Cassava Intercropping System of the Central Region, Ghana: Results and Implications of a Formal Survey.* Kumasi, Ghana: GGDP.

Perrin, R.K., Winkelmann, D.L., Moscardi, E.R. and Anderson, J.R. 1979. *From Agronomic Data to Farmer Recommendations: An Economics Training Manual.* El Batan, Mexico: CIMMYT.

Robertson, A.F. 1987. *The Dynamics of Productive Relationships.* Cambridge, UK: Cambridge University Press.

Southworth, V.R., Jones, W.O. and Pearson, S.R. 1979. Food crop marketing in Atebubu District, Ghana. *Food Research Institute Studies* 17(2): 157-95.

Tripp, R., Marfo, K., Dankyi, A.A. and Read, M. 1987. *Changing Maize Production Practices of Small-Scale Farmers in the Brong-Ahafo Region, Ghana.* Kumasi, Ghana: GGDP.

Twumasi-Afriyie, S. and Edmeades, G.O. 1983. *Ghana Grains Development Project Fourth Annual Report 1982. Part 2, Research Results.* Mimeograph. Ghana: CIDA.

Twumasi-Afriyie, S. and Edmeades, G.O. 1984. *Ghana Grains Development Project Fifth Annual Report 1983. Part 2, Research Results.* Mimeograph. Ghana: CIDA.

Twumasi-Afriyie, S. and Pratt, M.J. 1985. *Ghana Grains Development Project Sixth Annual Report 1984. Part 2, Research Results.* Mimeograph. Ghana: CIDA

Planned Change in Farming Systems
Edited by R. Tripp
© 1991 R. Tripp
A Wiley-Sayce Co-Publication

5

Alley Farming in South-Western Nigeria: The Role of Farming Systems Research in Technology Development

L. REYNOLDS, C. DI DOMENICO, A.N. ATTA-KRAH and J. COBBINA

In 1978 the International Livestock Centre for Africa (ILCA) set up a research programme aimed at improving the living standards of smallholder farmers in the humid zone of West Africa by raising the productivity of their sheep and goats. The initial work, based on the assumption that the whole farming system would have to change, focused on pasture and improved pasture species that could be used for animal fodder. A new approach was adopted 2 years later, when it became clear that farmers were unwilling to divert resources towards livestock production and away from their primary objective, crop production. Livestock are a minor component of the farming system in the humid zone. Usually, they are left to scavenge for themselves and are kept with minimal inputs of time and other resources from farmers. Consequently, researchers began to seek technologies that built on and worked within the farming system. Their methods of diagnosing the main constraints to crop and livestock production, as well as a description of the research area and the farming system, are presented here.

DIAGNOSING CONSTRAINTS TO CROP AND LIVESTOCK PRODUCTION IN THE HUMID ZONE

To diagnose the livestock production constraints, ILCA used a combination of methods similar to those described by Chambers and Jiggins (1987) and Farrington and Martin (1988). Chambers and Jiggins argued that the large interdisciplinary teams, extensive surveys and cumbersome, multidimensional data analysis that sometimes characterize the diagnostic phase of farming systems research may be avoided by encouraging resource-poor farmers to discuss and articulate their problems at the start of a research programme. They added that farmers should determine research priorities and then design, manage and evaluate on-farm trials. However, this recommendation makes unrealistic assumptions about farmers' breadth of knowledge about various options to existing technologies (Farrington and Martin, 1988). A more practical approach would be for researchers to draw on field observations, discussions with farmers and reports in the literature to suggest possible solutions which may be outside farmers' experiences. Other methods which address the problems identified by Chambers and Jiggins through involving farmers in the diagnostic work and limiting the size of survey teams include the 'design and diagnosis' technique developed by the International Council for Research in Agroforestry (ICRAF) (Raintree, 1986) and the 'rapid rural appraisal' technique (Khon Kaen University, 1987).

The ILCA team consisted of a social scientist, an agronomist and an animal scientist, assisted by two research technicians (agronomist and socioeconomist) based in the area and two enumerators. Although the research issues were identified through farmer surveys, researchers developed their own set of priority research topics through observations of participating farmers. The researchers' experience, together with farmers' comments on the design of alternative technologies, determined which technology would be selected for on-farm testing.

The full ILCA team undertook a rapid appraisal exercise that provided the framework for developing a single-visit questionnaire. This exercise involved the technicians and enumerators, under the supervision of the team leader, a socioeconomist. After the villages that constituted the main on-farm research (OFR) site had been selected, the technicians lived there and established a rapport with the inhabitants that made it possible to record details of household and farm activities daily and provide valuable information to the research team.

The research area and farming system

In selecting the research area, the ILCA team was assisted by a senior extension officer from the Ministry of Agriculture and Natural Resources, Oyo State. The team used aerial photographs, ground data, existing information and field visits to select an area encompassing the villages of Owu-Ile and Iwo-Ate, about 20 km north-east of Oyo town (*see* Figure 5.1). There were both forest and savanna farms in this area and it was far enough from the nearest major town for the inhabitants to be full-time agriculturalists rather than commuters. Despite the rural setting, the villages were served by a regular local transport service and the laterite roads were passable throughout the year.

The population density of the research area is about 190 persons/km^2 (Francis, 1987a). Demographic surveys showed that, together, Owu-Ile and Iwo-Ate consist of 172 households averaging 3.6 adults per household. Women account for 60% of the adult population. In the questionnaire, farming was given as the main occupation of 95% of the men and 29% of the women. Many of the other women gave farming as their secondary occupation, especially when they farmed for subsistence; their primary occupations ranged from processing palm oil and cassava to trading (Francis and Atta-Krah, 1988). Some of the women farmed on their own account, while others worked on their husbands' farms. A typical smallholder in the research area cultivates 2 ha of land by hand, maintains 6 ha under fallow, owns four small ruminants and supports a family of seven people (Okali and Cassaday, 1985).

Rainfall in the area is about 1 300 mm/yr. The rainy season extends from April to October; rainfall distribution is bimodal and allows two cropping seasons each year. The three major staple food crops are yam (*Dioscorea* spp.), cassava (*Manihot esculenta*) and maize (*Zea mays* L.); they are interplanted with numerous minor crops such as cocoyam, pepper, cowpea, tomato, pigeonpea, banana and okra. Tree crops, including oil palm (*Elaeis guineensis*) and cocoa (*Theobroma cacao*), are also common.

Figure 5.1   Agroecological zones of West Africa and the study area

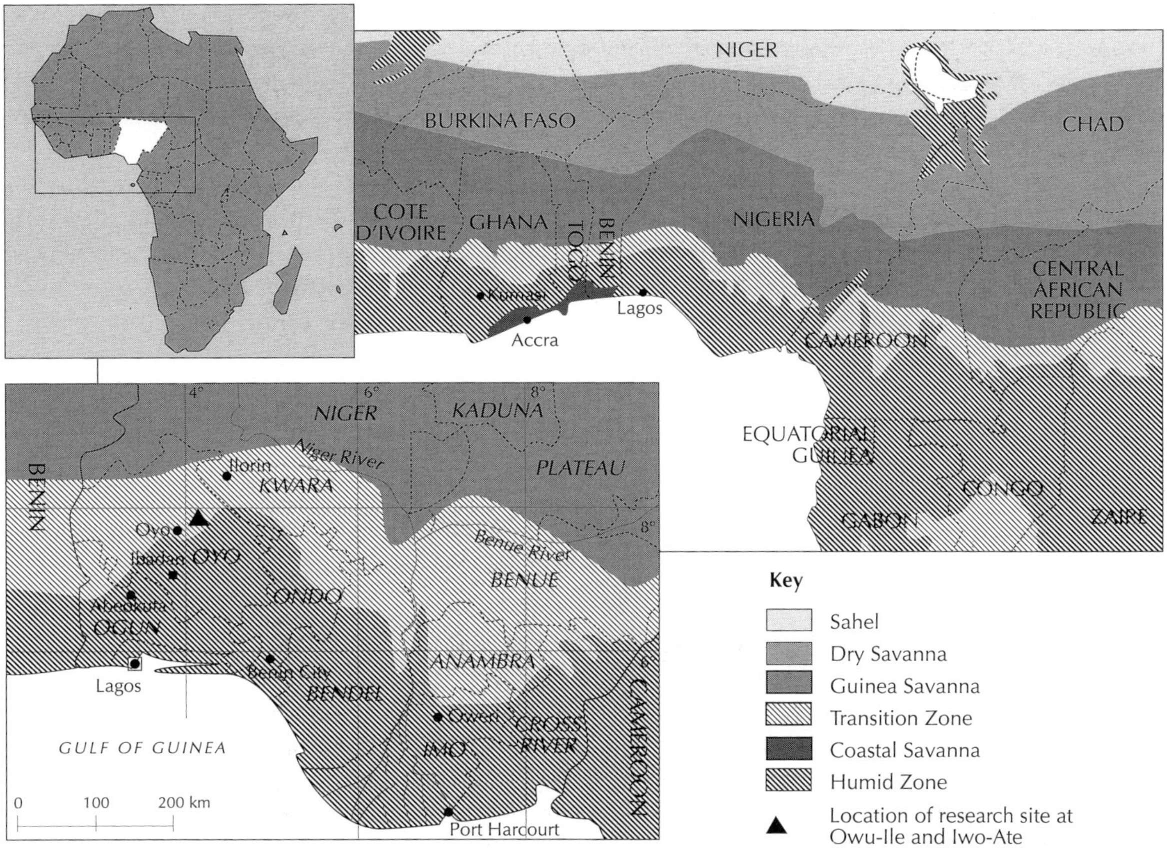

Source: Adapted from *IITA Annual Report and Research Highlights 1987/88*

Mixed cropping predominates, with yam/maize and cassava/maize as the primary combinations. Yam is planted in December and harvested the following September/October. Cassava is planted in June and harvested 18 to 24 months later. Maize is planted in late April/May and harvested in July/August, and a second maize crop is planted to harvest dry in November/December. Newly cleared land will often be planted first to maize, and the maize stover is left in the field as stakes or trellises for yams. Once the yam crop is harvested, cassava is planted and the land left as a 'cassava fallow' that is given minimal attention in the following year.

In south-western Nigeria individual farmers have long-term usufructuary rights, but ownership of land remains vested in the family. Lack of land is not a serious constraint to agriculture. Strangers in an area can be allocated land to farm upon payment of rent. Generally, women obtain the land they farm from their husbands, and widows retain usufructuary rights in the land allocated to them. Although women normally do not inherit land in this region, they may rent and occasionally purchase land for farming (Francis and Atta-Krah, 1988). There are some differences in agricultural practices and conditions between the forest and derived savanna zones. Farm plots in the forest zone tend to be smaller and more scattered than those in the derived savanna.

Fallowing is always practised in the forest zone, but in the savanna some plots are cultivated continuously for many years. Although considerable labour is needed to bring forest land back into cultivation from fallow, fallowing is unavoidable because of the decline in soil fertility under cropping. Farmers say that crop yields are declining and that fertilizer, where it is available, tends to be applied to savanna plots. Spear grass (*Imperata cylindrica*), a rhizomatous weed that thrives in soil that is low in nitrogen and subject to regular burning, often severely affects fallow land in the savanna. Labour for weeding is a greater problem in the savanna than in the forest during the cropping season.

Slash-and-burn farming predominates in the research area, but as human population density increases, the amount of land available for farming diminishes and thus the length of the fallow period decreases as well. The recycling of nutrients during the fallow period is essential to maintaining soil fertility in slash-and-burn agriculture. Leaf litter from the natural regeneration of bushes and trees on fallow land decays to provide soil nutrients and raise soil fertility levels for the following crop cycle. As a result, shorter fallow periods reduce soil fertility from one cycle to another.

The livestock production system

South-western Nigeria suffers from tsetse fly, the vector of trypanosomiasis. Although the incidence of trypanosomiasis has declined as the human population and the amount of cultivated land have increased, indigenous livestock breeds that are tolerant of the disease, such as West African dwarf sheep and goats, predominate. (Cattle are kept only rarely by the indigenous population in the research area.) Some 90% of the households in the research area own small ruminants; over half the owners of small ruminants are women. Herds number from three to five animals. Goats outnumber sheep by about 3:1; farmers say goats are less trouble to manage than sheep, as sheep tend wander away from the village area unless they are watched (Mack et al., 1985).

People keep small ruminants mainly as a flexible financial reserve (sales account for 40% of offtake from the herd). In south-western Nigeria, owners of small ruminants are obligated, if asked, to help relatives or friends establish herds by loaning animals to them in exchange for maintaining the animals. Such loans can account for 50% of offtake (Reynolds and Francis, 1988). Only rarely is an animal slaughtered in the household just for its meat. Ceremonial slaughter of animals at festivals or celebrations, after which the meat is eaten, accounts for 6% of offtake. Livestock are estimated to provide only 5% of farm income; the main income for 90% of the farmers in south-western Nigeria derives from crop production, with trading as the second most important source of income.

Purchased feed and veterinary care are rarely part of small ruminant production in south-western Nigeria (Sempeho, 1985). The conventional livestock system is truly a low input system, in which farmers rely on grasses, bushes and trees to provide forage for free-roaming animals (ILCA, 1988). Household wastes, such as maize chaff and the yam and cassava peels that remain after food processing, are also available for scavenging animals to feed upon. However, in response to increasing pressure on the land and to prevent free-roaming animals from damaging growing crops, local government authorities have started to pass by-laws similar to regulations already enforced in south-eastern Nigeria, which require animals to be tethered or confined, either in the growing season or all year round (Okali and Sumberg, 1985). Farmers therefore have to provide feed for confined animals. Apart from the use of bushes and trees on uncultivated land as a major source of feed, farmers also leave some indigenous tree species on fallow land to help regenerate soil fertility and to provide browse for confined animals (Wahua and Oji, 1987; Kang et al., 1989).

Growth rates are lower and mortality rates higher in young confined stock than in free-roaming animals (Mack, 1983; Reynolds and Adeoye, 1989). The dry matter intake has been estimated to be about 25% lower among confined village animals than among small ruminants on the experiment station, which have growth rates similar to those of free-roaming village animals. Confined animals that rely on forage will consume an inferior diet compared to free-roaming animals, because the latter can select the most palatable species and most nutritious plant parts, whereas the diets of confined animals are limited to what the farmer brings.

Results of initial surveys

The initial surveys revealed that farmers perceived the major constraints to crop production as declining soil fertility, a decrease in land available per capita and the high cost and scarcity of inputs such as fertilizer and pesticide (Francis and Atta-Krah, 1988). Farmers' priorities were clearly linked with crop production rather than livestock production. Previous attempts in the humid zone to interest smallholder farmers in planting herbaceous legumes to regenerate soil fertility on fallow land or to plant grass pasture to provide animal feed had not led to adoption. Labour supplies were insufficient to allow farmers to give much attention to maintaining plants on fallow land, and they did not consider the economics of using that same labour and productive (non-fallow) land for pasture as favourable.

A suitable alternative solution to farmers' problems would incorporate the benefits of fallowing into the cropping period, so that land use intensity could be increased in a sustainable fashion. Multipurpose leguminous trees were considered by the research team to meet those requirements. The trees would improve soil fertility and crop yields and would provide forage for animals. Their management would be similar to accepted farming practices (Okali and Sumberg, 1985) and, like the naturally occurring trees and bushes, they would regenerate soil fertility during fallow periods. For the trees to be acceptable to farmers, it was essential that they could be established in the cropping season as part of normal farming procedures, without reducing crop yields in the establishment year. Any additional labour required to manage them would have to be limited to what might be available from within the normal farm workforce. The trees would then have to remain productive without needing external inputs of fertilizer, while producing nitrogen-rich foliage that would, when used as mulch, also reduce the need for applying inorganic fertilizer to food crops. These conditions, if met, would fulfill the farmers' demands with respect to their major objective — subsistence food production. It was felt that if this objective was addressed, farmers would be more willing to give attention to meeting their secondary objective of improving livestock production.

DEVELOPMENT OF THE ALLEY FARMING TECHNOLOGY

Alley farming is the cultivation of rows of trees and arable crops; the trees are usually leguminous and the crops are planted between the tree rows. The trees are pruned regularly to prevent them from casting too much shade on the crops. The pruned foliage can be used as mulch and organic manure to improve soil fertility and as cut-and-carry forage for ruminant livestock. In 1975 the International Institute of Tropical Agriculture (IITA) began studying tree/crop interactions in alley farming, and in 1980 ILCA initiated research on livestock as a component of alley farming, with the idea that farmers could use a portion of the tree foliage for animal forage. The research effort described in this chapter was able to draw on the much of the work which had been undertaken, as well as on the findings of concurrent work.

It was necessary at the outset of the ILCA programme to determine the biological parameters of the proposed system, in order to ensure that targets could be met for soil fertility, crop yield, tree production, fodder palatability, livestock production and sustainability. This phase of research consisted of trials on the experiment station, conducted partly in concurrence with on-farm studies.

The tree species *Leucaena leucocephala* had been shown by IITA to be suitable for inclusion in arable areas (Kang et al., 1984) and its potential for animal feed was well established (NAS, 1977). Apart from working with *Leucaena*, ILCA also investigated *Gliricidia sepium*, for two reasons. First, there was some concern about the effects that consuming high levels of *Leucaena* might have on small ruminants, because it contains the toxin mimosine; second, researchers did not wish to rely on only one tree species and sought the additional flexibility that an alternative would provide (Sumberg, 1985). Trees suitable for alley farming must establish easily, have a deep rooting system, grow fast, tolerate frequent pruning, have a high biomass productivity and be palatable to livestock. *Gliricidia*, which had been used in West Africa as a shade tree for cocoa and was familiar to farmers, appeared to meet these criteria. A number of new *Gliricidia* accessions were collected in Central America and their productivity under frequent pruning was tested at Ibadan, Nigeria. The most productive line, which yielded 40% more than the local variety, was selected for inclusion in the alley farming trials (Atta-Krah et al., 1986; Atta-Krah and Sumberg, 1987).

At the start of the on-farm trials, alley farms were established using *Gliricidia* stakes, but it was clear that for most of the trials the bulk, weight and quantity of stakes needed would be a serious constraint. As a result, researchers directed their efforts towards establishing the trees from seed and began to collect the seed they would need for their trials. *Gliricidia* flowers on mature tree branches in the middle of the dry season, and seed is available from February to the start of the rains in late March. Pods must be collected when they are still green, as they shatter when dry. Establishment from seed is now the recommended practice for farmers.

Experiment station trials have clearly demonstrated the fertilizer substitution effect of trees in alley farms. At a spacing of 25 cm within rows and 4 m between rows, *Leucaena*, if pruned every 6-8 weeks in the cropping season, produces 6-8 t of leaf and soft stem dry matter per hectare, whereas *Gliricidia* yields 5-7 t/ha. If 75% of the prunings are used for mulch on the soil surface, maize grain yield can rise by 40%, from 1.5 t/ha to 2.2 t/ha. Incorporating tree prunings in the soil increases the efficiency of inorganic fertilizer, doubling crop response (Atta-Krah et al., 1986; Sumberg et al., 1987). Foliage from the first pruning, applied as mulch just before a maize crop is planted, has the greatest effect on crop yield, with mulch from subsequent prunings having a progressively smaller effect as the maize crop matures. Clearly, if tree foliage is used as animal feed it cannot be used for mulch, but second and third prunings would have a lower opportunity cost and, in the short term, would be more valuable as feed (ILCA, 1989).

The beneficial long-term effect on soil nutrient status of tree prunings applied as mulch has been demonstrated (Kang et al., 1984; Atta-Krah et al., 1986). The microclimate is also modified by the hedgerows and the mulch applications, which reduce soil temperature and increase moisture infiltration and retention (Lal, 1975). As a result of improved soil organic matter levels, the activity of microbes and earthworms is greater in alley farm plots (Yamoah et al., 1986; Kang et al., 1989).

Most work on the experiment station to date has concentrated on maize as the companion food crop in alley farming, but trials with yam, cassava, cowpea and rice have also been conducted (Kang et al., 1984; Budelman and Pinners, 1987; Ngambeki and Wilson, n.d.). Sole cropping has generally been practised in experiment station trials, although mixed cropping is the conventional practice in farmers' fields.

In a long-term trial that has continued since 1982, alley farm plots, with or without a 2-year fallow after 4 years of cropping, have been compared with conventional treeless plots (Atta-Krah, 1990). On the alley-farm plots maize yields were maintained during the cropping period, while yields on the conventional plots tended to decrease with time (*see* Table 5.1). In the first year after fallow, maize yields on the alley farms were 40% higher than those on continuously cropped alley plots, 20% higher in the second year, and not significantly different in the third or fourth years.

Table 5.1 **Maize grain yield (t/ha) from *Leucaena* alley farms with and without fallow: totals for two cropping seasons per year[1]**

	1983[2]	1984	1985	1986	1987	1988	1989[2]
Continuous alley farming	2.26	4.60	4.21	3.36	4.13	2.90	3.02
Alley farming with fallow 1983-84	—	—	5.54	4.15	4.81	3.24	—
Alley farming with fallow 1985-86	2.29	4.16	—	—	6.14	3.35	3.03
Continuous no-tree farming with fallow 1987-88	1.96	3.28	3.06	2.10	—	—	4.20

Notes: 1. Adapted from ILCA (1989, 1988, 1986) and ILCA, Ibadan (unpubl. data). Inorganic fertilizer (45 kg N/ha/yr) was given as a basal treatment during cropping years, except in 1984 when 60 kg N/ha was applied.
2. First season crop yield only.

The ILCA team also explored options for farmers who owned small ruminants and wished to place more emphasis on their livestock enterprise. The team developed tree/grass and tree-only systems that could produce fodder and supplement the feed obtained from alley farms. A hedgerow spacing of 4 m and four rows of guinea grass (*Panicum maximum*) within the *Leucaena* alleys produced 20 t forage DM/ha, but over 3 years, with no manure or inorganic fertilizer input, the yield declined (Atta-Krah and Reynolds, 1987). Tree-only plots of *Leucaena* without grass produced up to 37 t DM/ha in the first year of pruning, depending on inter-row spacing and cutting frequency, but again the forage yield declined to about 15 t DM/ha in the absence of returned nutrients. A cutting cycle of 12 weeks and an inter-row spacing of 1 m was recommended as the most suitable arrangement for farmers (ILCA, 1988).

Leucaena and *Gliricidia* browse (1:1, w/w) was given to pregnant and lactating sheep and goats at levels of inclusion ranging from 0% to 50% of the daily dry matter intake, as supplements to a basic diet of guinea grass ad-libitum and cassava peels (50 g DM/day). The major effect of this diet was seen in the survival rates of the offspring, which ranged from 40% without browse to 100% at the highest level of inclusion. In addition, growth rates up to weaning and 6 months of age improved as the quantity of browse in the diet increased (*see* Table 5.2 *overleaf*), and parturition intervals declined. The response of sheep to supplementary feeding was almost twice as great as that of goats (Reynolds and Adediran, 1988; Reynolds, 1989a).

When the sheep and goats grazed directly on grass/tree areas, they stripped the bark from the trees, reducing the productivity of the trees and killing some of them. Therefore, a cut-and-carry feeding system was recommended for smallholder farmers. Researchers also recommended that trees in feed gardens (plots of trees alone or trees and grass) should be protected from free-roaming animals, especially in the first year of growth.

Table 5.2   **Research station trial results on effects of supplementary *Leucaena* and *Gliricidia* browse on the growth and survival rates of West African dwarf goats and sheep**

| Species | Browse intake (g DM/day) | | Growth rate (g/day) | | Survival to 24 weeks |
	Dam[1]	Offspring[2]	Weaning[3]	24 weeks	
Goats	143	39	17.4	14.0	0.36
	254	83	28.7	20.1	0.46
	554	160	25.9	20.9	0.82
	719	246	31.9	28.3	0.94
Sheep	0	0	39.0	25.4	0.50
	120	34	46.7	30.7	0.62
	239	77	57.2	34.0	0.70
	441	136	66.3	44.5	0.89
	741	250	84.0	50.3	1.00

Notes: 1. During the final 2 months of pregnancy up to weaning.
        2. From weaning to 24 weeks.
        3. Weaning at 12 weeks for lambs and 16 weeks for kids.

Source: ILCA, 1988

In an economic assessment of alley farming, Ngambeki (1985) showed that *Leucaena* alleys could increase maize yields by over 60%, reduce the need for nitrogenous fertilizer, and give an attractive net income and marginal rate of return per unit cost, with cost-benefit ratios between 1.23 and 1.32. A preliminary economic assessment by Sumberg et al. (1987) indicated that alley farming with maize was more profitable than the conventional production system incorporating a 3-year fallow, but that the advantage of alley farming decreased as the price of maize rose relative to the cost of labour. They also concluded that the inclusion of small ruminants in alley farming would be profitable if the output of livestock products rose by 25-30% as a result of supplementary feeding. Additional economic analyses are in progress, taking into consideration new data from trials on the experiment station and in farmers' fields.

IMPLEMENTATION OF ON-FARM TRIALS

In 1981, ILCA staff established two alley farms on farmers' fields near Badeku, 30 km east of Ibadan. In 1982 they helped two farmers in Fashola, 60 km north of Ibadan, to establish alley farms (one each in the forest and savanna zones); a farmer near Badeku also established an alley farm that year with help from ILCA. One of these early farms was planted on fallow and no food crops were included. It was worked collectively, with the labour organized by an established farmers' cooperative. However, it was difficult to mobilize labour effectively as the idea of improved fallow was unfamiliar to farmers and they were unwilling to invest labour in the enterprise. These exploratory on-farm trials indicated that the most effective management unit was the household and that farmers preferred to integrate browse trees into arable crop farms (Atta-Krah and Francis, 1987).

In 1983, 12 alley farms were established on farmers' fields in the Fashola and Badeku areas with joint researcher-farmer participation. Although the farmers managed the alley farms imaginatively and successfully, other farmers in the area did not adopt the system. Atta-Krah and Francis (1987) suggest three reasons for this. First, ILCA's continued, if reduced, involvement in managing and monitoring the trials did not encourage other farmers to regard the technology as one to be adopted without similar assistance. Second, the programme was oriented toward individual farmers rather than the community. Third, there was no extension involvement.

Barlow et al. (1986) argue that progressive farmers are more likely than other farmers to accept new technologies and will act as role models for the rest of the community. ILCA's experience of working with individual progressive farmers provided no indication that the rest of the community was likely to follow the example set by a few persons and adopt a complex technology such as alley farming. The researchers concluded that a community approach was needed and that, to be effective, it must be accompanied by an extension effort.

Establishing the pilot research project

The pilot research project established in 1984 gave farmers control of and responsibility for their alley farms. ILCA's input was limited to providing tree seed and giving advice on planting and management. The research area was chosen to avoid the possibility that participants in the pilot project would have close contact with farmers who had alley farms at Fashola or Badeku and would thus expect more input from researchers. Owu-Ile village was selected after discussions with the chief and the village elders, followed by a general meeting involving the whole village population. The adjacent village of Iwo-Ate asked to be included in the pilot project at an early stage and the ILCA team agreed. This community approach was strengthened by the presence of an extension agent, posted to the villages from the local State Ministry of Agriculture to assist the ILCA technician based in the villages (Atta-Krah, 1985; Atta-Krah and Francis, 1987).

At the first explanatory meeting, farmers were asked to state their problems, and the team members explained how alley farming could help overcome those problems. In some areas farmers recalled unfortunate experiences with government-sponsored forestry projects in which land that farmers had allocated to plant trees was lost to cropping once the trees matured and their canopies closed. It was important to assure farmers that the land was still theirs and would remain theirs. This question surfaced at every meeting and farmers demanded continual reassurance. Although this initial suspicion, which persisted for several years, seems now to have been allayed, the point must be made that researchers should be continually receptive to farmers' concerns.

The farmers were then taken to visit some of the alley farms established in 1982-83 at Fashola and to talk to the individuals who had worked them. It was felt that farmers should not be brought to see alley farms on the research station at this stage, but should rather see farms similar to their own. After these visits, 78 farmers indicated their willingness to establish alley farms. The ILCA team then visited the plots these farmers intended using to assess the suitability of the land; some farms were rejected on the basis of poor farm management or intense shading from existing tree crops, leaving 68 farmers involved in the project. The ILCA technicians arranged discussions and demonstrations of tree planting on farmers' fields. These activities gave farmers practical experience in laying out and establishing alley farms. Issues such as planting depth, seeding rate and seed spacing were emphasized and seed was distributed. The extension agent and research technician were available to give advice on planting, but the work was the farmers' responsibility. Although 12 farmers failed to establish alley farms, during the year 12 new farmers volunteered to participate, so that by the end of a year 68 alley farms were in existence.

During the establishment year, senior technicians visited the alley farms and farmers once a fortnight. An ILCA postdoctoral agronomist coordinated monitoring and evaluation to determine which activities had occurred since the previous visit. Farms were scored on the basis of tree condition, weediness, the type and condition of the companion food crop and the general tidiness of the farm. Management practices were also recorded. The team used these scores to form conclusions about the suitability of specific crops with which the trees could be integrated and about management practices.

It was evident that a mature cassava crop reduced the intensity of light available to the trees, diminishing their growth, and that the creeping vines of melons or yams could smother tree seedlings. However, monitoring revealed that, despite a slow start, trees planted with mature cassava or yam did survive and gradually reached a productive stage. Direct comparisons between farmers were difficult because all farmers practised mixed cropping, with any permutation of two or three out of six crops planted with the trees. Not surprisingly, the other critical factor in the success of the alley farms was weeding. Although the food crop was weeded first, the tree seedlings benefited as well. However, hired labourers were unfamiliar with the tree seedlings and on a number of occasions farmers found that many of the trees had been cleared out together with the weeds. Researchers recommended that the labourers should be closely supervised or, alternatively, that farmers should weed along the tree rows before hiring labour to weed the rest of the field. In general, the better farms in terms of condition and food crop yield had better tree establishment (Atta-Krah and Francis, 1987). The entire research team visited all 68 alley farms at 6-month intervals to look mainly at tree establishment and the general condition of the farms.

During the year in which the pilot alley farms were being established, ILCA staff and the local extension agent gave demonstrations of weeding around tree rows, pruning and mulching to groups of about 12 farmers on the farmers' own fields. To understand the farmers' views on the technology, throughout this period researchers held discussions with them, observed their activities, noted any modifications in the recommended management method and closely watched the effects. It was observed that enthusiasm for alley farming was generally higher in Iwo-Ate, where the farmers themselves had initiated contact with ILCA.

As with the experiment station trials, in the on-farm trials *Gliricidia* initially grew faster than *Leucaena* and was less prone to attack by termites and rodents. When *Gliricidia* was planted in the early part of the first rains, it was necessary to prune the trees by the time the second season food crop was planted to prevent them from shading the food crop. *Leucaena* did not require pruning until it was 12 months old.

In the second year, researchers organized additional pruning and mulching demonstrations for the farmers, who could now expect to see some benefits to food crops on the more mature and better established alley farms. In addition, foliage was now available for supplementary feeding of livestock. Because some farmers were over enthusiastic about animal feeding, possibly because of their perceptions of researchers' objectives, they pruned trees too frequently, using small branches rather than waiting for branches to reach 1.5 m in length. However, in the second year only a few farmers used any foliage from alley farms to feed animals.

During 1984 it became apparent that women farmers were under-represented in the alley farming group when compared to the farming community as a whole. Although only 29% of the women had said their primary occupation was farming, most of them were actively engaged in working on the land, often on land allocated to them by their husbands, who expected the women to produce food for their families. An investigation was started, with a female researcher based full time in the village, to find out why women were not planting alley farms and to encourage their participation (Cashman, 1986; Youdeowei, 1986). It appeared that, in general, decisions to plant trees were taken by men, and thus only widowed or divorced women planted without referring to men. It also emerged that the village meetings, demonstrations and discussions were dominated by and oriented towards men. Women did not approach the male research and extension staff for advice but had fewer inhibitions about consulting another woman. By the end of the second year, the proportion of new plantings by women reflected the proportion of women in the farming community. Some of the best farms were managed by women, but overall there was little difference between men's and women's farms. This response by women indicated that alley farming was a technology suitable for men or women to adopt (Francis and Atta-Krah, 1989).

Progress of the research project

As the above discussion shows, during the first 2 years research efforts concentrated on studying farmers' attitudes and adaptations, rather than on collecting biological data of the kind already available from experiment station trials. In subsequent years, more emphasis was placed on obtaining the biological data needed to determine the economic impact of alley farming at the village level.

Farmers' degree of involvement in making decisions on how to manage the trials affected the results of the economic analysis. Ashby (1987) has shown that when farmer participation involved autonomous decision making in managing trials, crop yields under identical experimental conditions were lower than when farmers and researchers consulted together. Conclusions on the economic viability of a technology will therefore depend upon the degree of farmers' involvement in the test. Ideally, to provide an accurate picture of the impact of a technology on productivity, evaluations should be made under the conditions that will prevail after researchers have withdrawn from an area. This requirement creates difficulty in obtaining data on which to base statistical comparisons. Quantitative yield data on arable crops were therefore obtained from on-farm plots where researchers had worked closely with farmers, so that crop patterns and work activities could be standardized.

In 1986-87 standardized trial plots were superimposed on 10 farms to determine tree productivity and on 15 farms to study effects on soil fertility and crop yields. Labour requirements were measured on the basis of the farmer's own time and that of the farm family, together with hired labourers supplied by ILCA. Data were collected from farms established in 1984 and from newly established farms which were laid out to allow statistical comparisons to be made. The proportions and quantities of *Leucaena* and *Gliricidia* foliage used for feed and the effects of supplementary feeding on livestock are now being studied with 20 alley farmers and 20 conventional farmers. As with all trials where control rests with the farmer, sometimes the sequence of events must be modified to meet farmers' priorities and timing.

Field days were organized by ILCA and the local extension agent so that participating farmers could see what was possible on the better managed farms and see the work on the experiment station. During one field day, participants visited trial farms where 100%, 50% or 0% of the tree foliage was being used for mulch. This visit gave an excellent demonstration of the benefits of high-nitrogen mulch on maize, for the crops showed clear gradations in height and leaf colour. Benefits to animal performance were more noticeable in the experiment station trials, in which the amount of browse fed to animals varied more widely than in the villages.

In both Owu-Ile and Iwo-Ate, the alley farms had been established in forest and savanna areas. The benefits of alley farming were clearly greater in the savanna areas, where hedgerow establishment was better and tree biomass production 90% higher than in the forest areas, despite a lower soil nutrient status. The most likely reason for this was the amount of shade on the *Leucaena* or *Gliricidia* trees cast by the tall forest trees. Mean alley farm size was 0.13 ha, and each farm required 1.9 h of labour to plant from seed. Management time on all alley farms in the year of establishment was low, as weeding was part of the normal operations for the companion food crop.

Over a 2-year period a study was carried out on the labour requirements of four mature alley farms that had produced crops for several years before the start of the study. These farms were compared in an on-farm trial with four adjacent conventional farms without leguminous trees. The conventional farms had been under cultivation for only 1 or 2 years before data collection started. Weeding time was not recorded as farmers performed this task at varying times in the absence of researchers. Excluding the labour required for weeding, the labour needed to produce a maize crop on an alley farm was 285 h/ha compared to 260 h/ha for a conventional farm, but the variability between farms was such that the differences were not significant. The additional time required for pruning (21 h/ha) and for spreading the mulch (12 h/ha) was compensated for partly by a reduction in the time needed for land preparation (clearing and ridging) before the crop could be planted (*see* Table 5.3). Farmers reported that fewer weeds were found on alley farms, indicating that less labour would be required for weeding than on conventional farms. The better farmers were weeding twice in the first season, at 4-5 and 8-10 weeks after planting and were saving labour on each occasion. During the second season, one weeding was the norm, unless rainfall was exceptionally high. The less conscientious farmers weeded only once, irrespective of season.

Table 5.3   **Labour requirements (h/ha) for maize cultivation in farmer-managed trials in south-western (values shown are means for 4 cropping seasons on 5 farms)**

	Alley farm	Conventional farm
Land clearing and ridging	191.7	234.9
Pruning and mulching	67.5	—
Planting	25.4	25.1
Total[1]	284.6	260.0

Note: 1. Data were not collected on labour used for weeding.
Source: Reynolds, 1989b

These labour requirement figures contrast with those reported by Ngambeki (1985) from trials on the experiment station at higher tree and maize planting densities, using casual labour. Ngambeki found that alley farming required 50% more labour than conventional farming to produce a maize crop, with weeding taking 175 h/ha in conventional plots compared to 105 h/ha in alley plots. This finding confirms farmers' reports to the ILCA team on how trees in alley farms affected weed growth. On the experiment station, the use of hired labour for pruning took twice as long per tree than when farmers in Owu-Ile or Iwo-Ate did the work themselves. The economic assessment reported by Ngambeki therefore tends to undervalue the advantage of alley farming.

In savanna areas, where spear grass becomes a severe problem on fallow land, the growth of this grass is more limited on land fallowed after alley farming than on land fallowed after conventional farming. Data are available from only two pairs of farms, but 46% less time was needed to clear land on an alley farm in the savanna after a 2-year fallow than on an adjacent conventional farm. In an area where spear grass was less abundant, the alley farm took 21% less time to clear than the conventional farm after a 2-year fallow. When weeding time is included, the labour requirements for crop production are no greater, and possibly less, on an alley farm than on a conventional farm (ILCA, 1987a).

Tree productivity (leaf and soft stem) on farmers' plots was 5.3 t DM/ha/yr, a high value because farmers had not cut the trees for forage during the preceding dry season. In a study of four alley farms that had been established for 5 years, the effects of using different proportions of tree foliage for mulch or forage were measured. Without mulch, maize during the first season yielded 2.1 t/ha compared to 3.1 t/ha with a full application of mulch (*see* Table 5.4), a response similar to that recorded on the experiment station. The best farm was producing 7.3 t of tree foliage DM/ha during the period of the trial, which was a higher yield than had been achieved from any alley farm trials on the experiment station. Response to mulch at the 50% application level was greater than the incremental effect observed with the additional mulch on the 100% application plots. This finding is at variance with the linear response to mulch application reported by ILCA (1987b) in experiment station trials at three levels of inorganic fertilizer and indicates that using some foliage for animal feed need not drastically reduce crop response. Experiment station trials have shown that mulch applied at the time of crop planting has the largest effect on crop yields; however, similar studies have not been done on farmers' fields.

Table 5.4   **Effect of mulch application on maize grain yield in researcher-managed, farmer-executed alley farm trials (first season 1989), in the fifth cropping year**

Farm	Tree foliage yield (t/ha)	Maize grain yield (t/ha)		
		0% mulch	50% mulch	100% mulch
Farm no. 1	5.78	1.75	3.15	3.19
Farm no. 63	2.80	2.71	2.54	2.59
Farm no. 70	5.50	1.53	2.22	2.97
Farm no. 72	7.28	2.57	3.49	3.47
Mean	5.34	2.14	2.85	3.06

Source: ILCA, Ibadan (unpubl. data)

Data collected in 1987, after only 3 years' experience with the technology, indicated that 12% of the tree production was being given to livestock and the rest used for mulch. Cut-and-carry browse was given to animals on 8.8 days/mo, at a rate of 128 g edible dry matter per animal per occasion (ILCA, 1988). In 1989, a study of supplementary feeding among alley farmers and conventional farmers showed that the most common supplement provided by conventional farmers was cassava peels, followed by household wastes (*see* Table 5.5 *overleaf*). The use of *Leucaena* and *Gliricidia* foliage by conventional farmers occurred mainly in the first month of the trial during the dry season but decreased to less than 1 day/mo for the remainder of the year. The community as a whole traditionally has access to land under fallow, and anyone can harvest tree or bush regrowth for wood or forage. Preliminary results indicate that the growth rate of goat kids belonging to alley farmers is higher than the growth rate of kids belonging to conventional farmers. The picture is complicated because alley farmers have larger herds and each animal receives smaller quantities of cassava peel and household wastes than animals of conventional farmers.

Table 5.5 **Quantity (kg/occasion) and frequency (days/month) of supplementary feed given over 12 months to goats by alley farmers and conventional farmers in south-western Nigeria**

Supplements	Owu-Ile village		Iwo-Ate village	
	Alley farmers	**Conventional farmers**	**Alley farmers**	**Conventional farmers**
Leucaena and *Gliricidia*	2.90[1]	5.05	4.72	4.10
	(5.3)[2]	(2.5)	(10.1)	(2.5)
Grass	0.60	3.20	0.39	0.0
	(<0.1)	(0.1)	(0.2)	(0.0)
Indigenous browse	1.76	0.77	3.72	4.30
	(0.8)	(0.1)	(0.8)	(0.1)
Cassava peel	2.08	2.02	3.83	2.91
	(13.2)	(17.2)	(13.8)	(18.4)
Household wastes	2.09	2.39	3.63	3.36
	(12.0)	(14.3)	(7.8)	(11.8)

Notes: 1. Quantity offered.
             2. Values in parenthesis show frequency of feeding.
Source: ILCA, Ibadan (unpubl. data)

FEEDBACK FROM ON-FARM RESEARCH

The research described above clearly indicates ways in which the results of OFR helped refine the alley farming technology. For example, some of the farmers' practices were modifications of the technology recommended by researchers, who set up trials on the experiment station to investigate the effects of those modifications. Other aspects of farmer response to the new technology gave researchers a better idea of the social impact of the alley farming and the likelihood of adoption.

In the traditional OFR sequence, responsibility for trials passes in stages from researchers to farmers. In the initial stage of research, when researchers direct farmers' actions, only biological data can be collected. A technology's influence on the social organization of farm households cannot be determined until farmers are free to organize cropping and input patterns themselves. Trials on the experiment station or researcher-managed trials in farmers' fields will not provide information about the social impact of a technology, as such trials tend to focus only on a single portion of farmers' activities. When conventional experimental designs are used to produce results that show statistical differences between treatments, the constraints placed upon farmers will invalidate any social interpretation of the trial results. Specific investigations are necessary to study the effects of interventions on the social life of a village, especially in the case of a technology as complex as alley farming.

Land tenure and availability

One of the most urgent issues that the ILCA team needed to investigate was how alley farming might be affected by land tenure arrangements and agreements about using the trees on alley farms. Although land is not scarce in south-western Nigeria, at an early stage of research the team raised the question of the effect of land rights and the land tenure system on the adoption of alley farming. Planting trees has long-term implications for land use. If trees are regarded as permanent crops, people who have only temporary rights to land may be less enthusiastic about planting trees or may even be prohibited from doing so by landlords. A prospective alley farmer requires access to land on which he or she has the right to plant trees, and the farmer's rights to the land must be sufficiently secure to justify the effort of planting such a long-term crop. Furthermore, the farmer must be able to plant arable crops with the trees to benefit from the system's ability to maintain or improve soil fertility and must have the right to harvest and use foliage from the trees to obtain an adequate return on the investment (Francis 1987a, 1987b).

Noronha (1985) reports that, as land becomes increasingly scarce in Africa, people sometimes plant tree crops to affirm title to land and to lengthen the period of control over land. Introducing tree crops, such as cocoa, results in a significant change in the individual appropriation of land. However, there is no evidence yet that such a change is happening in south-western Nigeria as a result of planting leguminous trees in alley farms, possibly because the trees are pruned regularly and so take on the characteristics of shrubs rather than trees.

The first on-farm trials in south-western Nigeria were planted on land obtained mainly through inheritance, although a few plots had been obtained through purchases, gifts, leases or loans. Farmers who rented land had to ask their landlords' permission before planting trees, but Francis (1987a) found no cases of landlords refusing permission. Thus access to land does not appear to be a constraint to the adoption of alley farming in south-western Nigeria. A study of the impact of tenure systems on the adoption of tree-based technologies in Nigeria, Cameroon and Togo is in progress.

Participation by women

Another social issue in alley farming is the participation of women, many of whom, in addition to engaging in agricultural work, also own small ruminants. Women farming on their husbands' land usually sought their spouses' permission before planting alley farms. Widows established farms on the land of their late husbands. In general, however, women's participation in alley farming in south-western Nigeria seems to have been determined less by land tenure factors than by competing demands on women's time and the initial participation of village-based male technicians and extension agents in promoting the technology. In 1984, 12 women planted alley farms compared to 56 men, although more women than men were actively involved in farming.

The presence of a female researcher in the village for one year (1985) to study women and alley farming increased women's participation. Encouraged by the researcher, women planted 23 alley farms in 1985; in all, women planted 50% of the alley farms established that year. However, when the female researcher left the village at the end of 1985, most of these adopters could not identify with the broader group of alley farmers and eventually stopped alley farming. By 1989 only three women from the 1985 group were cultivating alley farms.

Targeting women as a special group to the exclusion of other members of the community is therefore not a long-term solution to persuading women to adopt the technology, unless women farmers receive specific attention from the establishment period until the benefits of alley farming become apparent. On the other hand, all 15 women alley farmers who, with male farmers, responded to the community approach in 1984-86 still cultivate their alley farms. The lesson to be learned is that although a female extension agent may be better able to influence women farmers, a more effective strategy in the long term is to integrate the activities of both male and female extensionists through a community approach to men and women farmers.

Agronomic practices

Feedback from biological studies was also valuable to the development of alley farming. As *Gliricidia* stakes had been found to be unsuitable for establishing trees on a wide scale, researchers investigated the collection and use of *Gliricidia* seed on the farm. A method of establishing *Gliricidia* from seed was introduced to the farmers and extension agents are now developing seed production plots to increase seed availability.

Smallholder farmers plant most crops at a lower density than researchers use in on-farm trials. Trials on the experiment station over a 2-year period showed that the highest tree foliage yield was obtained by spacing trees 8 cm apart. Yield per tree declined with close spacing but yield per hectare increased (Atta-Krah and Sumberg, 1988). However, researchers considered that tree mortalities would be higher if trees were planted so closely, and thus a spacing of 25 cm was recommended to farmers. In practice, however, many farmers often plant trees 30-60 cm apart, with a subsequent reduction in the potential foliage yield.

Spacing between tree rows was very variable on farmers' fields (2.5 m to 6 m). A space of 4 m was recommended on plots cultivated by hand, but for farmers in Fashola who hired tractors for ploughing, this gap was found to be too narrow. They were advised to increase the space to 5 m.

The traditional method of clearing land at the end of the dry season involves starting fires to clear crop residues and trash, but it was not certain how the farmers would modify this practice on the alley farms. The practice had not been tested on the experiment station and researchers viewed with apprehension the activities of one farmer who moved dry material to the middle of the alleys and set it alight. The leaves and soft stems of the trees suffered some scorching but the woody material was unharmed. It was later observed (unintentionally) on the experiment station that fire could clear a *Leucaena* and *Gliricidia* feed garden of weeds, leaves and soft stems, so that the entire plant above ground was scorched, but that within 3 weeks fresh green shoots reappeared. The trees were therefore capable of withstanding the fires used by farmers to clear land before the start of the rainy season.

Livestock feeding

Researchers paid considerable attention to how the farmers allocated foliage for mulch and supplementary feed. It was originally believed that farmers would use most of the foliage pruned from cultivated trees for mulch in the wet season and forage in the dry season. In practice, however, farmers used varying proportions of foliage throughout the year for feeding. This observation led to the experiment station studies (described above) on the effect on the food crop yield of removing all or part of the mulch at different pruning times. These studies gave an early indication of the opportunity cost of using tree foliage as feed compared to applying foliage as mulch.

As animal confinement becomes more common in south-western Nigeria, farmers are realizing that they cannot satisfy the demand for mulch and animal feed from the relatively small alley farms (0.1-0.2 ha) established so far. Some animal owners have responded to legislation on livestock confinement by disposing of their animals (Mutsaers, pers. comm.). Others have accepted the new situation and rely on material from fallow and spare land, together with household waste, as the main sources of animal feed. This new situation offers an entry point for improving forage production by planting feed gardens.

Feed gardens are designed to provide the maximum quantity of animal feed from the minimum amount of land. The loss of soil nutrients from the feed gardens could be partially balanced by returning manure from confined animals to the plots, but in practice farmers have not yet used the manure. In Owu-Ile, Iwo-Ate and surrounding villages some farmers have resorted to partially confining their animals, building pens in which to keep them overnight. In 1988 ILCA technicians planted a demonstration feed garden in Iwo-Ate and in 1989 farmers planted eight tree-only feed gardens to supplement the forage obtained from alley farms. One large feed garden of about 0.1 ha was established by a cooperative farmers' group in a village near Iwo-Ate. This is an indication that farmers have started to take the livestock component of the farming system more seriously.

Another response to the limited amount of feed available from alley farms when most of the tree foliage was applied as mulch was to give supplementary browse to animals sporadically, as and when it became available. Experiment station trials using this system indicated that, at the level of supplementation used on the farm, there was no significant difference between sporadic and continuous feeding, although animal performance continued to be lower in the sporadic treatment group (ILCA, unpubl.). Another limited feeding system tested on the experiment station involved providing supplementary browse to female animals at different stages of the reproductive cycle. Supplements given to the dam during lactation tended to have a greater effect on kid growth rates than supplements given during pregnancy, but the differences were not significant at the level of feeding used on-farm (ILCA, 1989). Farmers could supplement particular animals in a partial confinement system by allowing those animals into the pens first and giving them first access to supplements.

Tree management

Some of the earliest alley farms were planted on land that was reaching the end of the normal cropping period and would have gone into fallow the following year, because farmers were uncertain of the benefits of the technology and were 'hedging their bets'. When little effect on crop yield was seen in the establishment year, a few farmers allowed the alley farms to remain fallow for the following 2 years. It was observed that the trunk girth of trees on the fallow land was thicker than the girth of trees that had been pruned regularly. On farms where trees were pruned too frequently, trunk girth was small and tree productivity reduced. A few farmers used tree foliage from fallow alley farms to feed livestock. As a result of these observations, an experimentation station trial was started to study the effect of pruning intervals on tree foliage yields in the absence of a companion food crop. Pruning intervals ranged from 3 to 12 months, periods that would be relevant only to trees on fallow land as the shade produced from a 3-month regrowth would reduce crop yield. The highest forage yield was obtained with a first pruning for the year after a coppice growth period of 9 months (January to October), followed by a further 3 months' regrowth from October to January before the second pruning (ILCA, 1989).

An exception to this approach was provided by a farmer in Fashola. In 1982 he planted an alley farm using only *Leucaena* on 1.5 ha (*Gliricidia* was locally abundant and he was already using the foliage as supplementary feed for small ruminants). The *Leucaena* alleys were cultivated by tractor. In the dry season the farmer's cattle grazed on the crop residues in the alley farmed plot and browsed on the *Leucaena*. To reduce the growth of spear grass the farm was left in fallow for 1 year (1984), after which the land was returned to cultivation. This farmer obtained seed to plant another large alley farm once the rains started in 1990, again mainly for providing forage.

MODIFICATIONS IN FARMERS' BEHAVIOUR AND LESSONS FOR RESEARCH

In 1989, 134 household heads with alley farms and 169 who had not adopted alley farming were interviewed. These interviews covered almost all household heads who had worked directly with ILCA and many of those in villages adjacent to the pilot study area. The non-adopters were selected randomly from the same villages as the adopters. The response rate was 90% for the alley farmers and 80% for non-adopters. The two main reasons for failure to respond were that male heads of households were absent or that household members were too busy.

The distribution of respondents is shown in Table 5.6. Some 56% (75) of the alley farmers lived in the pilot villages of Owu-Ile and Iwo-Ate, and most of these farmers claimed to have been alley farming for 4-5 years. In three other villages — Badeku, Fashola and Ikire — where ILCA had introduced alley farming to individuals or had contributed inputs to other projects, 18 farmers who still practised alley farming were interviewed. In villages within a 20-km radius of the pilot study area, where spontaneous adoption had occurred, 41 adopters were interviewed.

Table 5.6   Alley farmers and conventional farmers interviewed in south-western Nigeria

Location	Alley farming households	Conventional farming households	Total households
Pilot/research villages	75 (56%)	80 (47%)	155 (51%)
Villages where ILCA personnel had input	18 (13%)	49 (29%)	67 (22%)
Villages adjacent to research sites with less ILCA input	41 (31%)	40 (24%)	81 (27%)
Total	134	169	303

Source: ILCA, Ibadan (unpubl. data)

In socioeconomic terms, the alley farmers resembled conventional farmers in a number of ways (*see* Table 5.7). Both groups were likely to have received no formal education (81% of alley farmers, 72% of conventional farmers), but alley farmers were more likely to have been involved in a training course (31%, as opposed to 12%, mainly because of the training ILCA organized for alley farming). More male alley farmers owned the land they farm than was the case with conventional farmers (87%, as opposed to 76%), and 70% of both groups had inherited rights to the land they used. The position of women farmers (11% of heads of households) was rather different in that they had been allocated land by their husbands or by the community. Some 10% of alley farmers used land allocated by the *Baale* (chief), compared to 18% of conventional farmers, and the latter were twice as likely to be making monetary rental payments for land. Alley farming households were also more likely to be involved with a cooperative farmers' group than were conventional farming households (82%, as opposed to 66%).

The survey recorded sex of household head rather than sex of the alley farmer. In most cases, the household head and alley farmer were the same person, although sometimes husbands responded on behalf of their wives. Nevertheless, the proportion of women alley farmers in the sample would not differ much from the 11% of female respondents in the survey. This finding confirms earlier comments in this chapter about the need for female extension agents to work alongside their male counterparts so that the benefits of alley farming are also accessible to women.

Table 5.7   **Characteristics of alley farmers and conventional farmers surveyed in south-western Nigeria, 1989**

Characteristic	Alley farmers		Conventional farmers	
	No.	%	No.	%
Total number	134	100	169	100
Male-headed households	114	85	161	95
Education:				
none	109	81	122	72
primary/Koranic	20	15	39	23
above primary	5	4	8	5
Other training	42	31	21	12
Main source of income from crops	121	90	154	91
Farming own land	116	87	127	75
Land inherited	96	72	120	71
Land owner if not the farmer:				
Baale (chief)	13	10	31	18
other	5	4	11	7
Rental payments:				
monetary	11	8	31	18
in kind	7	5	11	7
Household head/wife is member of farmers' co-operative	110	82	112	66

Source: ILCA, Ibadan (unpubl. data)

The main characteristics of alley farms and conventional farms were somewhat different (*see* Table 5.8 *overleaf*). One-third of the alley farms were in fallow or cassava fallow at the time of the survey. As noted earlier, cassava fallow occurs when cassava is planted on land approaching the end of the cropping cycle; normally, in the year after planting, cassava is not weeded and the farm receives no other attention, so the land acquires the appearance of a fallow plot. Cassava may be harvested during the year or may remain in the ground until the following year. Mature cassava extracts few nutrients from the soil.

All alley farmers cultivated conventional plots as well, but those interviewed cultivated fewer of them than conventional farmers. However, the number of conventional farms in fallow or cassava fallow was rather higher among alley farmers (2.52, as opposed to 2.02). A surprising number of both conventional farmers (14%) and alley farmers (10%) stated that they had no conventional plots in fallow. Most successful alley farmers (77%) had plots within easy walking distance from the village, closer to home than the conventional plots.

Alley farmers tended to plant fewer crop species on their farms. Some 19% had planted only tubers, compared to less than 1% in the case of conventional farmers. The alley farmers who grew only cereals (7%)

Table 5.8   **Characteristics of farms belonging to alley farmers and conventional farmers**

Characteristic	Alley farmers	Conventional farmers
Total farmers	134	169
Number of alley farms per farmer:		
cultivated	0.80	—
fallow/cassava fallow	0.39	—
Number of conventional farms per farmer:		
cultivated	1.42	1.78
fallow/cassava fallow	2.52	2.02
Alley farm located less than 3 km from house	105 (78%)	—
Conventional farm located less than 3 km from house	36 (27%)	65 (38%)
Major crops:		
tubers only	26 (19%)	1 (<1%)
cereals only[1]	10 (7%)	—
tuber + cereal	56 (42%)	25 (15%)
tuber + cereal + other	18 (13%)	54 (32%)
Fertilizer applied to:		
alley farm only	4 (3%)	—
conventional farm only	116 (87%)	162 (96%)
both types of farm	5 (4%)	—
neither	9 (7%)	7 (4%)

Note: 1. Researcher-managed, farmer-executed trials.

Source: ILCA, Ibadan (unpubl. data)

were collaborating with ILCA in quantitative crop yield studies. Data were not available on whether alley farmers cropped their alley farms differently from their conventional farms. Of the conventional farmers interviewed, 32% grew tuber/cereal/other crop mixtures, compared with 13% of alley farmers.

There were considerable differences among alley farmers in the use of inorganic fertilizer. Overall, 96% of alley farmers applied fertilizer to their conventionally farmed plots, but only 9% still applied inorganic fertilizer to the alley plots, with little difference between forest and savanna areas. In recent years, the cost of inorganic fertilizer has risen dramatically in Nigeria, and alley farming is recognized by farmers as an innovation that helps reduce input costs.

Almost all alley farmers (90%) reported that they used tree foliage mainly for mulch and feed, although measurements of foliage use on-farm show that 85-90% is used by farmers for mulch. Some 78% of the alley farmers also stated that they had enough browse for their animals, and only 11% were willing to consider planting a feed garden. Apart from mulch and feed, additional uses of *Leucaena* and *Gliricidia* included the production of stakes to support crops such as yam (49%) and firewood (28%). The last point is rather surprising as wood is not scarce in the rural humid zone. Furthermore, after *Leucaena* and *Gliricidia* leaves have dried and dropped for mulch or been eaten by livestock, the wood that remains is rather spindly. Substantial amounts of wood can be obtained only after the trees have grown uncut for 9 months or more, and this is practicable only if the trees are growing on fallow land. Four relatively new alley farmers who took part in the interviews were still waiting for the benefits to materialize once the trees were sufficiently mature to start pruning, and so were unable to comment.

Farmers perceived the main benefits of alley farming to be the provision of forage and mulch, apparently placing rather more emphasis on the forage (*see* Table 5.9). However, that response contrasts with the reported use of tree foliage and probably reflects what farmers believed enumerators from a livestock organization wished to hear. Some 17% of alley farmers had more than one alley farm, and a further 51% said that they intended to plant another or to extend the size of their alley farms in the 'near future'. Only 19% of the alley

Table 5.9   Alley farmers' perceptions and uses of alley farms

	Alley farmers	
	No.	**%**
Number of alley farmer respondents	134	100
Trees used for:		
mulch only	4	3
feed only	7	5
mulch + feed	47	35
mulch + feed + other	69	51
Perceived benefits:		
better soil fertility/crops	39	29
better animal performance	55	41
better soil/crops/animals	31	23
Intend to plant more alley farms	69	51
Willing to plant a feed garden:		
alley farmers	15	11
conventional farmers (N = 169)	9	5
Sufficient browse available for livestock	105	78
Cultivated tree species preferred by farmer:		
Leucaena	88	66
Gliricidia	3	2
Leucaena + Gliricidia	32	24
Reason for preference:		
liked by animals	101	75
grows fast	7	5
other	7	5

Source: ILCA, Ibadan (unpubl. data)

farmers were satisfied with the amount of land under alley farming at present, and 9% said they would have increased the number of alley farms if more land had been available. This reaction contrasts with that of the 18 farmers in various villages who once practised alley farming but had stopped. The main reasons they gave for stopping were that they had not been able to establish the trees successfully (56%) or had lost interest (33%).

Some 15% of the alley farmers did not wish to use the technique on their other farms, mostly because of previous failures and the amount and cost of labour required. Those who complained of the amount of labour involved in alley farming (9%) tended to have alley farms at some distance from the village, some more than 4 km away. Almost 50% of the alley farmers reported that the labour needed for pruning was the most difficult aspect of alley farming, although direct measurements indicate that the total time needed for cropping on an alley farm may be less than on a conventional farm. However, with the presence of trees the distribution of labour requirements changes. Less time is needed to prepare land and to weed, but extra labour is required to prune the trees on two or three occasions during the life of a maize crop. Trees are pruned once before ridging, so that the prunings can be incorporated in the soil for maximum effect. If the farmer follows the recommended agronomic practices and weeds 3-4 weeks after planting the maize crop, with a second weeding 6-8 weeks later, the second pruning occurs between those two dates. If the farmer does only one weeding 6-8 weeks after planting the food crop, as often happens, the weeding will coincide with the need to prune the trees to prevent them from shading the growing food crop.

Thus, a farmer following the recommended practices should not have a labour problem. In the pilot study area, where farms were close to the homestead it was easier to make use of family labour, especially labour provided by children after school. The use of family labour has increased in recent years as the cost of hiring labourers and tractors has risen. Women farmers, who were often involved in trading as well, said they could not use the profits from trading to cover the costs of hired labour and so had to rely increasingly on their families for labour.

CONCLUSION

Planting leguminous trees in alley farms to maintain and improve soil fertility has proved to be biologically viable and socially acceptable to smallholder farmers in south-western Nigeria. Farmers' involvement at early stages in the development of the technology was particularly important for ensuring that researchers did not specify inappropriate input requirements. Moreover, on-farm trials with farmer management were essential for assessing the social acceptability and suitability of this complex technology.

Within the geographic limits of the pilot study area, land tenure has not proved to be an obstacle to the adoption of alley farming (investigations are under way to broaden the base of information on land tenure). The adoption of alley farming by women would certainly be facilitated by including female extension staff to work alongside their male counterparts, dealing with the village community together. Alley farming requires sustained input from extension to ensure that farmers remain interested in the technology until the biological benefits become apparent — until hedgerows mature and foliage for mulch or feeding becomes available in sufficient quantities to affect crop yields and livestock performance.

Alley farming helps farmers to overcome what they perceive as the major biological constraint — maintaining soil fertility and crop yields without expensive external inputs. Alley farmers have minimized the use of inorganic fertilizer on alley farms while still applying fertilizer to conventional plots. After 5 years' experience with alley farming, a number of farmers have accepted the need for forage plots to augment the animal feed obtained from their alley farms. A major change in attitude has thus occurred among these farmers, for now both the crop and livestock components of their farm enterprises benefit from the inclusion of leguminous trees in the farming system.

References

Ashby, J.A. 1987. The effects of different types of farmer participation on the management of on-farm trials. *Agricultural Administration and Extension* 25: 235-52.

Atta-Krah, A.N. 1985. Developmental approach to on-farm research. In Norblom, T.L., Ahmed, A. El K.H. and Potts, G.R. (eds) *Research Methodology for Livestock On-Farm Trials*. Ottawa, Canada: IDRC.

Atta-Krah, A.N. 1990. Alley farming with *Leucaena*: Effect of short grazed fallows on soil fertility and crop yields. *Experimental Agriculture* 26: 1-10.

Atta-Krah, A.N. and Francis, P.A. 1987. The role of on-farm trials in the evaluation of composite technologies. *Agricultural Systems* 23: 133-52.

Atta-Krah, A.N. and Reynolds, L. 1987. Utilisation of pasture and fodder shrubs in the humid tropics of West Africa. In *Sheep and Goat Meat Production in the Humid Tropics of West Africa*. FAO Animal Health and Production Paper No. 70. Rome, Italy: FAO.

Atta-Krah, A.N. and Sumberg, J.E. 1987. Studies with *Gliricidia sepium* for crop-livestock production in West Africa. In Withington, D., Glover, N. and Brewbaker, J.L. (eds) Gliricidia sepium *(Jacq). Walp.: Management and Improvement*. Hawaii, USA: NFTA.

Atta-Krah, A.N. and Sumberg, J.E. 1988. Studies with *Gliricidia sepium* for crop-livestock production systems in West Africa. *Agroforestry Systems* 6: 97-118.

Atta-Krah, A.N., Sumberg, J.E. and Reynolds, L. 1986. Leguminous fodder trees in the farming system. In Haque, I., Jutzi, S. and Neate, P.J.H. (eds) *Potential of Forage Legumes in Farming Systems of Sub-Saharan Africa*. Addis Ababa, Ethiopia: ILCA.

Barlow, C., Jayasuriya, S.K., Price, E., Maranam, C. and Roxas, N. 1986. Improving the economic impact of farming systems research. *Agricultural Systems* 22: 109-25.

Budelman, A. and Pinners, E.C.M. 1987. The value of *Cassia siamea* and *Gliricidia sepium* as *in situ* support systems in yam cultivation. In Withington, D., Glover, N. and Brewbaker, J.L. (eds) Gliricidia sepium *(Jacq). Walp.: Management and Improvement*. Hawaii, USA: NFTA.

Cashman, K. 1986. *Women's Activities and the Potential of Alley Farming in Southwest Nigeria*. Report to the Ford Foundation. Ibadan, Nigeria: Humid Zone Programme, ILCA.

Chambers, R. and Jiggins, J. 1987. Agricultural research for resource-poor farmers. Part II: A parsimonious paradigm. *Agricultural Administration and Extension* 27: 109-28.

Farrington, J. and Martin, A.M. 1988. Farmer participatory research: A review of concepts and recent fieldwork. *Agricultural Administration and Extension* 29: 247-64.

Francis, P.A. 1987a. Land tenure and agricultural innovation. *Land Policy Use* 4: 305-19.

Francis, P.A. 1987b. Land tenure and the adoption of alley farming in southern Nigeria. In Raintree, J.B. (ed) *Land, Trees, and Tenure*. Madison, Wisconsin, USA: ICRAF/Land Tenure Center, University of Wisconsin.

Francis, P.A. and Atta-Krah, A.N. 1988. Institutions, resources and land management in southern Nigeria. Draft paper submitted for publication.

ILCA. 1986. *Annual Report, 1985/1986*. Addis Ababa, Ethiopia: ILCA.

ILCA. 1987a. *Final Report to the Federal Livestock Department of the Federal Military Government of Nigeria*. Ibadan, Nigeria: Humid Zone Programme, ILCA.

ILCA. 1987b. *Annual Report, 1986/1987*. Addis Ababa, Ethiopia: ILCA.

ILCA. 1988. *Annual Report, 1987*. Addis Ababa, Ethiopia: ILCA.

ILCA. 1989. *Annual Report, 1988.* Addis Ababa, Ethiopia: ILCA.

Kang, B.T., Reynolds, L. and Atta-Krah, A.N. 1989. Alley farming. *Advances in Agronomy* 43: 315-59.

Kang, B.T., Wilson, G.F. and Lawson, T.L. 1984. *Alley Cropping: A Stable Alternative to Shifting Cultivation.* Ibadan, Nigeria: IITA.

Khon Kaen University. 1987. *Proc. of 1985 Conference on Rapid Rural Appraisal.* Khon Kaen, Thailand: Rural Systems Research and Farming Systems Research Project, Khon Kaen University.

Lal, R. 1975. *Role of Mulching Techniques in Tropical Soil and Water Management.* Technical Bulletin No. 1. Ibadan, Nigeria: IITA.

Mack, S.D. 1983. *Evaluation of the Productivities of West African Dwarf Sheep and Goats in Southwest Nigeria.* Humid Zone Programme Document No. 7. Ibadan, Nigeria: ILCA.

Mack, S.D., Sumberg, J.E. and Okali, C. 1985. Small ruminant production under pressure: The example of goats in southeast Nigeria. In Sumberg, J.E. and Cassaday, K. (eds) *Sheep and Goats in Humid West Africa.* Addis Ababa, Ethiopia: ILCA.

NAS. 1977. *Leucaena: Promising Forage and Tree Crop for the Tropics.* Washington DC, USA: National Academy Press.

Ngambeki, D.S. 1985. Economic evaluation of alley cropping leucaena with maize-maize and maize-cowpea in southern Nigeria. *Agricultural Systems* 17: 243-58.

Ngambeki, D.S. and Wilson, G.F. n.d. *Economic and On-Farm Evaluation of Alley Cropping with* Leucaena leucocephala. Ibadan, Nigeria: IITA.

Noronha, R. 1985. *A Review of the Literature on Land Tenure Systems in Sub-Saharan Africa.* Discussion Paper ARU 43. Washington DC, USA: World Bank.

Okali, C. and Cassaday, K. 1985. *Community Response to a Pilot Farming Project in Nigeria.* Discussion Paper No. 10. Boston, USA: African-American Issues Center, Boston University.

Okali, C. and Sumberg, J.E. 1985. Sheep and goats, men and women: Household relations and small ruminant development in southwest Nigeria. *Agricultural Systems* 18: 39-59.

Raintree, J.B. (ed) 1986. *An Introduction to Agroforestry Diagnosis and Design.* Nairobi, Kenya: ICRAF.

Reynolds, L. 1989a. Effects of browse supplementation on the productivity of West African dwarf goats. In Wilson, R.T. and Melaku, A. (eds) *African Small Ruminant Research and Development.* Addis Ababa, Ethiopia: ILCA.

Reynolds, L. 1989b. Involving farmers in integrated crop-livestock research: Lessons from alley farming. Paper presented at the Winrock/SADCC Workshop on Integrating Research to Improve Agricultural Productivity and Natural Resource Management, November 1989, Lilongwe, Malawi.

Reynolds, L. and Adediran, S.O. 1988. The effects of browse supplementation on the productivity of West African dwarf sheep over two reproductive cycles. In Smith, O.B. and Bosman, H.G. (eds) *Goat Production in the Humid Tropics.* Wageningen, Netherlands: PUDOC.

Reynolds, L. and Francis, P.A. 1988. The effects of PPR control and dipping on village goat populations in southwest Nigeria. *ILCA Bulletin* 32: 22-27.

Reynolds, L. and Adeoye, S.A.O. 1989. Planted leguminous browse and livestock production. In Kang, B.T. and Reynolds, L. (eds) *Alley Farming in Humid and Sub-Humid Regions of Tropical Africa.* Ottawa, Canada: IDRC.

Sempeho, G. 1985. Sheep and goats in Nigerian agriculture. In *Socio-Economic Studies on Rural Development.* Vol. 59. Gottingen, F.R. Germany: Edition Herodot.

Sumberg, J.E. 1985. Collection and initial evaluation of *Gliricidia sepium* from Costa Rica. *Agroforestry Systems* 3: 357-61.

Sumberg, J.E., McIntire, J., Okali, C. and Atta-Krah, A.N. 1987. Economic analysis of alley farming with small ruminants. *ILCA Bulletin* 28: 2-6.

Yamoah, C.F., Agboola, A.A. and Mulongoy, K. 1986. Soil properties as affected by the use of leguminous shrubs for alley cropping with maize. *Agricultural Ecosystems and Environment* 18: 167-77.

Youdeowei, D.F. 1986. *Farming System, Household Economy, and Browse Utilisation: A Comparative Study of Southwest and Southeast Nigeria.* Humid Zone Programme Document. Ibadan, Nigeria: ILCA.

Wahua, T.A.T. and Oji, U.I. 1987. Survey of browse plants in upland areas of Rivers State. In Reynolds, L. and Atta-Krah, A.N. (eds) *Browse Use and Small Ruminant Production in Southeast Nigeria.* Ibadan, Nigeria: Humid Zone Programme, ILCA.

6

Intensifying Rice-Based Cropping Systems in the Rainfed Lowlands of Iloilo, Philippines: Results and Implications

G.L. DENNING

Rainfed lowland rice is grown on about 40 million ha, representing 29% of the global rice area. This category of rice culture covers a wide range of growing conditions in terms of rainfall, hydrology, soil and topography. As defined by the International Rice Research Institute (IRRI), rainfed lowland rice 'grows in bunded fields that are flooded for at least part of the cropping season at water depths that do not exceed 50 cm for more than 10 consecutive days' (IRRI, 1989a). During the past two decades, increasing efforts have been made to raise the productivity of rainfed areas, recognizing that the Green Revolution had favoured farmers in areas with access to irrigation and had largely bypassed farmers in rainfed areas. In the case of the Philippines, 78% of the rainfed lowlands had been planted to modern varieties by 1980, compared with only 2% in 1966 (Herdt and Capule, 1983).

With the availability of early maturing varieties, IRRI began on-farm cropping systems research in suitable rainfed lowland areas in the mid-1970s in order to identify opportunities for crop intensification. Initially, the research was undertaken in two areas of the Philippines, the Central Luzon province of Pangasinan and Iloilo province in the Western Visayas region (*see* Figure 6.1 *overleaf*); the Western Visayas region encompasses Panay Island and Negros Occidental. On-farm testing of modern rainfed rice technologies was later carried out in Mindanao and several other parts of the country in cooperation with the Philippine Department of Agriculture.

Based on cropping systems research and multilocational trials which had been conducted throughout the Philippines, the rainfed lowlands of Iloilo were identified as an area with high potential for crop intensification through the following measures: the use of early maturing varieties, dry seeding of the first rice crop at the beginning of the wet season (rather than waiting to puddle the fields), a short turnaround time between the crops and the introduction of a second rice crop. Variants of this technology included the substitution of the second

Figure 6.1 The Philippines, showing the Iloilo study area

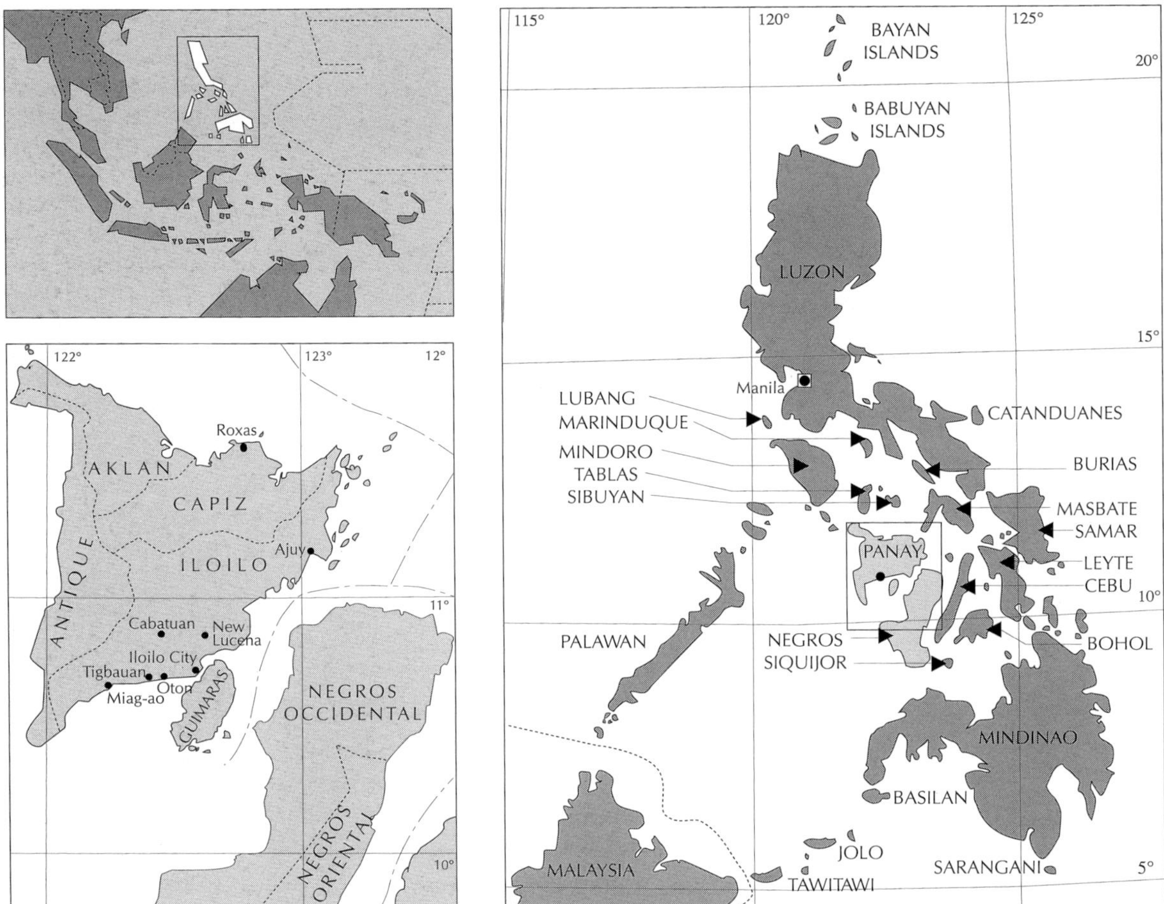

rice crop with an upland crop in drier areas, and the addition of an upland crop to a two-rice crop pattern where moisture was adequate.

The major objectives of the IRRI study were:

- to describe the development and dissemination of dry-seeded rice (DSR) technology in Iloilo;

- to determine the comparative productivity of DSR cropping patterns and non-DSR cropping patterns in Iloilo;

- to identify the factors influencing the adoption of DSR cropping patterns in Iloilo;

- to use the Iloilo case as a basis for an alternative model for the generation, adaptation and adoption of appropriate technology among small farmers.

THE STUDY AREAS

Within the Philippines, rainfed lowland rice is grown under a great diversity of conditions. This study was restricted to areas free of serious floods and droughts. Although Iloilo province falls into the relatively favourable category of rainfed lowlands, it exhibits a wide range of rainfed lowland environments. The classification of the rice growing environments in the study areas is given Table 6.1.

Table 6.1   Classification of rice growing environments in the study areas

Zone/rainfall pattern	Landscape position	Soil type[1]	Municipality
North	Sideslope	Sara sandy loam	Ajuy
(wet zone)	Plain	Sara sandy loam	Ajuy
Central	Plateau	Barotac loam	Cabatuan
(intermediate zone)	Plain	Santa Rita clay	New Lucena
South	Plateau	Santa Rita clay	Miag-ao
(dry zone)	Plain	Santa Rita clay	Miag-ao

Note: 1. Soil types based on Alicante et al., 1947

Physical environment

Iloilo is the largest of four provinces on the island of Panay, covering an area of 532 400 ha. There are 46 municipalities and 1785 *barangays* (villages) in the province. Of the total population of 1.5 million, 72% live in the rural areas (World Bank, 1979). About 319 000 ha, representing 60% of Iloilo's land area, is under cultivation. Rice is the main crop grown, accounting for 175 000 ha, and Iloilo is one of the major rice producing provinces in the Philippines. In 1982, almost 75% of the rice in the province was grown under rainfed lowland conditions.

The rainfall patterns are influenced mainly by the south-west and north-east monsoons. The length of the wet season increases along an axis running from the south to the north-east. Wetter conditions are also experienced closer to Panay's main mountain range, the Western Cordilleras, which runs roughly north-south. For this study, locations in southern ('dry zone'), central ('intermediate zone') and northern ('wet zone') parts of Iloilo province were identified as representative of the province's main rainfall zones (*see* Figure 6.2 *overleaf*). Iloilo City is typical of the drier southern coastal area. Long-term rainfall records for the city (1951-70) showed an annual rainfall of 1900 mm, with 5-6 months exceeding 200 mm/month and 4-5 months having less than 100 mm/month. Records from Miag-ao, about 20 km south of Iloilo City, reveal a similar annual rainfall figure, but the wet and dry seasons are more pronounced. Cabatuan and New Lucena, in the central area, have generally more favourable rainfall patterns. In the north is Ajuy, where the rainfall is typical of the wet zone, with a less pronounced wet season peak and a less distinct dry season. Records from Roxas in neighbouring Capiz province resemble the rainfall pattern in Ajuy.

Figure 6.2　Rainfall distribution in Panay Island: a) Iloilo City (1951-70), b) Miag-ao (1972-79), c) Roxas (1951-74)

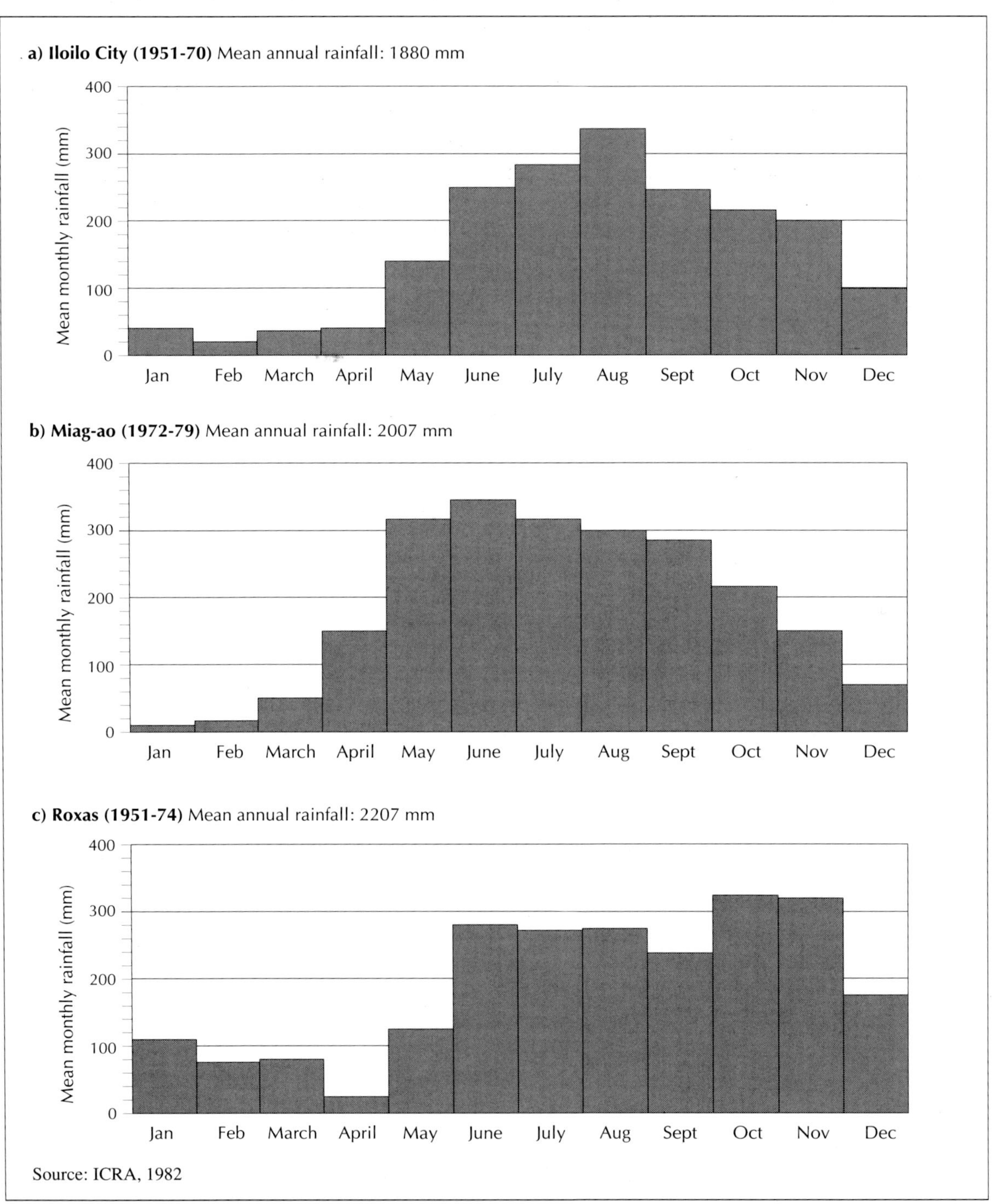

The soil type in two of the research areas, Miag-ao and New Lucena, is Santa Rita clay, which is the most important rice soil in Iloilo (Alicante et al., 1947). The wet zone research area, Ajuy, is situated on a Sara sandy loam, an important soil for lowland rice in northern Iloilo. In the Cabatuan research area in the intermediate zone, the soil type is a Barotac loam; it is acidic, red to reddish brown, and very stony.

The landscape varies throughout the province. Rainfed lowland rice areas are generally classified as plateau, sideslope, plain or bottomland (*see* Figure 6.3). In central Iloilo, the plateau and plain locations are several kilometres apart and thus there are variations in soil type and rainfall. For this reason, the two central locations —Cabatuan (plateau) and New Lucena (plain)—are discussed separately, although they were initially selected to represent the intermediate rainfall zone. In the south, although the plateau and plain locations in Miag-ao are only 2 km apart, there were soil and rainfall differences between these two locations.

Figure 6.3   Schematic representation of geomorphic and pedologic conditions in the rice-growing areas of Iloilo

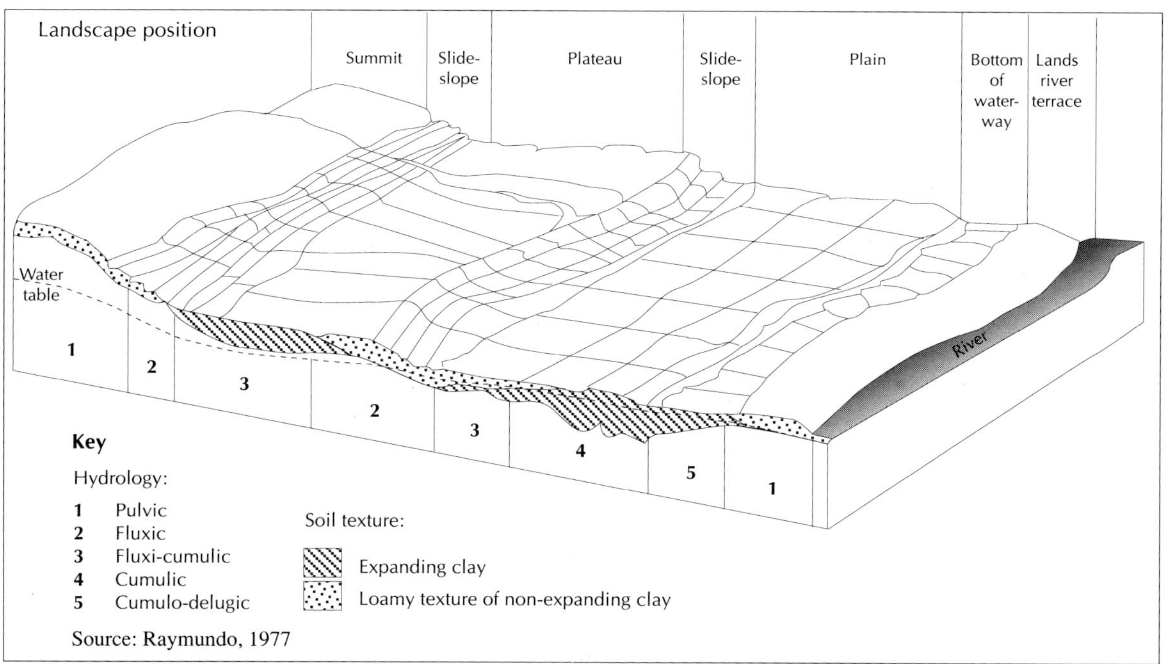

Farming systems and the socioeconomic environment

The following outline of the farming systems and some aspects of the socioeconomic environment of Iloilo is derived from surveys undertaken by the IRRI Cropping Systems Program in the late 1970s (Barlow et al., 1983; Chapman, 1983).

Rainfed rice farms in Iloilo average about 1.5 ha and are subsistence oriented, although most farmers sell rice and other farm produce to meet cash needs. Most farmers own all or part of their land, but a significant minority are share-tenants. Landlords and tenants usually share crop production expenses and receive equal shares of the crop produce. Farms are often fragmented, with farmers owning or renting fields which are not necessarily contiguous. Most farmers use the *carabao* (water buffalo) as the source of animal power for land preparation and transport. Pigs and poultry are reared to supplement cash income.

The dominant traditional cropping pattern varies according to the rainfall pattern. In the far north, a two-rice crop sequence has predominated since the introduction of modern (early maturing) varieties; in the rainfed areas of central and southern Iloilo a single rice crop pattern is more common. Throughout the province, upland crops such as mungbean and cowpea are sown after rice if there is adequate moisture. In the drier southern areas, farmers often plant a very early maturing maize crop at the onset of the wet season prior to transplanting rice in the same fields. Modern rice varieties from the University of the Philippines at Los Baños (UPLB) and IRRI were introduced into Iloilo in the 1960s, and are now widely grown on both irrigated and rainfed lowlands. A photoperiod-sensitive variety, BE 3, was introduced from Burma (Myanmar) in the late 1950s but by the late 1970s it had been largely replaced by modern photoperiod-insensitive varieties, in particular IR36.

The other major change in Iloilo during the late 1970s was the shift in the method of rice crop establishment. Wet-seeding of rice (WSR), the sowing of pre-germinated rice seed in a puddled soil, widely replaced transplanting (TPR) in both irrigated and rainfed areas. This shift was mainly in response to increases in labour costs associated with TPR and the availability of cheap and effective herbicides for weed control in WSR.

TECHNOLOGY DEVELOPMENT AND DISSEMINATION

The description given here of the efforts to introduce and promote the use of modern rainfed technology draws on several reports (World Bank, 1979; Bolton, 1980; Cuyno and Lambert, 1980; IRRI/PCARRD, 1980; Nicolas et al., 1980; Cardenas et al., 1981; ICRA, 1981). Discussions with project staff and observation findings were used to validate the reports.

During the 1970s, IRRI was involved in two separate but related efforts in Iloilo to evaluate and introduce technologies for the rainfed lowlands. The approaches were described as 'applied research' (Haws, 1977) and 'cropping systems research' (Zandstra et al., 1981).

Applied research

The applied research approach involved the multilocational testing and evaluation of a technology package which showed promise in the target area. This approach rested on the premise that viable technology existed but was not reaching farmers (Haws, 1977). Researchers identified rice production technologies that were successful in areas with similar climate and soils. Data sources included soil and rainfall maps and the results of on-station and on-farm trials conducted over a wide range of environments throughout the Philippines.

In the case of Iloilo, the potential for DSR and the two-rice crop pattern was established through adaptive research trials conducted in farmers' fields in the Bulacan and Nueva Ecija provinces in the early 1970s (Cardenas et al., 1981). These trials were usually managed by extension agents who had been trained in adaptive research methods. Few treatments or replicates were used at the sites, but identical trials were conducted at multiple locations throughout Bulacan and Nueva Ecija; in 1971, for example, 116 on-farm trials were conducted in the rainfed lowlands of the two provinces (IRRI, 1972). The aim of the trials, carried out over a 4-year period, was to evaluate rice varieties, establishment methods, fertilizer rates and timing, and the effectiveness of herbicides and insecticides (IRRI, 1972, 1973, 1974, 1975). The trial results provided the basis for the technology that was to be tested and evaluated in Iloilo.

In 1974, multilocational testing using the applied research approach began in Santa Barbara municipality in central Iloilo. The aim of these tests, carried out by IRRI and the Philippine Bureau of Agricultural Extension (BAEx), was to evaluate the potential for double-cropping in the rainfed lowlands. The main components evaluated were early land preparation, dry seeding of rice at the onset of the wet season (rather than waiting for sufficient water to puddle the soil), use of pre-emergence herbicide to control weeds, use of early maturing rice varieties, effectiveness of a short turnaround period (about 15 days) between the first and second rice crops, and planting an upland crop after the second rice crop. Research and extension staff considered this cropping pattern and its main components to be suitable for the Iloilo environment. Initial field trials, designed to verify a predesigned technology package, were conducted on farmers' fields in Santa Barbara, using 1000 m² plots. Farmers were able to evaluate the package with little risk, as IRRI covered the cost of cash inputs.

Two farmers grew the new cropping pattern in 1974 in place of their traditional practice of growing one TPR or WSR crop. With close supervision from research and extension staff, average yields of 3 t/ha and 4 t/ha were achieved for the first and second rice crops, respectively. In the second year, nine farmers, on an area of 25 ha, participated; by the third year, 54 farmers were participating, on an area of 89 ha, and the average yield rose to 9 t/ha. At this stage, farmers in Santa Barbara were adopting the technology with intensive extension supervision and support. Project-financed multilocational test plots, each 1000 m², were located in strategic areas throughout the province to evaluate technology adaptation and to demonstrate the technology to farmers.

The rapid increase in the number of adopters was supported by the formal launching in late 1976 of a project called Kabsaka (from *Kabusugan sa Kaumahan*, which in Ilonggo means 'abundance on the farm'). Extension workers and village leaders were trained at IRRI and in Iloilo, and a vigorous information and extension campaign made use of various media, farmer field days and farmers' classes to promote the project. In the first year of the formal Kabsaka project (1977), 88 farmers dry-seeded their first rice crop on 149 ha of land and obtained an average yield of about 5 t/ha, but drought affected the subsequent wet-seeded crop, resulting in crop failure for many farmers. Contrary to expectation, the number of adopters in 1978 increased to 276, covering 477 ha. For the first time, following excellent results from the multilocational test plots, the adoption of DSR spread beyond the Santa Barbara pilot area to include (among others) the areas around Cabatuan and Ajuy. A two-crop yield of about 9 t/ha was reported for the 1978 season. In 1979, the project expanded to include farms on the adjacent island of Negros and in the following year 1040 farmers were participating, on an area covering 1500 ha in 20 municipalities in provinces in both Panay and Negros (*see* Figure 6.4 *overleaf*).

Cropping systems research

Concurrent with the on-farm testing and extension activities in Santa Barbara, IRRI was also working with the Bureau of Plant Industry (BPI) in undertaking cropping systems research in the municipalities of Oton and Tigbauan from 1975 to 1980. The main objective of this research project was to evaluate proposed cropping systems research methods (based largely on earlier work in another Philippine province, Batangas); a secondary objective was to develop more productive but economically viable cropping patterns for small farmers in rainfed and partially irrigated areas (Morris et al., 1982).

Cropping systems research was conceived initially to increase the benefits for crop production from available physical resources (Zandstra, 1977). IRRI's cropping systems research methodology, described by

Figure 6.4   Participation of farmers in the Kabsaka project, Iloilo 1974-80

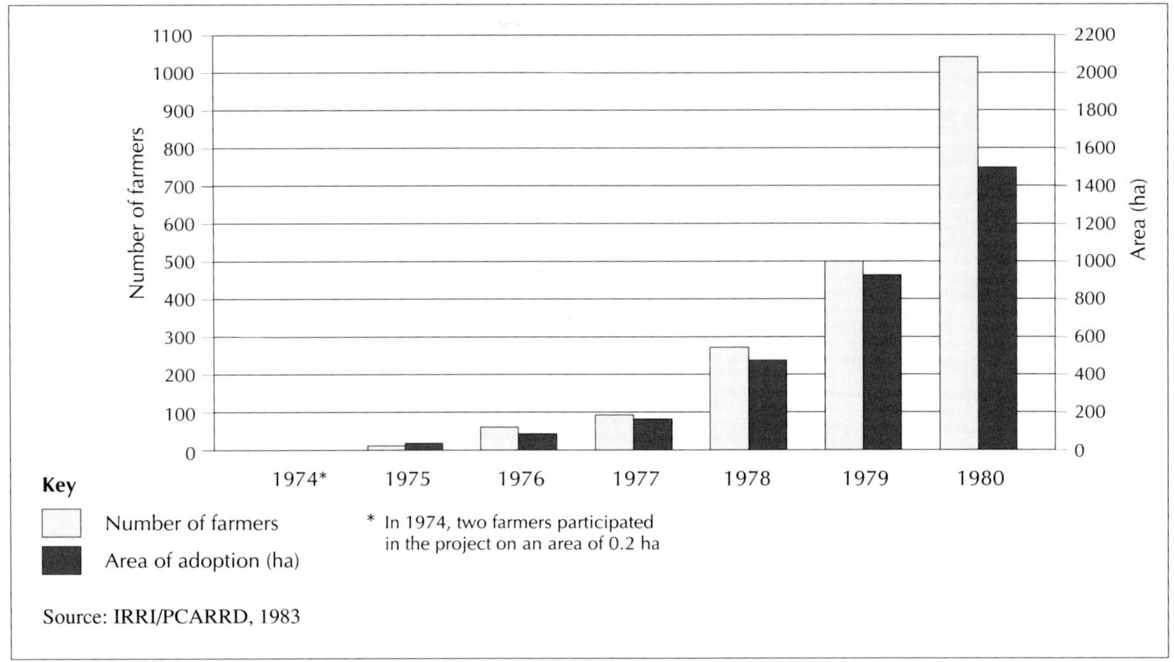

Zandstra et al. (1981) and elaborated further by Denning (1985a), involves a step-by-step procedure of technology design, testing, evaluation and transfer. The target area is identified before the site is selected and described, and it is further defined after subsequent technology testing and evaluation. IRRI's use of the term 'cropping systems' rather than 'farming systems' reflects its mandate to focus on the crop component of rice-based farming systems. However, the methodology does take account of the importance of non-crop enterprises (such as livestock, off-farm employment and cottage industries) in the description, design, testing and evaluation phases.

After 4 years of testing in Oton and Tigbauan, several cropping patterns and component technologies were identified as agronomically suited to Iloilo (*see* Table 6.2). From these tests, WSR emerged as superior to DSR in all sites except those on sideslopes and light-texture plateau soils. This finding appeared to conflict with the results from Santa Barbara (on a plain landscape), where DSR was identified as suitable for Iloilo rainfed areas.

By the time the Kabsaka project was implemented, WSR was being spontaneously adopted by farmers throughout the country as a means of saving labour associated with preparation of seedbeds, pulling of seedlings and transplanting. The effect of WSR on intensifying rice-based farming systems is limited, as farmers still have to wait for the soil to be puddled before planting. On the other hand, DSR, which involves sowing the rice at the beginning of the wet season, showed clear potential for increasing cropping intensity by earlier planting of the first and subsequent crops. In addition, it was clear that researchers and extension staff from BAEx and BPI, in cooperation with IRRI, were directly responsible for the introduction of DSR into Iloilo. For these reasons, DSR and DSR-based cropping patterns, rather than WSR, became the focus of the study, and thus this chapter describes the impact of the applied research approach rather than the cropping systems research approach.

Table 6.2   Suggested cropping patterns for specified field conditions, IRRI-BPI project, Iloilo[1]

Field conditions	Suggested cropping patterns
Sideslope (all rainfed):	
Medium texture	DSR/TPR-MG/CP[2]
Light texture	DSR-P
Plateau:	
Rainfed, medium texture	WSR/TPR-MG/P
Rainfed, light texture	DSR/TPR-P
Partially irrigated[3], medium texture	WSR-TPR-CP
Partially irrigated, light texture	DSR-TPR-CP
Plain:	
Rainfed, medium texture	WSR-ratoon/fallow-CP; WSR-TPR-CP
Rainfed, heavy texture	WSR-TPR/WSR-CP
Partially irrigated, medium texture	WSR-WSR-CP
Partially irrigated, heavy texture	WSR-WSR-CP
Bottomland:	
Rainfed, heavy texture	WSR-WSR-CP
Partially irrigated, heavy texture	WSR-WSR-CP

Notes: 1. DSR = dry-seeded rice; WSR = wet-seeded rice; TPR = transplanted rice; MG = mungbean; CP = cowpea; P = potato.
        2. DSR/TPR means DSR or TPR; MG/CP means mungbean or cowpea.
        3. Partially irrigated assumes water availability late May/early June to early December, but decreasing frequency from
           late October.
Source: Morris et al., 1982

Project expansion

To support widespread dissemination of what became known as the 'Kabsaka technology', between 1978 and 1980 the Philippine Government negotiated a US$ 12 million loan from the World Bank. The designers of the Rainfed Agricultural Development (Iloilo) Project (RADIP) were impressed by the potential of the Kabsaka technology and its implications for rainfed area development. A World Bank Staff Appraisal Report on RADIP in 1979 proposed that the project should 'draw on the experience of the highly successful and innovative Kabsaka project and seek to extend the technology to all suitable areas of the province and, in the long run, to other provinces in the Philippines which may have comparable conditions' (World Bank, 1979).

RADIP was designed as an integrated rural development project. Its aim was to promote a 'comprehensive and balanced development in rainfed areas and facilitate the realization of a broad range of opportunities rather than merely pursuing the narrow focus of having new technology accepted by the farmers' (World Bank, 1979). It was also concerned with backyard livestock enterprises, the construction of small-scale water impounding schemes, the improvement of marketing facilities (through the construction of grain warehouses and farm-to-market roads), human nutrition and health services, strengthening institutions and manpower resources, assisting adaptive research and future project preparation. However, crop intensification through DSR remained a central thrust of RADIP. The target of the extension component was to intensify crop production on 60 000 ha through DSR and other means over a 5-year period. In the first full year of the implementation of RADIP (1981), 4600 farmers covering 6300 ha in Iloilo participated in the project (IRRI, 1982).

COMPARATIVE PRODUCTIVITY OF RICE CROPPING PATTERNS

Dry-seeding of rice at the onset of the wet season was the main technical component advocated by the Kabsaka and RADIP projects. The perceived advantage of DSR over TPR or WSR was that the first rice crop could be established earlier, thus producing a higher and more reliable second crop yield or, alternatively, enabling farmers to plant an additional crop to the one or two usually grown. These changes were feasible since DSR allows crop establishment without puddling. It was considered that the shift to DSR would benefit the whole cropping pattern. The discussion that follows seeks to establish whether these advantages are, in fact, evident upon field implementation of DSR by farmers in the study areas and whether productivity from farmers' DSR patterns matched the potential shown by experiments in the study areas.

As rice dominates Iloilo cropping patterns in terms of production and income, the biological productivity of cropping patterns was evaluated using rice yield as the unit of measurement. This measure of productivity, however, may be less relevant in environments where other crops contribute significantly to farm production and income. To reflect the contribution of non-rice crops, productivity is also measured in economic terms. The data collected for this analysis cover the 1982 and 1983 growing seasons (a 'growing season' refers to the year when the first crop of a pattern is established). The 1982 growing season corresponded to the 1982 wet season, although some crops in the pattern may have been harvested or established in early 1983. As the study areas are rainfed, it was important to take into account the rainfall patterns in the 1982 and 1983 growing seasons.

Throughout the Philippines, the 1982 growing season was one of the driest ever recorded. The wet season ended 2-3 months earlier than usual in many areas, and the ensuing dry season lasted 7-8 months in some parts of the country. In Iloilo, the start of the 1983 wet season was delayed by about 1 month. However, this second growing season subsequently proved to be about normal or slightly wetter than average. These quite different rainfall patterns experienced in 1982 and 1983, coupled with local variations, provided an excellent opportunity to study cropping pattern performance over a wide range of growing conditions.

Yield performance

DSR patterns under recommended management and input levels

Rice yields were measured from DSR cropping patterns in researcher-managed production plots (1000 m² each) grown under recommended management and input levels (*see* Table 6.3). The highest yields in 1982 were obtained in the wet zone (Ajuy), followed by the intermediate zone (Cabatuan and New Lucena), with lowest yields recorded in the dry zone (Miag-ao) (*see* Table 6.4 *overleaf*). These results were largely as expected, based on the long-term rainfall patterns for these areas, despite a drier than average 1982 wet season.

In Ajuy, yields exceeding 8 t/ha were obtained from the two-crop patterns in 1982. First crop yields in northern Iloilo were higher than those obtained elsewhere in the province. This result was probably related to the ease with which thorough land preparation can be accomplished prior to DSR establishment in northern Iloilo. In Ajuy, farmers can take advantage of light rainfall during February and March to plough and harrow the relatively light-textured soils, prior to the onset of the main monsoon season. Second crop yields in Ajuy also exceeded those obtained elsewhere in Iloilo, but were 22% lower, as well as being more variable, than the

Table 6.3   Component technology used for production plots during 1982 and 1983

Component technology	First rice crop	Second rice crop
Method of establishment	Dry-seeded	Wet-seeded
Timing of establishment	As soon as possible after first monsoon rains	15 days after harvest of first crop, or as soon as possible thereafter
Variety	IR36	IR36
Seeding rate (kg/ha)	100	100
Nitrogen application: rate (kg N/ha) timing	90 1/3 basal 1/3 at 20-25 DAS[1] 1/3 at 30-35 DAS	90 1/3 basal 1/3 at 15-20 DAS 1/3 at 30-35 DAS
Phosphorus application: rate (kg P_2O_5/ha) timing	30 basal	0 —
Potassium application: rate (kg K_2O/ha) timing	30 basal	0 —
Weed control	Butachlor at 2.0 kg a.i./ha, applied to moist soil after seeding; 1 handweeding, if necessary	Butachlor at 1.5 kg a.i./ha, applied 6-8 DAS
Pest control	Carbofuran granules at 1 kg a.i./ha, applied and incorporated before seeding ; Mono-crotophos at 0.75 kg a.i./ha at about 15 DAS and at 30-35 DAS; other applications per economic threshold levels; no fungicide used	As for first rice crop

Note: 1. DAS = days after seeding.

first crop. This finding was probably because of greater exposure of the second rice crop to drought. There were negligible differences between yields from sideslope and plain landscape positions in Ajuy during 1982.

The early dry season of 1982 had a marked effect on the second crop yields in Cabatuan and New Lucena. Second crop yields were 82% and 60% lower than first crop yields, respectively, and tended to be more variable, as in Ajuy. In Miag-ao, only one rice crop was established in each field in 1982. Yields were comparable to those obtained for DSR in the intermediate zone, but were lower than wet zone DSR yields. The difficulty of land preparation and less effective weed control in Miag-ao than elsewhere are likely to have been contributing factors. However, the relatively low yield in the plain location, which occurred again in 1983, is believed to have been caused by zinc deficiency (Denning, 1985b).

The first rice crops of 1983 were established after the long dry spell. Average DSR yields in all locations except the Miag-ao plain exceeded 5 t/ha. The yields exceeded the corresponding 1982 yield levels on Ajuy sideslope (by 50%), Ajuy plain (38%), Cabatuan (132%), New Lucena (30%), Miag-ao plateau (48%) and Miag-ao plain (9%). These high first crop yields could be attributed partly to the reduction of the weed and

Table 6.4   **Grain yield of rice from 1000 m² production plots at various locations in Iloilo, 1982 and 1983**

Rainfall	Location		1982 yield (t/ha)			1983 yield (t/ha)		
			1st crop	2nd crop	Total	1st crop	2nd crop	Total
North (wet zone)	Ajuy	sideslope	4.53 (6.1)[1]	3.97 (40.0)	8.50	6.78 (6.0)	4.76 (16.0)	11.54
		plain	4.80 (14.9)	3.26 (39.7)	8.06	6.60 (15.4)	4.11 (13.8)	11.00[2]
Central (intermediate zone)	Cabatuan		3.54 (24.6)	0.62 (144.2)	4.16	7.85 (13.7)	4.35 (17.7)	12.20
	New Lucena		4.05 (27.5)	1.63 (123.5)	5.68	5.28 (16.9)	4.53 (10.0)	9.81
South (dry zone)	Miag-ao	plateau	3.91 (32.7)	—[3]	3.91	5.80 (10.8)	—	5.80
		plain	3.49 (26.7)	—	3.49	3.79 (23.1)	—	3.79

Notes: 1. c.v. % given in parenthesis.
   2. The computation of total yield excluded the first crop yield of one farmer who did not participate in the second cropping.
   3. Not planted to a second rice crop.

insect pest population during the long dry spell. Another contributing factor could have been the 'Birch Effect' — the rapid release of nitrate and other nutrients associated with the first rains of the dry season (Birch, 1958; Nye and Greenland, 1960); this hypothesis is now being tested in Iloilo through more intensive field research.

Second crop yields in 1983 averaged over 4 t/ha in Ajuy, Cabatuan and New Lucena, resulting in exceptionally high total rice yields across the wet and intermediate zones. Nevertheless, they were again lower than the first crop yields. These crops presumably would not have benefited from the rapid organic matter mineralization at the beginning of the wet season. The lower yield variability of the second crop, relative to that found in 1982, was probably because of the absence of excessively dry weather during the 1983 growing period.

The contrast between the 1982 and 1983 growing seasons enabled an analysis to be carried out of DSR cropping pattern performance in different rainfall zones under both favourable and unfavourable growing conditions. The analysis showed that productivity potential in northern Iloilo is relatively high even in a less favourable year such as 1982, and demonstrated the stability and high adoption potential of the DSR-WSR pattern in this environment. It also showed the susceptibility of a DSR-WSR pattern to an early dry season in central Iloilo. Simulated growing season analyses by Bolton and Zandstra (1981) predicted that the western part of Iloilo, which includes Cabatuan, would have unstable second crop yields because of erratic late season rains. Although results obtained for Cabatuan in 1982 seem to support this prediction, the 1983 yield of over 12 t/ha suggests that the two-crop pattern in this environment should be further explored. Recognition of the sensitivity of second crop yields to drought conditions in this environment led to an evaluation of the performance of an early maturing rice variety, IR58. This variety matures up to 13 days earlier than the existing variety, IR36, although its potential yield is lower than that of IR36 under favourable rainfall conditions (IRRI, 1984).

In the drier southern part of Iloilo province, the potential of a single DSR crop is questionable. There are difficulties in taking advantage of early upland crop establishment in this environment because the soils are very heavy and are subject to waterlogging during the monsoon season. Bolton and Zandstra (1981) also concluded that there were few opportunities for planting a second rice crop in Miag-ao. They advocated that in most years a single rice crop should be followed by an upland crop or ratoon rice. A DSR crop would appear to be best suited for ratoon cropping because of its earlier establishment. The inability to grow two rice crops in this environment, however, suggests that alternative cropping patterns involving WSR, TPR and upland crops should be explored.

DSR and non-DSR patterns under farmers' management and input levels

The yields obtained from DSR and non-DSR cropping patterns grown and managed by farmers in various locations in Iloilo during 1982 are shown in Table 6.5.

Table 6.5   **Grain yields of rice from DSR and non-DSR cropping patterns grown by farmers at various locations in Iloilo, 1982**

Rainfall zone	Location	Yield (t/ha) from DSR pattern			Yield (t/ha) from non-DSR pattern			Difference (t/ha)		
		1st cp[1]	2nd cp	Total	1st cp	2nd cp	Total	1st cp	2nd cp	Total
North (wet zone)	Ajuy sideslope	3.27 (28.6)[2]	2.88 (37.7)	6.15 (30.5)	3.43 (40.0)	1.64 (86.4)	5.07 (42.4)	-0.16[3] (0.46)	1.24[4] (0.56)	1.08[3] (0.93)
	plain	3.66 (27.1)	2.81 (36.8)	6.48 (28.6)	3.11 (38.2)	1.81 (57.8)	4.91 (29.9)	0.55[3] (0.48)	1.00[4] (0.47)	1.57[5] (0.77)
Central (intermediate zone)	Cabatuan	2.62 (38.2)	0.37 (190.5)	3.00 (36.6)	3.18 (37.5)	0.13 (332.3)	3.31 (37.6)	-0.56[1] (0.39)	0.24[3] (0.23)	-0.31[3] (0.42)
	New Lucena	3.60 (30.8)	1.74 (68.7)	5.34 (30.4)	3.53 (43.7)	0.84 (117.9)	4.37 (48.8)	0.07[3] (0.46)	0.90[4] (0.38)	0.97[3] (0.65)
South (dry zone)	Miag-ao plateau	2.27 (56.8)	0.002 (450.0)	2.27 (57.1)	1.98 (42.2)	0.23 (263.5)	2.21 (50.1)	0.29[3] (0.35)	—[6]	0.06[3] (0.39)
	plain	2.22 (45.9)	0.04 (282.5)	2.26 (44.6)	2.13 (68.6)	0.08 (06.3)	2.21 (66.7)	0.09[3] (0.55)	—[6]	0.05[3] (0.55)

Notes: 1. cp = crop.
     2. c.v. % given in parenthesis.
     3. Not significant at the 10% level.
     4. Significant difference at the 5% level, by t test.
     5. Significant difference at the 10% level, by t test.
     6. Not computed, as very few farmers established a second rice crop.

In all locations, it was found that DSR yields were not significantly different from first crop yields from non-DSR (WSR or TPR) patterns. However, on Ajuy sideslopes, Ajuy plains and New Lucena, the second crop yields were significantly higher ($p = 0.05$) by about 1 t/ha in fields where DSR was the first crop. This difference can be attributed to the earlier establishment of both the first and second crops in DSR patterns (*see* Table 6.6). For WSR and TPR, establishment of the first crop is always delayed to ensure that sufficient water is available for puddling; this, in turn, may delay second crop establishment and harvest as the varieties used were not photoperiod-sensitive. Because of the early dry season in 1982, the second crop yields in non-DSR patterns were affected more seriously than those in DSR patterns (although even second crop yields from DSR patterns were consistently lower than those obtained from the first crop).

Second crop yields in 1982 were generally more variable than first crop yields, and the variability of second crop yields was higher for non-DSR patterns than for DSR patterns. The second crop was clearly exposed to

Table 6.6   **Dates of first and second rice crop establishment (week number) from researcher-managed plots and from farmers' DSR and non-DSR patterns at various locations in Iloilo, 1982 and 1983**

Rainfall zone	Location		Field category[2]	Median establishment date (week number)[1]			
				1982		1983	
				1st crop	2nd crop	1st crop	2nd crop
North (wet zone)	Ajuy	sideslope	Researcher-managed	19	39	22	40
			DSR (F)	20	42	23	41
			non-DSR (F)	28	45	26	44
		plain	Researcher-managed	20	39	21	40
			DSR (F)	20	40	22	41
			non-DSR (F)	27	45	26	45
Central (intermediate zone)	Cabatuan		Researcher-managed	25	41	23	41
			DSR (F)	20	43	23	42
			non-DSR (F)	27	46	29	45
	New Lucena		Researcher-managed	22	39	27	43
		DSR (F)	21	42	26	43
			non-DSR (F)	25	44	33	49
South (dry zone)	Miag-ao	plateau	Researcher-managed	20	—	27	—
			DSR (F)	21	—	27	—
			non-DSR (F)	26	—	31	—
		plain	Researcher-managed	22	—	23	—
			DSR (F)	23	—	25	—
			non-DSR (F)	26	—	29	—

Notes: 1. For reference purposes, week 20 commences on 14 May, week 30 commences 23 July, week 40 commences on 1 October and week 50 commences on 10 December

         2. DSR (F) = farmers' DSR patterns; non-DSR (F) = farmers' non-DSR patterns

greater risk of drought when following WSR or TPR, rather than DSR. In southern Iloilo, where few second rice crops were established, there appeared to be no advantage of DSR over non-DSR patterns in terms of rice production.

Rice yields obtained by both adopters and non-adopters of DSR patterns were higher in all Iloilo research locations in the 1983 growing season than in 1982 (*see* Table 6.7). Although there was no significant advantage of DSR over WSR or TPR as a first crop, in all cases except Miag-ao plain the trend was in favour of DSR. The second crop yields, however, again showed the advantage of early establishment from DSR as a first crop. In Ajuy plain, Cabatuan and New Lucena, these yields were significantly higher for DSR patterns, compared with those for non-DSR patterns. In Cabatuan and New Lucena, second crop yield advantages of 2.03 t/ha and 2.87 t/ha respectively were obtained using DSR patterns, suggesting that even in a good year, there was a substantial yield advantage in using DSR in central Iloilo. The yield advantage of DSR patterns was also evident

Table 6.7 **Grain yields of rice from DSR and non-DSR cropping patterns grown by farmers at various locations in Iloilo, 1983**

Rainfall zone	Location	Yield (t/ha) from DSR pattern			Yield (t/ha) from non-DSR pattern			Difference (t/ha)		
		1st cp[1]	2nd cp	Total[2]	1st cp	2nd cp	Total[2]	1st cp	2nd cp	Total[2]
North (wet zone)	Ajuy sideslope	6.46 (14.7)[5]	5.26 (14.4)	12.32 (19.2)	5.67 (28.8)	5.13 (13.7)	10.80 (14.2)	0.79[3] (0.76)[5]	0.13[3] (0.46)	1.52[3] (1.33)
	plain	6.30 (15.1)	4.85 (12.9)	11.47 (16.1)	5.72 (19.9)	3.98 (20.9)	10.40 (18.1)	0.58[3] (0.46)	0.87[6] (0.32)	1.07[3] (0.82)
Central (intermediate zone)	Cabatuan	4.29 (20.5)	3.99 (28.7)	8.28 (22.0)	4.04 (22.6)	1.96 (86.3)	6.00 (33.3)	0.25[3] (0.39)	2.03[7] (0.63)	2.28[6] (0.83)
	New Lucena	3.97 (28.0)	3.90 (18.4)	7.87 (20.9)	3.51 (16.7)	1.03 (129.4)	4.53 (32.6)	0.44[3] (0.42)	2.87[8] (0.42)	3.34[8] (0.70)
South (dry zone)	Miag-ao plateau	4.31 (36.5)	—	4.31 (36.5)	3.91 (18.2)	0.05 (412.3)	3.96 (17.4)	0.40[3] (0.43)	—[9]	0.35[3] (0.42)
	plain	3.46 (23.9)	— —	3.46 (23.9)	4.02 (37.1)	0.08 (360.6)	4.10 (41.9)	0.56[3] (0.46)	—[9]	0.64[3] (0.51)

Notes: 1. cp = crop.
2. Totals do not necessarily equal the sum of first and second crop yields or yield differences because of a small number of third rice crop yields which were included in the total.
3. Not significant.
4. c.v. % given in parenthesis.
5. Standard error of difference (s.e.d.) given in parenthesis.
6. Significant difference at the 5% level, by t test.
7. Significant difference at the 1% level, by t test.
8. Significant difference at the 0% level, by t test.
9. Not computed, as very few farmers established a second rice crop.

in Ajuy in 1983, but this difference was significant only for the second crop in the plain location. On the sideslopes, part of the 1.52 t/ha advantage of DSR patterns was accomplished with some farmers growing a third rice crop. The results in Miag-ao were inconsistent. Even in 1983, DSR adopters and non-adopters could not establish a second rice crop in this area.

Results from farmers' fields in the 1982 and 1983 confirm the advantage of DSR in both wet and intermediate zones. The advantage is primarily from a better and less variable second crop yield, although there is no apparent yield loss from DSR as a first crop. The results from Miag-ao in the dry zone suggest that the advantages of DSR in this environment, if any, will come from its ability to create opportunities for the production of non-rice crops.

Comparison of results from researcher-managed and farmer-managed patterns

The yield results from researcher-managed production plots and farmers' DSR and non-DSR patterns are shown in Figures 6.5, 6.6, 6.7 and 6.8. The results from non-DSR plots are included for reference only.

In Ajuy, the yield advantage of crops from researcher-managed plots over that of farmers' DSR patterns was evident only in 1982; it was more pronounced in the first crop. In 1983, there was no apparent difference observed in either the first or second crop. Results from Cabatuan in 1982 also showed higher yields from researcher-managed production plots for the first crop and, to a lesser extent, for the second. In 1983, a similar result was obtained, although the yield difference was relatively less in the second crop. In New Lucena the

Figure 6.5   Comparison of rice yields from researcher-managed plots and from farmers' plots using DSR and non-DSR patterns in Ajuy, 1982 and 1983*

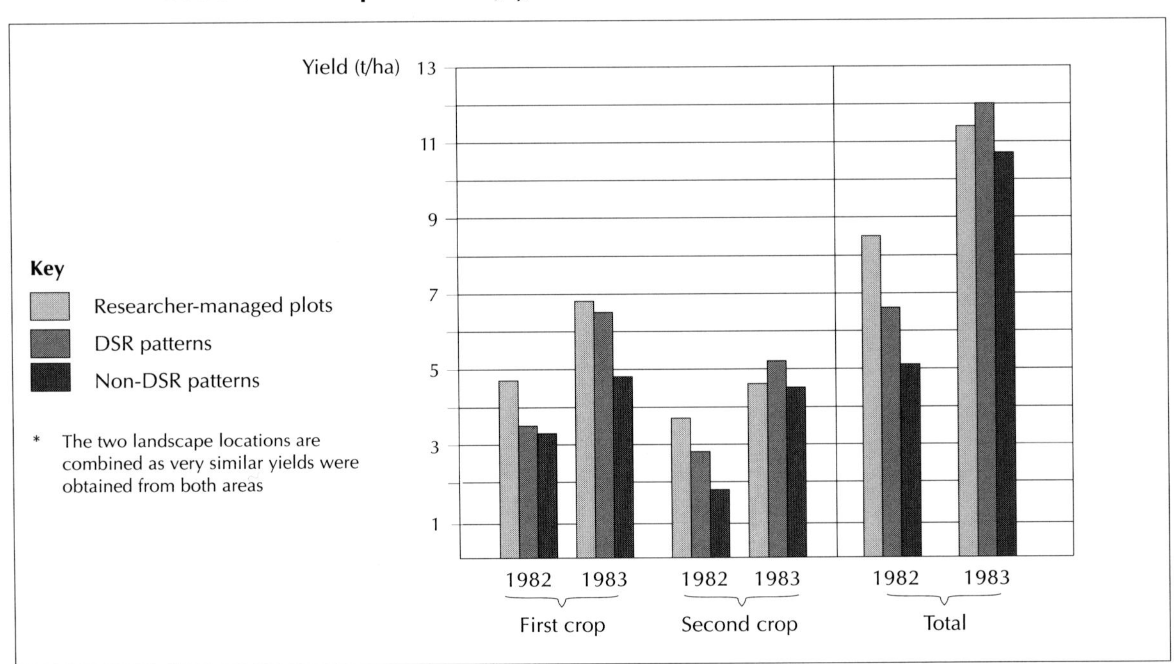

Figure 6.6 Comparison of rice yields from researcher-managed plots and from farmers' plots using DSR and non-DSR patterns in Cabatuan, 1982 and 1983*

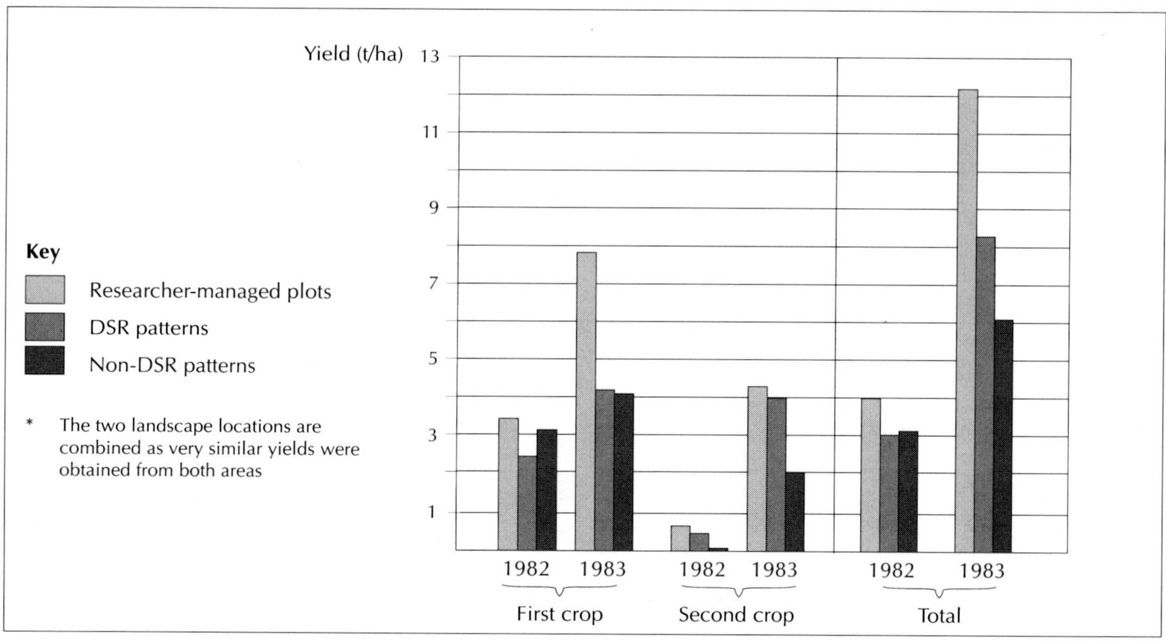

Figure 6.7 Comparison of rice yields from researcher-managed plots and from farmers' plots using DSR and non-DSR patterns in New Lucena, 1982 and 1983*

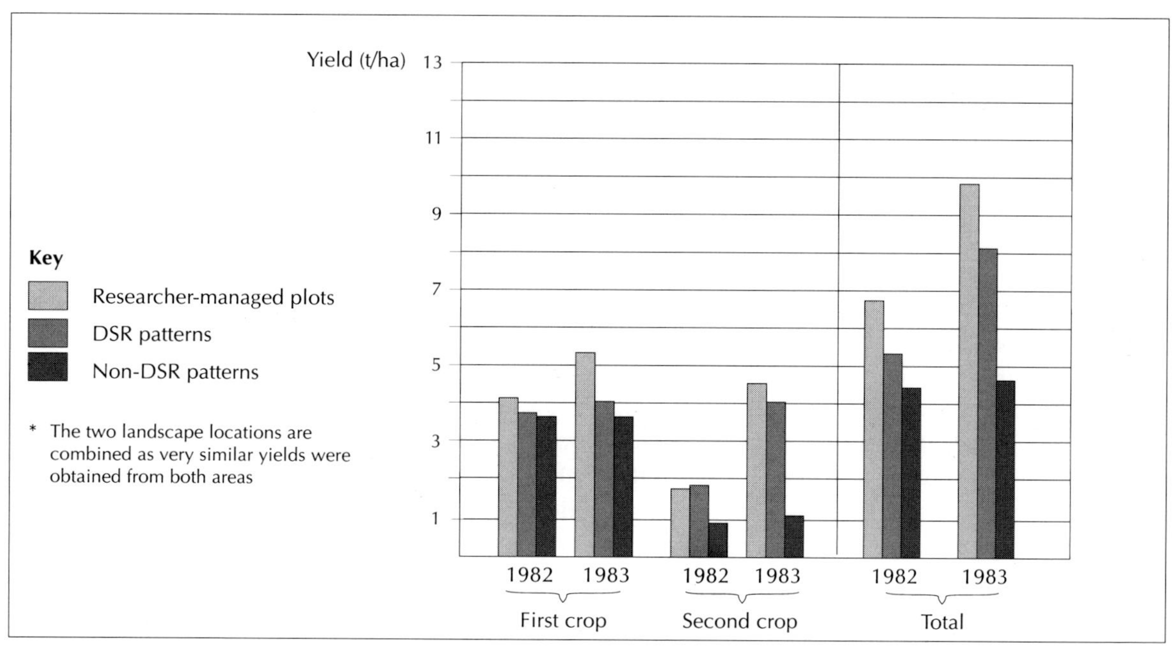

Figure 6.8   **Comparison of rice yields from researcher-managed plots and from farmers' plots using DSR and non-DSR patterns in Miag-ao, 1982 and 1983***

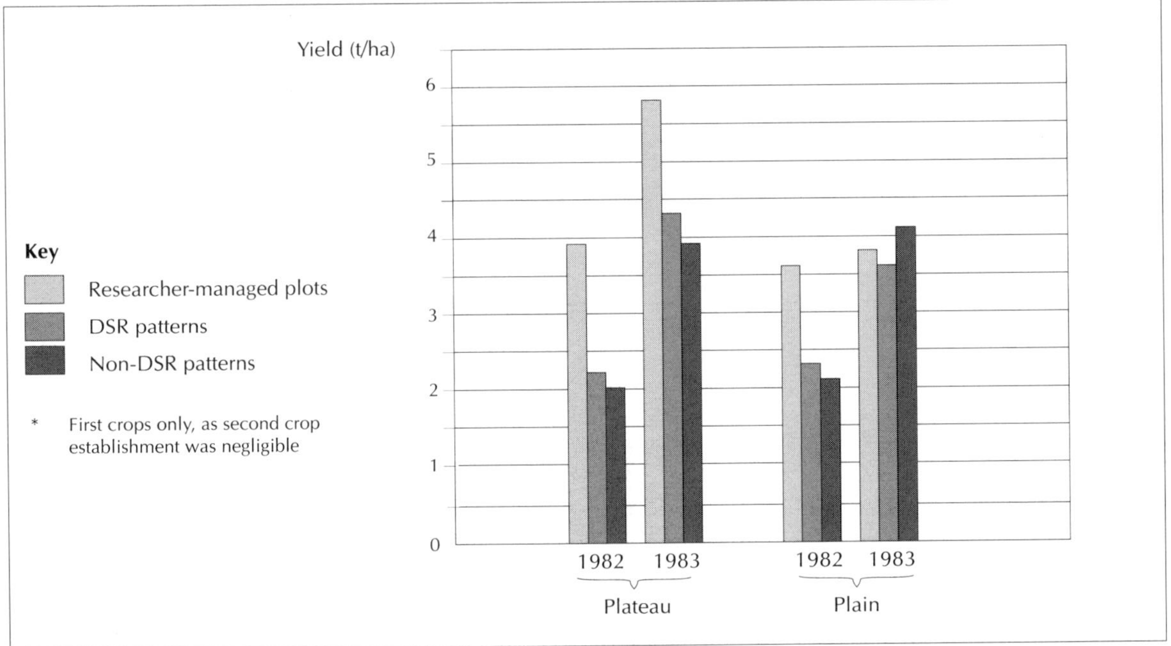

differences were far less pronounced, indicating that farmers were achieving yield levels from their DSR patterns comparable with those obtained by the researchers. There was, however, an advantage of about 1 t/ha from DSR growing in researcher-managed production plots over farmers' DSR crops during 1983. In Miag-ao, where a single rice crop was grown, the farmers' DSR yields were well below the researcher-managed plot levels, except on the plain location in 1983, when comparable yields were obtained.

The higher first crop yields from researcher-managed plots are probably related to higher input levels and more intensive management, relative to that found in the farmers' DSR fields. Farmers applied an average of 50-60 kg N/ha on DSR, compared with the recommended level of 90 kg N/ha. In addition, pre-emergence herbicide was seldom used by farmers, except in Ajuy. Other factors such as seed quality, pest control and timing of harvest might also have contributed. Where the yield from researcher-managed plots was not higher than that obtained from farmers' DSR patterns (as in Ajuy in 1983, Cabatuan in 1982, New Lucena in 1982, and Miag-ao plain in 1983), it seemed unlikely that farmers would obtain higher yields with available technology. This result suggests that further technical improvements need to be made to the DSR package.

In general, the higher second crop yield obtained from researcher-managed plots could be attributed to earlier establishment of the second crop. With 1000 m² plots, the researcher does not have the same difficulty as farmers in harvesting and preparing the land for establishment of a second rice crop. It appears that the long turnaround times experienced in farmer-managed fields mean that they lose much of the advantage obtained through dry seeding the first crop. Roxas et al. (1982), in another study of Iloilo rainfed farmers, using a simulation model, found that a shorter turnaround time between crops benefited production by offering opportunities for higher and more stable second crop yields. Thus, attention to ways of shortening the turnaround period appears to be justified if farmers are to obtain maximum advantage from using DSR cropping patterns.

There can be considerable difficulty in establishing a second rice crop quickly when it is surrounded by fields of standing crops in non-DSR patterns; the working animals lack access to the desired field and they cannot reach the area without damaging other farmers' crops. Small power tillers, carried along paddy bunds, can be used to overcome this constraint but the cost of these machines puts them beyond the reach of most farmers in Iloilo's rainfed areas.

Economic performance

The results from the economic analyses of farmers' DSR and non-DSR patterns in Iloilo province are given in Tables 6.8 and 6.9. The analyses involved a comparison of total net returns from DSR and non-DSR patterns in 1982 and 1983 and the contribution by crop type to the total net returns from DSR and non-DSR patterns in both years.

Comparison of total net returns from DSR and non-DSR patterns

The total net returns from both DSR and non-DSR patterns were considerably lower in 1982 than in 1983. This result reflected the differences in yields obtained in these two climatically contrasting seasons. Prices obtained

Table 6.8 **Total net returns (P/ha) from DSR and non-DSR cropping patterns grown by farmers at various locations in Iloilo, 1982 and 1983[1]**

Rainfall Zone	Location	1982 net returns (P/ha)			1983 net returns (P/ha)		
		DSR pattern	Non-DSR pattern	Difference[2]	DSR pattern	Non-DSR pattern	Difference
North (wet zone)	Ajuy	3676	2188	1488[3] (605.8)	9802	8647	1155[4] (686.6)
Central (intermediate zone)	Cabatuan	745	1867	-1122[3] (426.8)	4629	3532	1097[4] (660.8)
	New Lucena	2148	2087	61[4] (627.9)	5505	3334	2171[5] (643.9)
South (dry zone)	Miag-ao	1373	934	439[4] (432.7)	3682	3212	470[4] (462.1)

Note: 1. Exchange rates: US$ 1 = ₽ 8.47 (30 June 1982); US$ 1 = ₽ 11.00 (30 June 1983).
2. Standard error of difference (s.e.d.) given in parenthesis.
3. Significant difference at the 5% level, by t test.
4. Not significant.
5. Significant difference at the 1% level, by t test.

for rice in both years were similar. The difference between years, however, confirmed the sensitivity of production to the vagaries of seasonal weather, and may explain why few farmers choose to commit their entire rainfed lowland area to a single cropping pattern.

In Ajuy, the DSR patterns (DSR-WSR) appeared to be more profitable than the non-DSR patterns (mostly WSR-WSR) in both 1982 and 1983, reflecting the better yields obtained from the DSR patterns, particularly from the second crop. The advantages gained in terms of production and net income in both poor and relatively good seasons suggest that the DSR-WSR pattern is not a particularly risky innovation in this environment. It appears that the light soil texture, which facilitates effective chemical weed control, and the relatively long wet season combine to make the DSR pattern attractive to farmers in the northern areas of Iloilo. As a result, the spread of this innovation in northern Iloilo has probably been more rapid and uniform than elsewhere in the province.

The results from Cabatuan were less convincing. In 1982, the non-DSR patterns actually resulted in a significantly higher net income compared with that obtained from DSR patterns. Few non-adopters of DSR planted a second crop, whereas most DSR adopters established a second rice crop, and thereby incurred substantial losses because of the crop failure associated with the early dry season. In the more favourable 1983 season, the DSR patterns were more profitable than the non-DSR patterns by over P1000 (US$ 91) per hectare. In New Lucena, there was no significant difference in net income between DSR and non-DSR patterns in 1982. In the more favourable year, DSR patterns resulted in a significantly higher net return by over P2000 (US$ 182) per hectare. Thus it appears that for Iloilo's intermediate rainfall zone, DSR patterns lead to higher net returns in a relatively good year, but may be no more profitable or even less profitable than a non-DSR pattern in a year that is drier than normal.

In the drier location of Miag-ao, the results showed that there was no significant economic benefit, in terms of net return per hectare, from growing DSR. This finding reflects the inability to capitalize on early rice crop establishment.

Contribution by crop type to total net returns from DSR and non-DSR patterns

The contributions of various crops to net income indicate the relative unimportance of non-rice crops in Ajuy and Cabatuan. In New Lucena, 34% of the net income from non-DSR patterns came from mungbean in 1983. In that environment, farmers who dry-seeded their first crop opted to establish a second rice crop, while those who wet-seeded preferred to establish mungbean. Being drought-sensitive, several mungbean crops failed in 1982, resulting in little overall contribution to net income that year.

In Miag-ao, it appears that the establishment of DSR increased the opportunity to grow a successful mungbean crop, with contributions of 14% and 18% to the net income from DSR patterns in 1982 and 1983, respectively. These contributions, however, were insufficient to create any significant increase in overall net return from dry seeding in this environment.

The DSR crop generally costs 10-20% more to produce than a first WSR crop (Denning, 1985b). With regard to cash costs, the major inputs were fertilizer and seed. Fertilizer constituted 45-60% of total input costs in most areas and this appeared to be independent of crop establishment technique. Expenditure on herbicide was greatest in Ajuy, where it was twice as much for DSR than for WSR. This result probably reflected the greater effectiveness of herbicide in this environment, and thus greater reliance by farmers on this weed control technique.

Table 6.9 **Contribution by crop type to total net returns from DSR and non-DSR cropping patterns in Iloilo, 1982 and 1983**

		Contribution to total net return (%)					
	Cropping	1982			1983		
Location	pattern	Rice	Mungbean	Maize	Rice	Mungbean	Maize
Ajuy	DSR	96	4	0	100	0	0
	Non-DSR	100	0	0	100	0	0
Cabatuan	DSR	100	0	0	104	-2	-2
	Non-DSR	100	0	0	100	0	0
New Lucena	DSR	100	0	0	95	5	0
	Non-DSR	98	2	0	66	34	0
Miag-ao	DSR	86	14	0	77	18	5
	Non-DSR	98	2	0	94	1	3

ADOPTION OF DRY-SEEDED RICE CROPPING PATTERNS

By the 1981 growing season, the Kabsaka technology had been adopted by 4600 farmers, on an area covering 6300 ha. Closer examination of farmers' fields and discussions with farmers revealed a varied pattern of DSR adoption. Within apparently homogeneous physical environments, farmers exhibited individual preferences in their decision to adopt or not to adopt DSR.

Methods

Information about the socioeconomic characteristics of farmers in the study areas was obtained from interviews. At the beginning of the 1982 wet season, farmers were classified according to whether or not they had adopted DSR. A farmer was classified as an adopter if DSR was practised on at least some part of his/her farm; farmers who did not practise DSR on any part of their farms were classified as non-adopters. The respondents were selected randomly, with a more or less similar number of adopters and non-adopters in the sample. The study area was classified according to location in order to ensure that there was adequate representation from all three agroclimatic zones where the project was operating and where the researcher-managed plots were located.

After the draft questionnaire had been pre-tested, initial interviews were conducted during August 1982. Information on crop establishment and early crop management was collected, together with details on most of the socioeconomic factors which were later included in the analysis. Following the harvest of the first rice crop in each location (October-December), respondents were interviewed for the second time. Various analytical procedures to identify the determinants of adoption were reviewed (Denning, 1985b), and logit analysis (Goldberger, 1964) was selected as the most suitable.

Results

The sample characteristics of the independent variables used in the regression models are shown in Table 6.10. Continuous variables appear as mean values, while dichotomous variables are shown by frequency of occurrence of each discrete value.

The logit model describing DSR adoption behaviour by Iloilo farmers was significant (*see* Table 6.11). The independent variables, taken together, significantly influenced the adoption of DSR. This model correctly predicted the adoption status of 70.5% of 173 farmers in the sample from which the model was derived. Of the nine independent variables included in the model, three were found to be significant at the 5% level. The probability of DSR adoption increased with the area of rainfed lowland being farmed (p = 0.05), the availability of labour in the farm household (p = 0.05) and the farmer's knowledge of modern rainfed rice technology (p = 0.001) (*see* Table 6.12).

Over the range set for the explanatory variables given in Table 6.12, the three significant variables displayed fairly similar effects on the probability of adoption. The probability of adoption for a farmer with a large rainfed

Table 6.10 Characteristics of sampled farmers in Iloilo, as shown as by mean values for continuous variables and distribution of farmers for discrete variables

Variable	Units	All (N = 173)	Adopters (N = 94)	Non-adopters (N = 79)
Farmer's age	Years	48.52	47.85	49.32
Education	Years	6.47	6.38	6.57
Labour index	Index value	2.00	2.15	1.83
Draft index	Index value	1.01	1.05	0.95
Extension visits	Visits/year	26.63	30.26	22.32
Knowledge	Score/8	2.46	3.20	1.57
Tenancy	% land owned	30.14	28.50	32.09
Rainfed lowland area	m^2	15 162	17 430	12 463
Knowledge (herbicide only)	Score/3		1.34	
Area dry-seeded	m^2		12 544	
DSR seeding rate	kg/ha		115	
Credit	0 = no credit	124	67	57
	1 = credit	49	27	22
Tenancy dummy	0 = not owner	118	64	54
	1 = owner	55	30	25

Table 6.11   Logit regression analysis for adoption of DSR in Iloilo, 1982

	Explanatory variable	Coefficient	Standard error	Asymptotic t value	Significance level
b_0	Constant	-1.0154	1.1258	-0.902	n.s.[1]
x_1	Farmer's age	-0.0143	0.0160	-0.895	n.s.
x_2	Education	-0.0623	0.0539	-1.156	n.s.
x_3	Labour index	0.3132	0.1545	2.027	5%
x_4	Draft index	-0.2643	0.2331	-1.134	n.s.
x_5	Extension visits	0.00373	0.00829	0.451	n.s.
x_6	Knowledge	0.4922	0.1109	4.437	0.1%
x_7	Credit	-0.4662	0.4190	-1.113	n.s.
x_8	Tenancy	0.000130	0.00417	0.031	n.s.
x_9	Rainfed lowland area	0.0000558	0.0000227	2.455	5%

Log of the likelihood function for full model ($\log L_{max}$) = -96.96

Log of the likelihood function for constant only ($\log L_o$) = -119.26

Log likelihood ratio statistic [$-2(\log L_o - \log L_{max})$] = 44.6 (9 d.f.)

Cases predicted correctly = 70.5%

Sample size = 173

Note: 1. n.s. = not significant.

Table 6.12   Predicted probabilities of DSR adoption for an average Iloilo farmer, for different sizes of rainfed lowland area being farmed and different levels of labour availability and farmer knowledge[1]

	Small rainfed lowland area (1 ha)			Medium rainfed lowland area (2 ha)			Large rainfed lowland area (3 ha)		
	Labour index			Labour index			Labour index		
	1	2	4	1	2	4	1	2	4
Mean knowledge level (index = 2.46)	0.45	0.53	0.69	0.59	0.66	0.79	0.72	0.78	0.87
High knowledge level (index = 3.58)	0.59	0.66	0.79	0.71	0.77	0.86	0.81	0.86	0.92

Note: 1. The mean knowledge level (index = 2.46) was the average recorded for the sample; the high knowledge level (index = 3.58) corresponds to the value obtained in a similar study conducted by the author in South Cotabato, Philippines.

lowland area, a high level of knowledge (for Iloilo) and the equivalent of four full-time labourers in the household was twice that of a farmer with a relatively small rainfed lowland area, an average level of knowledge and the equivalent of only one full-time labourer in the household.

The model revealed relatively poor probability of adoption among the subset of Iloilo farmers in Cabatuan. The logit regression analysis was run again, with the same set of independent variables, but excluding data from Cabatuan, on the assumption that somewhat different determinants of adoption might be operating there. This reduced the number of observations from 173 to 141.

This second model (hereafter called the Iloilo-less-Cabatuan model) was also significant at the 1% level (*see* Table 6.13). However, the proportion of cases correctly predicted rose from 70.5% to 75.9%. Again, the availability of household labour and the farmer's knowledge level were significant ($p = 0.1$ and $p = 0.001$, respectively). The effect of area of rainfed lowland being farmed was now less apparent, being almost significant at the 10% level. The most striking difference in the Iloilo-less-Cabatuan model was that frequency of extension visits now became a significant explanatory variable ($p = 0.05$). This finding suggested that inclusion of Cabatuan accentuated the importance of area of rainfed lowland elsewhere in Iloilo, while

Table 6.13   Logit regression analysis for adoption of DSR in Iloilo, excluding farmers from Cabatuan, 1982

	Explanatory variable	Coefficient	Standard error	Asymptotic t value	Significance level
b_0	Constant	-1.664	1.4021	-1.187	n.s.[1]
x_1	Farmer's age	-0.0275	0.0206	-1.337	n.s.
x_2	Education	-0.0555	0.0663	-0.836	n.s.
x_3	Labour index	0.3178	0.1842	1.725	10%
x_4	Draft index	-0.1343	0.2685	-0.500	n.s.
x_5	Extension visits	0.0214	0.00999	2.139	5%
x_6	Knowledge	0.6766	0.1411	4.793	0.1%
x_7	Credit	-0.3687	0.4944	-0.746	n.s.
x_8	Tenancy	0.00101	0.00529	0.190	n.s.
x_9	Rainfed lowland area	0.0000417	0.0000256	1.631	n.s.

Log of the likelihood function for full model ($\log L_{max}$) = -68.87

Log of the likelihood function for constant only ($\log L_o$) = -97.56

Log likelihood ratio statistic $[-2(\log L_o - \log L_{max})]$ = 57.4 (9 d.f.)

Cases predicted correctly = 75.9%

Sample size = 141

Note: 1. n.s. = not significant.

simultaneously masking a positive influence of extension visits. This result is further illustrated by the statistics on rainfed lowland area and extension visits shown in Table 6.14.

According to the Iloilo-less-Cabatuan model, although extension visits significantly increased the probability of adoption, they were less significant than labour availability and farmer's knowledge level (*see* Table 6.15). From these data, a farmer with a high knowledge level, receiving fortnightly extension visits and with the equivalent of four full-time labourers in the household is almost twice as likely to adopt DSR as a farmer with an average knowledge level, receiving monthly visits from extension and with the equivalent of one full-time labourer in the household.

Table 6.14   Sample statistics from Iloilo for rainfed lowland area and frequency of extension visits by location and adoption status

Location	Rainfed lowland area (ha)		Frequency of extension contact (visits/yr)	
	Adopters	**Non-adopters**	**Adopters**	**Non-adopters**
Ajuy	2.03	1.71	39.97	12.86
Cabatuan	1.56	1.09	6.55	14.41
New Lucena	2.12	1.30	39.67	28.94
Miag-ao	1.32	1.26	31.74	25.59
Mean	1.74	1.25	30.26	22.32

Table 6.15   Predicted probabilities of DSR adoption for an average Iloilo farmer at different frequencies of extension visits and different levels of labour availability and farmer knowledge (excluding information from Cabatuan)

Farmer category	Mean knowledge level (index = 2.46)		High knowledge level (index = 3.58)	
	Monthly visits	**Fortnightly visits**	**Monthly visits**	**Fortnightly visits**
Labour index = 1	0.40	0.47	0.55	0.62
Labour index = 2	0.48	0.55	0.63	0.69
Labour index = 4	0.56	0.63	0.70	0.76

Adoption patterns

In both the Iloilo and Iloilo-less-Cabatuan models, there was a greater probability of farmers with more labour, larger areas of rainfed lowland and better knowledge of the Kabsaka technology adopting DSR in 1982.

Labour considerations

The importance of labour is likely to derive from the difficulty of weed control in DSR. Puddling the soil for WSR and TPR suppresses early weed competition, and because they are established later in the wet season, WSR and TPR crops are more commonly flooded, thereby further enhancing weed suppression. The adoption of DSR by farmers in Iloilo entails a significant increase in labour costs, particularly with regard to handweeding, and thus a farmer planning to adopt DSR must have either an adequate supply of farm labour to control weeds or sufficient cash reserves to employ outside labourers for handweeding.

It appears, therefore, that DSR may not be suited to poorer farmers with low labour availability in the household. In view of the high labour requirement for TPR, the main alternative for farmers in Iloilo is to wet-seed the rice crop and accept a higher probability of second crop failure. The widespread adoption of WSR in place of TPR in Iloilo shows that a large proportion of farmers are indeed opting for this alternative.

As DSR is relatively labour-intensive, there was the possibility of an inverse relationship between farmers' age and DSR adoption. However, results showed that age did not influence the decision to adopt. This conclusion was consistent with the findings of many recent studies (Herdt and Capule, 1983).

Farm size and related factors

The finding that farmers with a larger area of rainfed lowland tend to adopt DSR more readily than those with a smaller area is consistent with some past studies on technology adoption (Herdt and Capule, 1983). Reviews by Schutjer and Van der Veen (1977) and Feder et al. (1982), however, point out two important features of this generalization. First, although larger farmers tend to be the first to adopt modern rice technology, this relationship often diminishes over time; that is, smaller farmers tend to adopt the technology later than larger farmers. Second, farm size *per se* is often not the direct cause of rapid or slow adoption; it is often correlated to many other potentially important factors such as credit, wealth, access to inputs and access to information.

In Iloilo, the relationship between landholding and adoption was considerably weaker in the absence of Cabatuan data. DSR has recently been introduced and promoted among Cabatuan farmers. Although the observed effect of farm size (rainfed lowland area, in particular) may disappear over time, a follow-up survey of the same sample in 1988 continued to show a significant effect of farm size.

Considering that DSR in Iloilo is a new and thus potentially risky technology, it is likely that the larger farmers adopted earlier because of their capacity to absorb the risk of crop failure. In Ajuy, where the rainfall is more reliable and more conducive to the double cropping of rice, the difference in farm size between adopters and non-adopters was quite small. It appears that here even smaller farmers have recognized the value of DSR and have decided to adopt. The soundness of this decision is supported by the higher yields and incomes derived from DSR patterns compared with those obtained from non-DSR patterns in Ajuy. It is possible that the small farmers' initial caution could be overcome by further refinement in DSR technology, leading to greater confidence in technology performance. Feder and Slade (1984) point out that early adopters influence other farmers by disseminating information on their experiences. In essence, the early adopters, by experimenting with a new technology, are absorbing much of the risk associated with a new and largely unproven innovation.

Credit availability did not appear to inhibit adoption of DSR. However, results from a separate study in Santa Barbara indicated that although institutional credit was important in initially inducing farmers to adopt the Kabsaka technology, farmers continued to adopt the main components of the package (including DSR) some years later, without institutional credit support (IRRI, 1982). Had the adoption survey been undertaken in 1977 or 1978 instead of 1982, it is conceivable that credit might have been identified as an adoption determinant.

Tenancy status did not appear to influence the decision to adopt DSR. Although some farmers claimed that they were not allowed to adopt DSR because of their landlords' aversion to the risks inherent in the technology, it appears that such influence was not widespread. Landlords and tenants seem to have developed agreements on input and output sharing which made DSR adoption favourable to both parties.

There seemed to be no correlation between draught animal ownership and the decision to adopt DSR. It was observed that hiring tractors was a cost-effective substitute for draught animals in preparing fields for DSR establishment.

Farmers' knowledge level

Farmer knowledge of the Kabsaka technology was apparently an important factor influencing the decision to adopt. It is difficult to ascertain, however, to what extent greater technical knowledge by farmers actually influenced the decision to adopt because the acquisition of knowledge after adoption cannot be discounted.

Frequency of extension contact and the knowledge index were very weakly correlated in Iloilo. However, as extension was an important determinant of adoption in Iloilo (outside of Cabatuan), it appears that extension was contributing in ways not directly reflected in the knowledge index. Observations of extension activities in Iloilo revealed that the decision to adopt DSR could be encouraged by persuasive arguments on the potential benefits of the technology. These arguments were put forward by local extension agents, and reinforced by their supervisors who regularly attended meetings of farmer groups. There also appeared to be some group pressure applied to farmers to participate in the project.

The limited significance of extension contact in the adoption model which included Cabatuan reflects the low profile of extension agents in Cabatuan. That DSR adopters had a lower level of extension contact than non-adopters, however, cannot be explained on available information. Elsewhere in Iloilo, extension contact had a positive influence on DSR adoption. The complexities of the DSR technology appear to necessitate regular extension support, while simpler technologies, such as modern varieties, can spread rapidly with or without extension advice.

Formal education levels of Iloilo farmers also did not appear to influence DSR adoption. Although the results of previous studies have been inconclusive, some of these studies, including those by Mangahas (1970), Gerhart (1975) and Feder and Slade (1984), associated adoption with more literate and educated farmers.

ASSESSMENT OF THE ILOILO CASE

The introduction of DSR cropping systems in parts of Iloilo has been given widespread publicity in the Philippines and elsewhere in Asia as an example of the successful application of the farming systems approach to research and development. The introduction of double cropping of rice as a 'system' gives the impression that an entire cropping system, involving a cropping pattern and component technology, has superseded one previously practised by farmers. Research has shown, however, that the significant change which has taken place is that farmers in the study areas have been exposed to and have adopted DSR to varying degrees, for various reasons and with various results. In another component of the same study (not reported here), it was found that a substantial proportion of farmers did not use herbicide for weed control and, by and large, used fertilizer levels similar to those applied to rice established by other methods in the same areas (Denning, 1985b).

The main channel for the introduction of DSR to Iloilo appears to have been the IRRI and BAEx applied research programme, and observations indicate that this approach, involving the simultaneous testing and demonstration of innovations, has been successful in Iloilo. However, in several other Philippine provinces, including Pangasinan, Zamboanga del Sur and North Cotabato, DSR failed to spread significantly. Although the reasons for this have not been studied, the main reason appears to be the limited adaptation of DSR to some environments. Most of Pangasinan has a similar rainfall pattern to that of southern Iloilo and therefore offers little scope for double cropping of rice. In Zamboanga del Sur, heavy textured soils make land preparation and effective chemical weed control difficult.

This applied research approach to technology introduction could be viewed as almost the antithesis of the classical farming systems approach, which emphasizes the need to gain a comprehensive understanding of the farmers' resources, capabilities and objectives prior to designing and testing technical interventions (e.g., Byerlee et al., 1980; Zandstra et al., 1981). The main apparent limitation of the applied research approach in Iloilo is the inability to explain satisfactorily why innovations were not taken up by farmers in some locations. Such information was obtained only from the adoption surveys, described above, which were undertaken some years after the technology had been tested and introduced.

The 'adaptive research' methodology put forward by Haws (1977) and Cardenas et al. (1981) and the 'cropping systems research' methodology (Zandstra et al., 1981) have been viewed by many development practitioners in the Philippines as being alternative methodologies for introducing new technologies. However, rather than being alternatives, elements of both approaches have contributed to a more comprehensive and farmer-oriented research and development model.

A conceptual model for technology generation, adaptation and adoption

A conceptual model for the generation, adaptation and adoption of technology is given in Figure 6.9. A number of interrelated activities (A to F) are shown as part of a research and development process which is designed to lead to adoption of innovations by farmers.

Activity A: Technology generation at international and national research centres

This activity describes the type of research undertaken in international agricultural research centres (IARCs), national research centres and universities. Research in this category is conducted mainly in laboratories and on research stations, and tends to be commodity-oriented rather than systems-oriented. Technologies emerging from research undertaken at this level generally require further adaptation and modification to specific agroclimatic environments before adoption by farmers is likely to proceed.

Activity B: On-station technology generation and adaptation for specific regions and agroclimatic environments

The objective of this activity is to generate and adapt technology for specific agroclimatic environments. The main participants at this stage are researchers from national research systems who are working at regional

Figure 6.9   Conceptual model for technology generation, adaptation and adoption

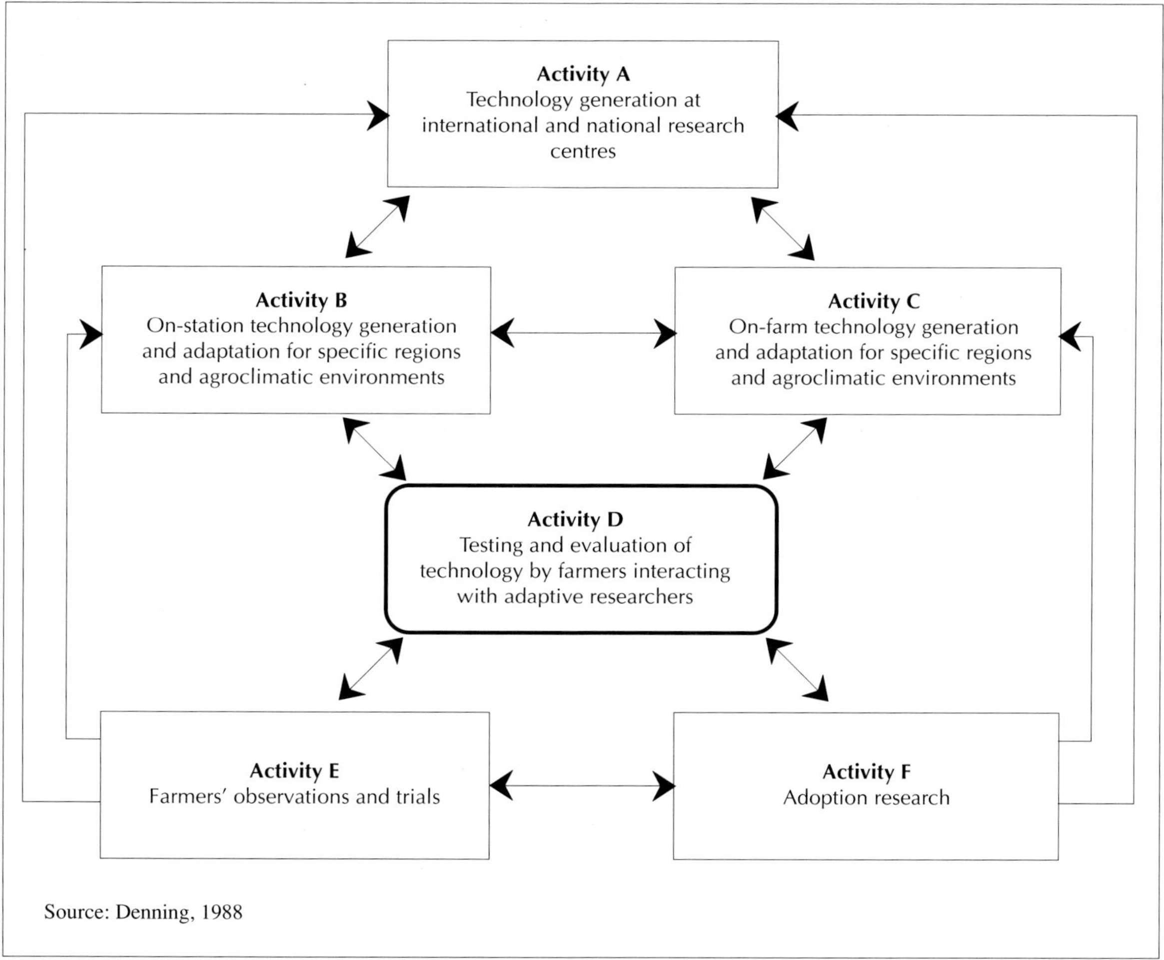

stations. Increasingly, certain aspects of IARC research are also being conducted at this level, in collaboration with national scientists. This research is often undertaken through IARC network activities; for example, testing and evaluation of rice germplasm in specific agroclimatic environments is supported by IRRI's International Network for Genetic Evaluation of Rice (INGER).

Activity C: On-farm technology generation and adaptation for specific regions and agroclimatic environments

The focus of this activity is also on technology development for specific areas and agroclimatic conditions, but although trials are conducted in farmers' fields, relatively conventional research techniques and experimental designs are employed. The main participants are national researchers, particularly those involved in national farming systems research (FSR) programmes. The on-farm cropping systems research programme conducted by IRRI throughout the 1980s falls within this research category (IRRI, 1983, 1986, 1989b). In addition to the involvement of researchers, whether from national institutions or IARCs, the on-farm nature of this research activity also requires participation by extension agents and farmers.

Historically, on-farm research (OFR) has been viewed as the link between research station activity and extension. However, it is suggested here that, whether or not by design, the IRRI cropping systems research undertaken in the Philippine provinces of Batangas, Pangasinan, Iloilo, Cagayan and Misamis Oriental is better viewed as a component of the technology generation process. It is further suggested that, rather than seeking to arrive at extension recommendations, a cropping systems research site (of the type described by Zandstra et al., 1981) could operate as a semi-permanent field location for testing new technologies, including those at an early stage of development. Information generated in this way would be of use primarily to researchers (involved in activities A and B) and not necessarily to extension agents or farmers in the short term. This concept is now applied by IRRI through the 'key site' model (IRRI, 1989c), whereby teams of national scientists work in collaboration with IRRI scientists at research sites representative of major rice ecosystems or sub-ecosystems. The key site model is a major element of IRRI's ecosystem-focused research programme. A role such as this for on-farm cropping (or farming) systems research conducted by IARCs is more consistent with their international mandates and more compatible with the emerging role and increasing capabilities of national FSR programmes.

The combination of Activities B and C, if closely linked and interactive, can provide the 'building blocks' for the improvement of existing farming systems.

Activity D: Technology testing and evaluation by farmers interacting with adaptive researchers

This activity is the focal point of the conceptual model. It is an integrating process that seeks to make the available technology (generated and adapted through Activities A, B and C) suit individual farming practices and production systems. It is concerned with the more direct and immediate goal of identifying opportunities to integrate 'outside' knowledge and skills (such as DSR) with those of the farmer, in a way that is compatible with the constraints on the farming system and the goals of the farmer. The main participants are the farmer and the adaptive researcher. The term 'adaptive researcher' is used here to describe an agriculturalist, possibly an extension agent, who has some training or experience in conducting on-farm trials.

There are two important requirements for an effective informal testing and evaluation component:

* *Identification, testing and evaluation of a range of technical options to cover a variety of farm conditions and farmer circumstances.* This procedure is designed to take account of and exploit the individuality of farmers, farms and farm plots within farms by testing and evaluating a number of potentially useful innovations. The adaptive researcher needs to be acquainted with the available technologies and the circumstances under which they are applicable. In addition, it is necessary to have sufficient knowledge of the farmers' environment to assess where the potential for various technologies exists. On-farm evaluation of innovations can then proceed, using multilocational testing procedures similar to those described by Denning (1985a). Concurrently, farmers should be encouraged to experiment with and evaluate the innovations. As such, this approach incorporates elements of the 'adaptive research' methodology of Haws (1977) and Cardenas et al. (1981).

- *Recognition of the farmers' role in technology evaluation adaptation.* A number of authors have illustrated the extent to which farmers actively experiment in order to improve their farming practices (Rhoades and Booth, 1982; Chambers, 1983; Goodell, 1984; Lightfoot, 1987; Fujisaka, 1989). However, farmers' experiments and innovations are rarely given recognition in FSR programmes. Instead, the onus is invariably on the researcher or the extension agent to prescribe what is best for farmers. This was the case in Iloilo where the Kabsaka technology was promoted actively, but it was clear that the farmers adjusted the recommendations to match their farm and field-specific conditions. Fertilizer rates were reduced by about half and pre-emergence herbicide was used only on the more light-textured soils of Ajuy (Denning, 1985b).

Activity E: Farmers' observations and trials

This activity represents the aspect of farmer innovation that is independent of the formal research and extension system; that is, the process through which farming practices evolved prior to the arrival of formalized research and extension efforts. Johnson (1981) cited several examples where farmers in traditional agricultural systems have been observed to conduct informal experiments. Most examples involved germplasm exchange and testing.

In the context of the overall model, it is important to recognize that Activity E, as well as formalized research efforts (A, B and C), contribute to Activity D which, in turn, may lead to wider and more long-term adoption of innovations.

Activity F: Adoption research

Adoption research can contribute to the development and dissemination of technology in various ways. First, it can provide those who are responsible for generating technology with important information on the agroecological and socioeconomic limitations to the adaptation of their technologies. This activity allows greater insight into the circumstances where innovations are appropriate and where alternatives are required. Second, adoption research may enable researchers to identify farmers' innovations, whether these are indigenous technologies or modifications of introduced technologies. The contributions of adoption research to technology generation and adaptation are illustrated in Figure 6.9 by the feedback loop between Activity F and Activities A, B and C.

A third purpose of adoption research is to identify institutional constraints to adoption. This knowledge may lead to institutional change to facilitate adoption (for example, through more liberal credit policies) or may point to the need for technical alternatives which are acceptable within existing institutional limitations (for example, low-cost innovations that do not require credit).

In Iloilo, an agroeconomic approach to the adoption study was employed, using a combination of field trials and surveys. It is believed that this represents an improvement over most earlier adoption study methodologies which have concentrated on the socioeconomic determinants of adoption. The field trials and farmers' yield data described earlier contributed to an understanding of adoption of DSR in Iloilo. Thus, this study has added to the growing body of literature which stresses the importance of agronomic adaptation of innovations as an essential prerequisite for adoption (e.g., Perrin and Winkelmann, 1976; Ashby, 1982; Horton, 1983).

CONCLUSION

From 'on-farm research' to 'cropping systems research' to 'farming systems research', IRRI has participated actively in the evolution of methods to make research more relevant to small rice farmers. In the past few years, farmer participation has played an increasingly important role at IRRI in the identification of research priorities and the evaluation of technology performance. This development has clearly strengthened IRRI's perception of the major research issues, especially as it takes on new challenges in the less favourable rice ecosystems. The various approaches that have been discussed in this chapter form part of a pragmatic evolution towards increasing the relevance of science to rice producers and, indirectly, to rice consumers. However, these approaches must be further improved to meet a predicted 60% higher global rice demand by the year 2020 (IRRI, 1989c). While it is recognized that most of this increase must come from the irrigated rice areas, significant increases must also be obtained in the rainfed lowlands, the deep water, tidal wetlands and the uplands where most of Asia's rural poor are located. To achieve sustainable increases in production without environmental deterioration remains the major challenge for FSR in the coming decade.

Notes

The author gratefully acknowledges the assistance of Arsenio R. Samiano, Jose S. Nicolas, German Turija and Apolinario Sotomil in the conduct of field trials and surveys in Iloilo. Dr Sam Fujisaka's critical review of an earlier draft of this chapter is also highly appreciated.

References

Alicante, M.M., Rosell, D.Z., Barrera, A. and Aristorenas, I. 1947. *Soil Survey of Iloilo Province, Philippines.* Soil Report No. 9. Manila, Philippines: Department of Agriculture and Natural Resources.

Ashby, J.A. 1982. Technology and ecology: Implications for innovation research in peasant agriculture. *Rural Sociology* 47: 234-50.

Barlow, C., Jayasuriya, S. and Price, E.C. 1983. *Evaluating Technology for New Farming Systems: Case Studies from Philippine Rice Farms.* Los Baños, Philippines: IRRI.

Birch, H.F. 1958. The effect of soil drying on humus decomposition and nitrogen availability. *Plant Soil* 10: 9-31.

Bolton, F.R. 1980. Double-cropping rainfed rice in Iloilo Province, Central Philippines. PhD thesis. Reading, UK: University of Reading.

Bolton, F.R. and Zandstra, H.G. 1981. *Evaluation of Double-Cropped Rainfed Wetland Rice.* IRRI Res. Paper Series 63. Los Baños, Philippines: IRRI.

Byerlee, D., Collinson, M. et al. 1980. *Planning Technologies Appropriate to Farmers: Concepts and Procedures.* El Batan, Mexico: CIMMYT.

Cardenas, A.C., Dilag, R.T., Pantastico, E.B. and Haws, L.D. 1981. *An Approach to Rainfed Farming: The Philippine Case.* Manila, Philippines: PCARRD/BAEx.

Chambers, R. 1983. *Rural Development: Putting the Last First.* London, UK: Longman.

Chapman, J.A. 1983. Design and analysis of appropriate technology for small farmers: Cropping systems research in the Philippines. PhD thesis. Michigan, USA: Michigan State University.

Cuyno, R.V. and Lambert, G. 1980. *The Role of Formal Organization in Agricultural Technology Utilization: The Kabsaka Case.* Research Report No. 1. Los Baños, Philippines: Management of Rural Development Program, UPLB.

Denning, G.L. 1985a. Integrating agricultural extension programs with farming systems research. In Cernea, M.M., Coulter, J.K. and Russell, J.F.A. (eds) *Research-Extension-Farmer: A Two-Way Continuum for Agricultural Development.* Washington DC, USA: World Bank.

Denning, G.L. 1985b. Adaptation and adoption of dry-seeded rice in the rainfed lowlands of Iloilo and South Cotabato, Philippines. PhD thesis. Reading, UK: University of Reading.

Denning, G.L. 1988. On-farm research to generate technology for rainfed rice farmers. *Entwicklung + Landlicher Raum* 3(88): 13-15.

Feder, G. and Slade, R. 1984. The acquisition of information and the adoption of new technology. *American Journal of Agricultural Economics* 66: 312-20.

Feder, G., Just, R.E. and Zilberman, D. 1982. *Adoption of Agricultural Innovations in Developing Countries: A Survey.* World Bank Staff Working Paper No. 542. Washington DC, USA: World Bank.

Fujisaka, J.S. (1989). A method for farmer-participatory research and technology transfer: Upland soil conservation in the Philippines. *Experimental Agriculture* 25: 423-33.

Gerhart, J. 1975. *The Diffusion of Hybrid Rice in Western Kenya.* El Batan, Mexico: CIMMYT.

Goldberger, A. S. 1964. *Econometric Theory.* New York, USA: John Wiley.

Goodell, G.E. 1984. Untying the HYV package: A Filipino farmer grapples with the new technology. *Journal of Peasant Studies* 11: 238-66.

Haws, L.D. 1977. Adaptive trials to determine production program feasibility. In *Proc. of Symposium on Cropping Systems Research and Development for the Asian Rice Farmer.* Manila, Philippines: IRRI.

Herdt, R.W. and Capule, C. 1983. *Adoption, Spread, and Production Impact on Modern Rice Varieties in Asia.* Manila, Philippines: IRRI.

Horton, D. 1983. Potato farming in the Andes: Some lessons from on-farm research in Peru's Mantaro Valley. *Agricultural Systems* 12: 171-84.

ICRA. 1981. *Agricultural Research and Development Related to the Kabsaka Project, Iloilo Province, Panay Island, the Philippines.* Wageningen, Netherlands: ICRA.

ICRA. 1982. Farming Systems in Miag-ao, Iloilo, Philippines. Wageningen, Netherlands: ICRA.

IRRI. 1972-75, 1982-84, 1986. *Annual Reports.* Los Baños, Philippines: IRRI.

IRRI. 1989a. *Implementing the Strategy: Work Plan for 1990-1994.* Los Baños, Philippines: IRRI.

IRRI. 1989b. *Annual Report for 1988.* Los Baños, Philippines: IRRI

IRRI. 1989c. *IRRI Toward 2000 and Beyond.* Los Baños, Philippines: IRRI

IRRI/PCARRD. 1980. *Annual Report for 1979-1980.* Series V, IRRI/PCARRD Cooperating Applied Research Project on Rainfed Rice. Los Baños, Philippines: IRRI/PCARRD.

IRRI/PCARRD. 1983. *Annual Report for 1981-82.* Series VI-VII, Cooperative Projects on Cropping Systems in Problem Areas. Los Baños, Philippines: IRRI/PCARRD.

Johnson, A.W. 1981. Individuality and experimentation in traditional agriculture. In Crouch, B.R. and Shankariah Chamala (eds) *Extension Education and Rural Development (Vol. 1): International Experience in Communication and Innovation.* Chichester, UK: John Wiley.

Lightfoot, C. 1987. Indigenous research and on-farm trials. *Agricultural Administration and Extension* 24: 79-89.

Mangahas, M. 1970. An economic analysis of the diffusion of new rice varieties in Central Luzon. PhD thesis. Chicago, USA: University of Chicago.

Morris, R.A., Gines, H.C., Magbanua, R.D., Torralba, R.I. and Sumido, A. 1982. The IRRI/BPI cropping systems research project in Pangasinan and Iloilo. In *Proc. of Workshop on Cropping Systems Research in Asia.* Los Baños, Philippines: IRRI

Nicolas, J. S., Price, E.C., Dilag, R.T. and Haws, L.D. 1980. An evaluation of the Iloilo Province Kabsaka Program for 1978-1979. Unpublished report. Los Baños, Philippines: RPTR Office, IRRI.

Nye, P.H. and Greenland, D.J. 1960. *The Soil under Shifting Cultivation.* Technical Communication No. 51. Harpenden, UK: Commonwealth Bureau of Soils.

Raymundo, M.E. 1977. Physical environment in rice-based cropping systems. Unpublished lecture for Cropping Systems Training Course, 22 December 1977. Los Baños, Philippines: IRRI.

Rhoades, R.E. and Booth, R.H. 1982. Farmer-back-to-farmer: A model for generating acceptable agricultural technology. *Agricultural Adminisitration* 11: 127-37.

Roxas, N.M., Angus, J.F. and Barlow, C. 1982. *The Economics of Timeliness in the Crop Intensification of Rainfed Farms in Iloilo, Central Philippines.* Department of Agricultural Economics Paper No. 82-01. Los Baños, Philippines: IRRI.

Schutjer, W.A. and Van der Veen, M.G. 1977. *Economic Constraints on Agricultural Technology Adoption in Developing Nations.* USAID Occasional Paper No. 5. Washington DC, USA: USAID.

World Bank. 1979. *Philippines: Rainfed Agricultural Development (Iloilo) Project.* Staff Appraisal Report. Washington DC, USA: World Bank.

Zandstra, H.G. 1977. Cropping systems research for the Asian rice farmer. In *Proc. of Symposium on Cropping Systems Research and Development for the Asian Rice Farmer.* Los Baños, Philippines: IRRI.

Zandstra, H.G., Price, E.C., Litsinger, J.A. and Morris, R.A. 1981. *A Methodology for On-Farm Cropping Systems Research.* Los Baños, Philippines: IRRI.

7

From Diagnosis to Farmer Adoption: MARIF's Maize On-Farm Research Programme in East Java, Indonesia

R. KRISDIANA, M. DAHLAN, HERIANTO, C. VAN SANTEN and L.W. HARRINGTON

The on-farm adaptive research programme described here was conducted by the Malang Research Institute for Food Crops (MARIF), located south of Malang in East Java, Indonesia. Founded in 1981, MARIF is one of six research institutes operated by the Central Research Institute for Food Crops (CRIFC), which is a unit of the Agency for Agricultural Research and Development (AARD). Each of the six research institutes has a specific mandate. MARIF's mandate (MARIF/RTI, 1989) is:

- to serve as a national centre for improving the yield and quality of maize, soybean, groundnut, mungbean, cassava and sweet potato;

- to develop the *palawija* (non-rice food crop) component of farming systems in order to help increase and stabilize agricultural production through more effective use of natural and human resources;

- to identify constraints in the agricultural development of *palawija* crops and evaluate ways of overcoming these constraints through technological and institutional changes;

- to assist in the transfer of technology to the farmer through cooperation with the agricultural extension services.

MARIF's responsibility for research on *palawija* crops includes plant breeding and disciplinary research, on-farm adaptive research and other approaches to farming systems research (FSR), and cooperation with extension in technology transfer. Applied research activities, especially plant breeding, have a national focus whereas adaptive research activities, including the on-farm research (OFR) programme, operate largely within East Java.

MARIF has a staff of 69 senior researchers (of whom six are PhD's), 106 research assistants and technical staff, 112 administrative staff and 134 permanent labourers. Only a small number of staff have been involved in the OFR programme at one time or another. At present, the programme is conducted by a core team of seven Indonesian researchers (one PhD) and two expatriate advisors (both PhDs), with other MARIF staff cooperating from time to time, especially from Agro-Economics, Crop Protection and Crop and Soil Management. In the past, researcher participation varied according to changes in the priorities assigned to different research themes, as well as individual interest.

In implementing the OFR programme, MARIF has received support and technical assistance from two non-Indonesian institutions. Project ATA-272 (Technical Cooperation Indonesia/The Netherlands: Strengthening MARIF) has provided capital equipment, training, expatriate staff and, more recently, additional operating funds.[1] Technical assistance on OFR procedures has been provided by the International Maize and Wheat Improvement Center (CIMMYT), largely through occasional visits from staff posted to the Asian Regional Office in Bangkok, Thailand.

THE STUDY AREA

MARIF has identified four different *palawija* production environments in East Java: the limestone hill area of south-eastern Java; the irrigated mid-altitude plains on alluvial and young volcanic soils; the rainfed mid-altitude areas on young volcanic soils; and the rainfed higher-altitude areas, on young volcanic soils and volcanic ash. Some characteristics of these production environments are shown in Table 7.1. MARIF has conducted adaptive research on *palawija* in two of these environments, the limestone hill area and the rainfed mid-altitude young volcanic soils area.

This report focuses on the rainfed mid-altitude young volcanic soils area in Malang District, where most of the early research was conducted (*see* Figure 7.1). The major soil categories in the district are young volcanic soils (36%), volcanic ash (18%), limestone and lithosols (37%) and alluvial soils (9%). The young volcanic soils area includes latosols (inceptisols; 60%), regosols (entisols; 26%) and other types (14%). These are known to be low in organic matter and deficient in phosphate and other plant nutrients, such as sulphur and zinc.

Table 7.1   Major *palawija* production environments in Malang District

Land type	Soils	Dominant crops	Altitude (m a.s.l.)	Physical area (%)
Rainfed	Limestone	Maize, cassava, grain legumes, sugar cane	< 600	43
Rainfed	Young volcanic	Maize, dry-seeded rice	400-700	37
Irrigated	Alluvial and young volcanic	Transplanted rice, maize	400-700	15
Rainfed	Young volcanic and volcanic ash	Maize, horticultural crops	400-1500	5

Source: Dahlan, 1987

Figure 7.1   East Java, showing Malang study area and MARIF headquarters

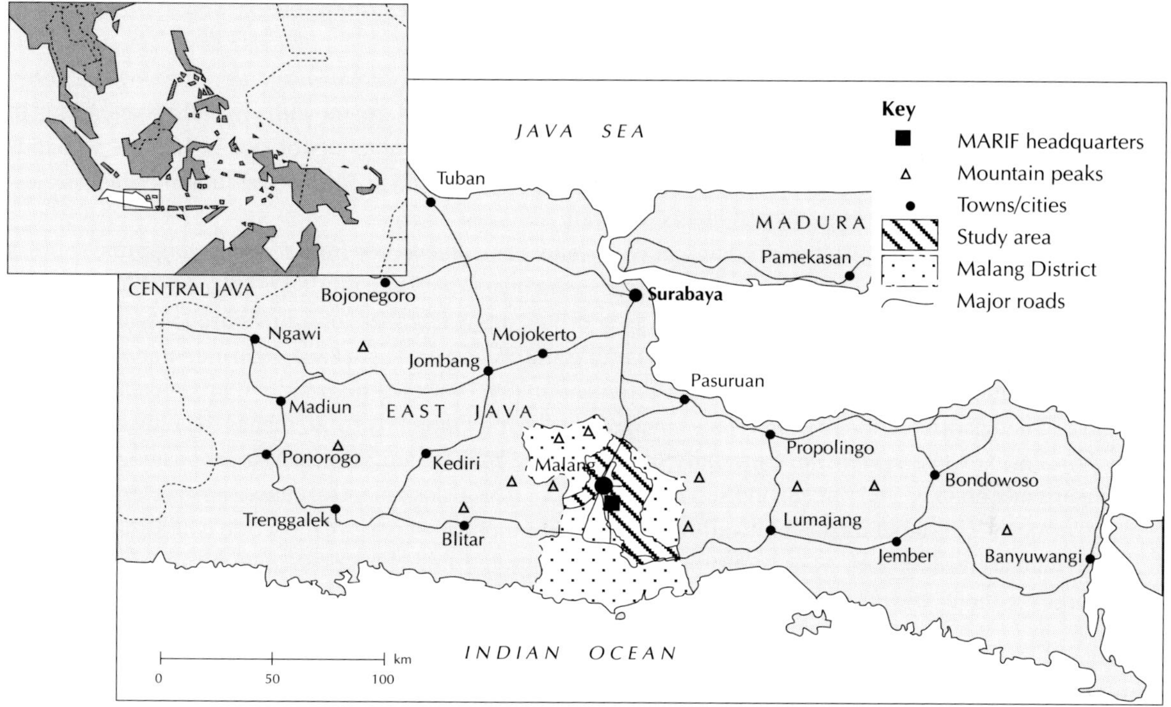

The study area in Malang District covers about 30 000 ha of physical area, or 60 000 ha of harvested area, as most fields are double-cropped; this would be roughly equivalent to 40 000 farms. The area is considered to be reasonably representative of an additional 150 000 ha of *palawija* area in East Java. The average annual rainfall in the study area is 2130 mm, most of which falls in 5-6 months (over 200 mm rain per month). The rainfall varies from year to year and from one location to another. The average annual temperature is 24°C, with an average minimum of 18.6°C and a maximum of 25.7°C.

Cropping patterns and sources of income

Given the relatively favourable rainfall conditions, two crops are normally grown per year. The dominant crops in the study area are maize and upland rice; cassava, sugar cane and grain legumes are also grown. Common cropping patterns include maize followed by maize, or upland rice followed by maize. Cassava is often restricted to field borders or to an occasional row within the field.[2] Grain legumes are occasionally substituted for maize in the second crop.

The first (rainy season) crop, whether maize or upland rice, is normally planted between September and November, depending on the onset of the rains. Rainy season maize is usually harvested after 90-100 days, between December and early March; upland rice takes longer to mature and is harvested about a month later.

The second (post-rainy season) crop, usually maize, is ideally planted immediately after the harvest of the first crop but excess rain at this time frequently forces farmers to delay planting for up to several weeks. Post-rainy season maize planted after upland rice tends to be planted especially late, often several weeks after the main planting period.

Farms in East Java are very small by almost any standard. In the study area, an average farm includes only 0.8 ha of rainfed land suitable for field crops (*tegal*). Many study area farmers with *tegal* also have fields in other production environments. About a third of them produce small amounts of rice in bunded wetlands (*sawah*). Most farmers also have home garden plots (*pekarangan*), which are planted to numerous species, including perennials, and feature several layers of canopy.

Rice from *sawah* fields and citrus products, clove, bananas and other crops from the *pekarangan* are sources of income that can be as important as that derived from growing *palawija* on rainfed *tegal* fields. Most farmers also raise livestock, especially cattle and poultry. In addition, many members of farm households are engaged in off-farm and non-agricultural activities, such as cutting and processing sugar cane, light manufacturing work, small trading and retailing; and many off-farm activities involve seasonal migration to urban centres.

Thus, although production of *palawija* is a major source of income it is by no means the only source. For many families, however, *palawija*, especially maize, are the major food staples. When consumed directly, maize is usually cracked and boiled, and served as if it were rice. Consequently, consumers prefer flint-grain types. Curiously, consumers also prefer yellow maize, unlike most other parts of the world where maize is used for direct human consumption. Maize consumption is concentrated among lower-income families in rural areas and families without *sawah* suitable for rice production.

Input and product markets

On average, about 70% of the maize produced in the study area is used for direct human consumption.[3] The rest is sold, usually to private traders at harvest time. The proportion of maize sold for cash will probably increase rapidly in the future, however, as the demand for maize as livestock feed grows.

Maize markets in Malang are efficient (low unit costs for marketing services) and well integrated with markets in other parts of Indonesia. Apart from imports and exports, which have been negligible in the recent past, maize marketing is in the hands of private traders. Maize prices at the farm level in East Java are strongly influenced by several factors, including demand for maize in the Surabaya and Jakarta feed mills, which also receive supplies from other producing regions (Timmer, 1987).

Input markets in the study area also operate reasonably well. Fertilizer and pesticides are readily available at subsidized prices, although the level of subsidy is now lower than it was during early phases of the OFR programme. In 1984, when the programme was initiated, farmers in Malang could exchange 1 kg of nitrogen for about 1 kg of maize; this has since increased to 1 kg of nitrogen for about 2 kg of maize. Improved maize seed, hybrid and open-pollinated, is readily available from private sector sources.

RESEARCH ACTIVITIES

On-farm research on *palawija* has been conducted continuously by MARIF in the study area since the 1984 post-rainy season. Although maize, cassava and grain legumes have all been studied, most of the early work focused on maize. The content of this chapter reflects that focus. A total of 11 research cycles (two cycles per year) have been completed since 1984. Research activities have included exploratory and formal surveys, a wide range of on-farm trials, and laboratory tests. These activities have taken MARIF staff into dozens of villages and have led to cooperation with hundreds of farmers.

The sheer number of activities, especially the researcher-managed on-farm trials, makes it impossible to describe each one individually here. Instead, the following discussion on developing and testing hypotheses on different problems and the relevant system interactions draws on a synthesis of results from assorted research activities. A summary of the research activities conducted between 1984 and 1989 is presented in Table 7.2.

Table 7.2   On-farm research activities in the study area, cycles 1 - 10

Activity	1 1984 p-r[1]	2 1984-85 r	3 1985 p-r	4 1985-86 r	5 1986 p-r	6 1986-87 r	7 1987 p-r	8 1987-88 r	9 1988 p-r	10 1988-89 r
On-farm trials:										
Exploratory	*	—	—	—	—	—	—	—	—	—
Variety	*	*	*	*	*	—	*	—	—	—
Fertilizer	—	*	—	*	*	*	*	*	—	—
Crop protection	—	—	—	*	*	—	—	—	—	—
Verification	—	—	*	*	*	*	*	*	*	*
Joint extension	—	—	—	—	—	—	—	*	*	*
Surveys:										
Exploratory	*	—	—	—	—	—	—	—	—	—
Maize production	—	*	—	—	—	—	—	—	—	—
Seed quality	—	—	*	—	—	—	—	*	—	—
Pest and disease	—	—	—	*	*	*	*	*	*	*
Farmer inputs	—	—	—	—	—	—	*	*	*	*
Farmer assessment	—	—	—	—	—	—	—	—	*	*
Other activities:										
Pot experiments	—	*	—	*	*	*	—	—	—	—
Soil analysis	—	*	—	—	*	*	—	—	—	—
Field days	*	—	*	—	*	—	*	—	*	*

Note: 1. p-r = post-rainy season; r = rainy season.
Source: van Santen and Dahlan, 1989

Diagnosis and planning

The initial diagnosis

In the early days of the MARIF OFR programme, a decision was taken to restrict active research to maize, at least temporarily. As the programme developed, however, additional *palawija* crops in other study areas were included.

The main reasons for this early focus on maize were that:

- the OFR programme was a pilot effort, and MARIF was unsure as to whether it should therefore invest substantial resources in this type of research; by focusing on one enterprise it reduced the 'learning cost';

- maize was the major *palawija* crop grown in the largest study area which was easily accessible to MARIF staff;

- the early programme participants were maize specialists.

To a certain extent, the programme was initiated to answer a riddle: why were only a limited number of study area farmers using improved maize varieties? However, once the research team members began visiting farmers' fields and exchanging views with farmers the range of subjects the programme sought to address expanded swiftly.

Initial diagnostic work included an informal, exploratory survey and a formal verification survey of farmers' circumstances, practices and opinions. The first exploratory survey was conducted during the post-rainy season in 1984 by a small team composed of a local extension subject-matter specialist and researchers from MARIF, the ATA-272 project and CIMMYT. The maize crop had tasseled and was reaching maturity. Over about a week and a half, the team visited villages, talked to village authorities and interviewed farmers and their families. Team members made their way through small fields of maize and other *palawija*, each field enclosed by fruit and fuelwood trees, with *sawah* rice fields in the valley floor below and the volcanoes of East Java dominating the horizon.

By the end of the survey the riddle had still not been solved, but it had been modified (MARIF, 1985). The farmers appeared to be managing their maize fields fairly intensively, with adequate land preparation, row planting, excellent weed control and high levels of fertilizer and farmyard manure, and yet the maize plants, regardless of variety, had spindly stalks and discolored leaves, and in many fields lodging was a serious problem. The farmers reported low yields, normally less than 2 t/ha, consistent with researchers' observations and published data. On the experiment station, maize yields in excess of 5 t/ha were fairly common.

The survey results suggested that there were three factors which merited priority: plant population management, soil fertility management, and variety (it later became evident that other problems, especially control of early season insects, were equally important). The researchers hypothesized that the lodging problem was linked to the problem of poor stalk strength, and that this might be caused by excessive plant densities, hence the focus on plant population management. Stalk weakness and leaf discoloration were thought to be caused by imbalances in nutrient availability (published soil descriptions explicitly noted deficiencies of phosphate and sulphur), hence the focus on soil fertility management. And it was thought that overall plant vigour might be related to farmers' reliance on unimproved varieties, hence the focus on variety. In addition to these hypotheses, researchers thought that the excessive plant densities might be attributed to farmers' need for livestock fodder. The interactions between these hypotheses are shown in Figure 7.2.

Early diagnostic research

In the first cycles of research the emphasis was placed on further diagnosis. Trials and surveys were both used in this diagnosis. A formal survey was conducted in cycle 2, with the aim of quantifying farmers' practices and

Figure 7.2   Hypotheses on the problems and causes of low maize yields in the study area, after exploratory survey

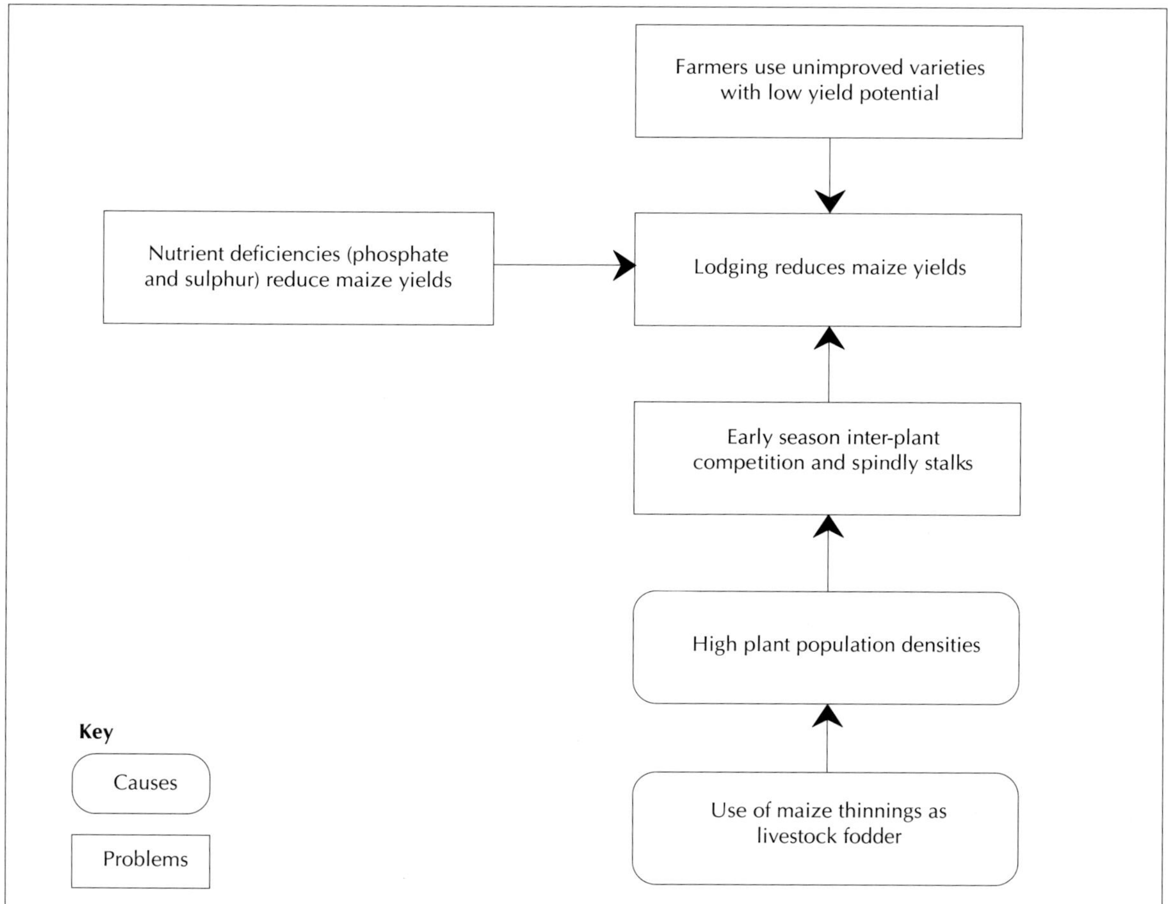

testing some of the hypotheses formed during the exploratory survey. This survey featured a questionnaire (which was written in Indonesian, and translated into Javanese and Madurese) and two-stage random list sampling.

The formal survey confirmed many of the initial hypotheses. It found that less than a quarter of the farmers were using modern varieties. All the farmers reported that they were using nitrogen fertilizer, at the astonishingly high average rate of 162 kg/ha N, but with little of it being applied at planting time. In contrast, less than a third of the farmers reported using phosphate and most of these were concentrated in one or two villages. Questions of nitrogen use efficiency arose, to complement questions on phosphate and sulphur deficiency.

The plant density of the maize fields was estimated at 150 000 plants/ha for traditional unimproved varieties and slightly less for improved materials, and most farmers reported either thinning or 'removing bad plants'.[4] The extensive lodging observed by researchers in the exploratory survey could easily be accounted for by this

Table 7.3   **Maize production practices in the study area**

Practice	Estimate unit
Plant traditional, unimproved varieties	48 %
Plant older, improved varieties	29 %
Plant modern, improved varieties	23 %
Use self-supplied seed	75 %
Average density at planting	150 000 plants/ha
Take out bad plants	83 %
Use thinnings as fodder	25 %
Applied manure last season	57 %
Applied nitrogen this season	100 %
Average nitrogen dose	162 kg/ha
Applied nitrogen at seeding	18 %
Applied first nitrogen after 3 weeks	78 %
Applied phosphate this season	30 %
Use pesticides	13 %
Average maize yield, local varieties	1.8 t/ha

Source: MARIF, 1985

combination of overplanting, extremely high nitrogen doses and little or no phosphate. Table 7.3 summarizes some of the results of the formal survey.

The formal survey, however, failed to confirm researchers' hypotheses on the causes of the high plant density. Only a quarter of the farmers reported that they were using maize thinnings as a major source of livestock fodder; other sources of fodder, including grasses and weeds, were reported as being more important, even during the maize growing season. Farmers with livestock indicated that fodder scarcity was not an issue.[5] Another explanation for the planting practice used by the farmers had to be found, and researchers considered two possibilities: that farmers were using high seed rates to compensate for a perceived risk of poor germination because of uncertain seed quality, or that they were using these high seed rates to compensate for expected pest damage.

Seed quality is easily measured through germination tests of farmer-stored seed at planting time. For various reasons, these tests were postponed for several cycles, delaying clarification of the causes of the farmers' planting practice. In the end, however, the tests showed very acceptable germination rates for farmers' seed. As to the pest damage explanation, farmers reported being familiar with numerous pests and diseases but said that none of these reduced maize yields significantly; only 13% of them reported using any pesticides on their maize. Thus, fodder requirements, seed quality and compensation for pest damage did not seem to account for the use of high plant densities.

During the first two cycles, however, farm surveys and laboratory tests were not the only diagnostic tools that were used. Exploratory trials and fertilizer and variety trials were also employed and they proved to be immensely useful, albeit in an unexpected way. Researcher-managed trial locations in the first two cycles were severely damaged by shootfly (*Atherigona* spp.), especially where planting had been a little late. Farmers indicated that this incidence of damage was normal and thus had not considered shootfly a particular problem when asked about pest damage during the survey.

After two cycles of diagnostic research, the hypotheses about problems and their causes evolved substantially. The list of problems now included: inefficient use of nitrogen fertilizer; phosphate and/or sulphur deficiencies; use of a maize variety with a low yield potential; damage from shootfly, especially in the post-rainy season; inter-plant competition, caused by high plant densities; and lodging. The revised set of hypotheses is given in Figure 7.3.

Figure 7.3   Hypotheses on the problems and causes of low maize yields in the study area, after two cycles of diagnostic research

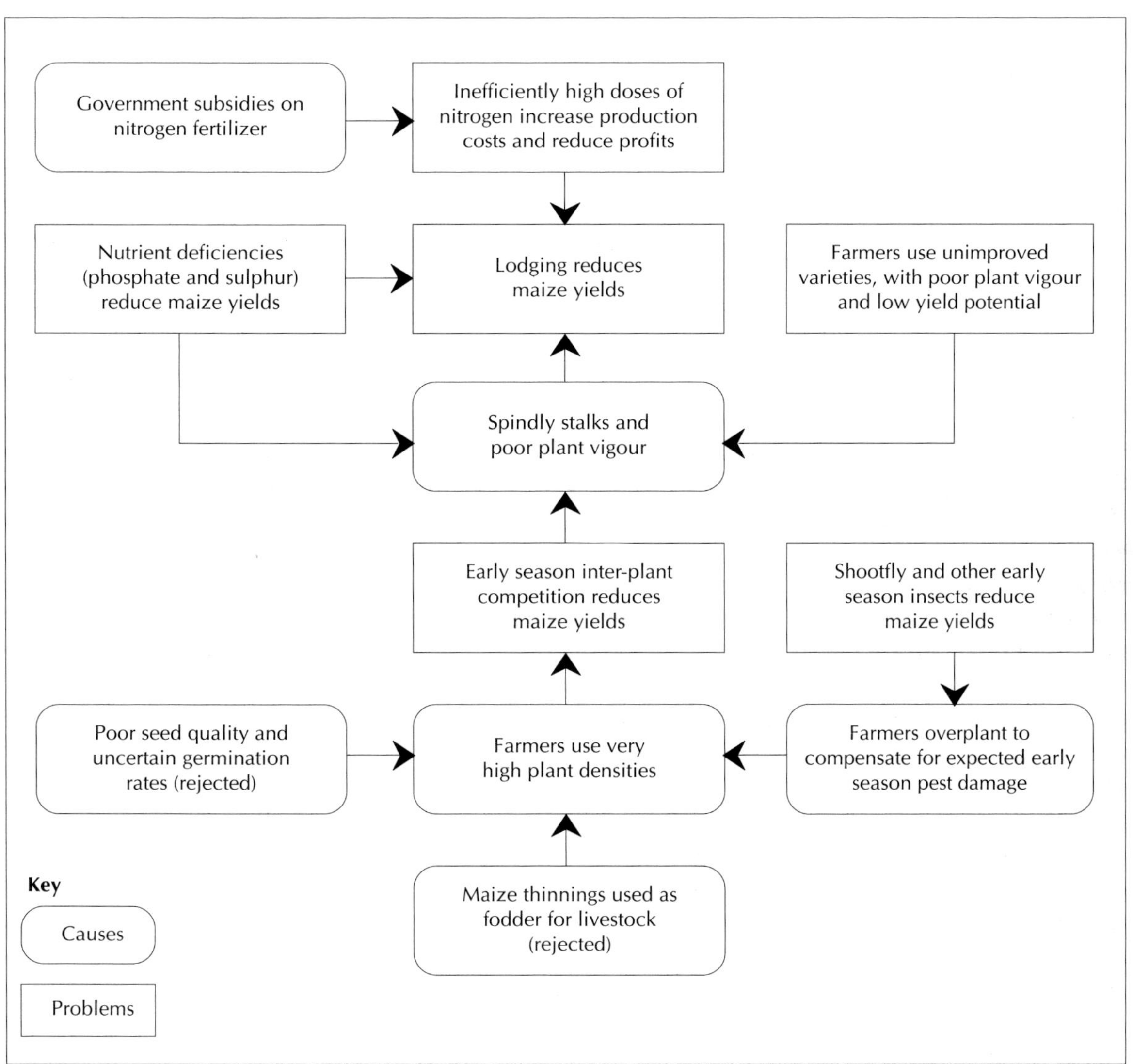

Setting research priorities

Initially, the question of setting research priorities was given little attention. Given the numerous interactions between problems and causes, it was felt that all the identified problems could be addressed by a relatively small number of research activities:

- early season insect control was expected to help solve three problems: yield loss from insect damage; yield loss from inter-plant competition (overplanting would no longer be required if insects were controlled); and yield loss from lodging (lodging would be reduced if plant density were reduced);

- research on nitrogen, phosphate, potassium and sulphur levels, and method and timing of application, was expected to help solve the following problems: nutrient deficiencies; the inefficient use of nitrogen; and lodging (stalk strength was expected to improve with lower rates of nitrogen combined with higher rates of phosphate);

- varietal research was expected to help solve problems associated with nitrogen efficiency and lodging, as well as to improve yields through higher yield potential.

This implied a rather modest research effort focusing on pest control, fertilizer management and variety. No scoring models were used to formally prioritize problems or possible solutions.

Research on plant protection

An emphasis on plant protection in the OFR programme was initially fairly controversial. Some researchers felt that methods of shootfly control were well described in the broader agronomic literature and that little was to be gained by pursuing the issue further. Shootfly damage was known to be typically associated with late planting, and control practices such as manipulation of planting time or host plant resistance were well known, as were a number of chemical control measures. Other researchers felt that further diagnostic studies were needed before proceeding with detailed work on alternative methods of control. They considered that there was a need for a better understanding of the causes and incidence of shootfly attack and the yield losses associated with shootfly damage (Supriatin et al., 1985).

Several kinds of research activities were undertaken in the following crop cycles to further define the pest control problem and assess possible solutions. Shootfly surveys were conducted, on-station crop protection trials were implemented and pest control was included in joint researcher-farmer verification trials. Survey data were reviewed and further informal surveys were conducted among collaborating farmers and their neighbours.

Field surveys to estimate the incidence of shootfly damage in a random sample of farmers' fields were conducted in cycles 4, 5 and 6 (the 1985-86 rainy season through to the 1986-87 rainy season). The average proportion of maize plants in farmers' fields damaged by shootfly was found to be about 10% in cycle 4 and about 18% in cycle 5, with considerable variation between fields and between villages; in some instances the proportion reached 50%. These surveys proved immensely frustrating, however, and they almost certainly underestimated the extent of the shootfly problem because of systematic measurement error. Shootfly damage proved difficult to observe in practice because most of the affected plants had died or had been removed by farmers before the survey team could count them. Eventually, this technique was abandoned.

Informal interaction with collaborating farmers and their neighbours led to useful information on the causes of the shootfly problem. The farmers confirmed that pest problems were more severe with late plantings than with early plantings, and were particularly severe in late plantings of post-rainy season maize. They gave a number of reasons why planting this second crop might be delayed (*see* Figure 7.4).

An on-station trial was conducted by a MARIF entomologist in cycle 4. The use of chemical pesticides[6] was compared with two untreated control treatments, one planted at a 'normal' density of about 80 000 plants/ha (two to three seeds per hill) and the other at the 'farmer' density of over 120 000 plants/ha (four to five seeds per hill). In general, pesticide use resulted in substantially higher yields than either of the unprotected treatments.[7] The entomologist concluded that the farmers' practice did provide some protection against shootfly (harvest density was as high as with chemical protection), but yields were nonetheless lower, probably because of inter-plant competition (Supriatin, 1986).

Figure 7.4   Hypotheses on the problems and causes of pest damage in post-rainy season maize

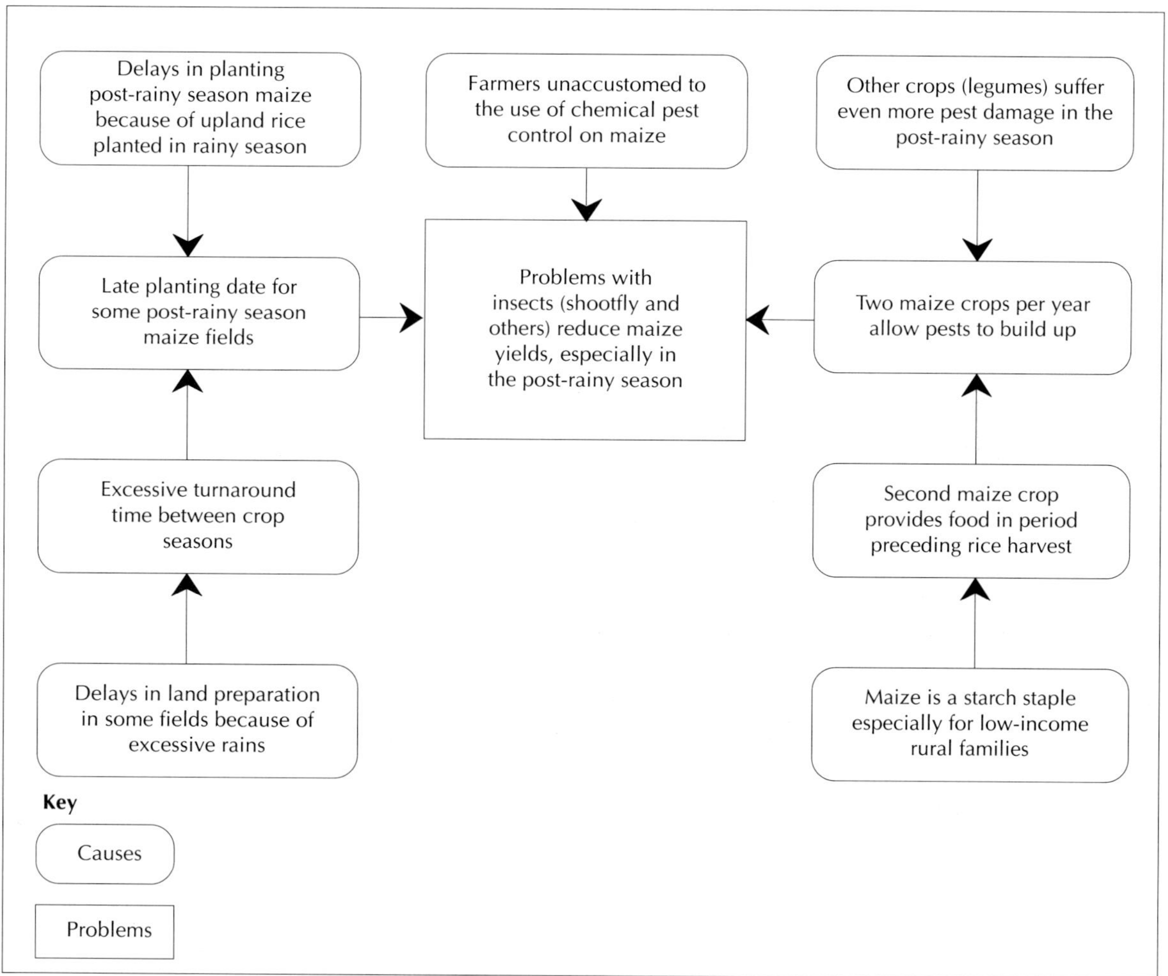

Some of the most interesting information on the effect of plant protection on grain yields, however, came from a series of simple on-farm verification trials. In these trials, comparisons were made between treatments that were identical (apart from plant protection) in four different research cycles.[8] In three of these cycles, comparisons were made between the unmodified farmer practice (planted and managed entirely by the farmer) and that same practice but with chemical pest control.[9] Some results of these trials are given in Table 7.4.

Table 7.4   Verification trial results with/without plant protection, cycles 3, 5, 6 and 7

Variable	Cycle 3 1985 post-rainy	Cycle 5 1986 post-rainy	Cycle 6 1986-87 rainy	Cycle 7 1987 post-rainy
Grain yield (kg/ha) (no plant protection [1])	2321	2887	4178	3019
Grain yield (kg/ha) (plant protection)	3171	3856	4953	3433
Yield difference (kg/ha)	850	969	775	414
Number of locations	8	3	7	16
Yield difference significant at 5%?	Yes	No	No	Yes

Note: 1. In cycles 3, 5 and 7, a comparison was made between the unmodified farmers' practice vs. that practice plus chemical pest control; in cycle 6, a comparison was made between the farmers' practice plus improved variety and planting method vs. those practices plus chemical pest control.

Sources: Dahlan, 1987; MARIF, 1987

The choice of pesticide for these verification trials (a low dose of Carbofuran 3% granules, applied in the hole with the seed at planting time) was made in the cycle 3 verification trial series and was never modified. The choice was based largely on existing knowledge ('on-the-shelf' technology). Although the technique was chosen so casually, the on-station trial conducted in cycle 4 confirmed that it was as effective as selected alternatives and subsquent work has shown it to be surprisingly attractive in many ways. First, it appears to be immensely profitable; breakeven yield increases were estimated at less than 100 kg/ha, negligible compared to the large yield increments usually observed in trials. Second, it is exceptionally divisible (can be tried on a small scale), is simple to implement and is easy to master. Third, reductions in government subsidies for Carbofuran appeared to have little impact on the profitability of the technique.

However, the OFR team has never been entirely happy with this choice because of the possible effects of Carbofuran on non-target insect species and because of concerns about farmer safety (although it should be noted that the 3% granular formulation was specifically chosen to minimize the danger to farmers handling this chemical). In several planning meetings it was recommended that safer control strategies be identified and examined, either by the research team or by MARIF entomologists. To date, however, researchers have been unable to identify an alternative control practice using readily available inputs that is equally profitable.

Research on plant population management

After early diagnostic activities showed a strong correlation between early season insect damage and the farmers' overplanting practice, the themes of insect control and plant population management were seen by most OFR staff as inextricably linked (*see* Figure 7.4). They assumed that solving the problem of early season pest damage would foster spontaneous adjustments by farmers in plant stand management — that is, reductions in seed rates.[10] Thus, relatively little research was done on the theme of plant population management *per se*.

One research activity that did examine the effects of plant and harvest densities on maize grain yields was the set of plant population trials conducted in cycle 5 (1986 post-rainy season). These trials compared five different combinations of spatial distribution and plant density. The results suggested that maize grain yields would be reduced by either excessively low or excessively high harvest densities. Harvested densities of 70 000 plants/ha gave the highest yields (*see* Table 7.5). These trials were somewhat inconclusive, however, in that yield differences were not statistically significant.

Table 7.5   **Results from plant population trials, cycle 5 (post-rainy season, 1986), three locations**

Distance between rows (cm)	Distance between hills (cm)	Seeds per hill (units)	Plant density (000 pl/ha)	Approximate harvest density (000 pl/ha)	Yield (kg/ha)
80	25	5	250	106	4.0
80	35	3	107	71	4.3
80	35	2.5	89	60	4.0
80	40	2.5	78	55	3.8

Source: MARIF, 1987

Verification trials, the major source of data generated by the OFR programme, were of little use in formally assessing plant population management issues. This was because plant population tended to be closely associated with other factors: with variety in cycles 5, 6 and 7; with soil fertility management in cycles 7 and 8; and with plant protection in cycle 8. Indirectly, however, these trials provided further support for the hypothesized link between early season insect control and seed rates. Among the farmers who collaborated in these trials, those who adopted chemical insect control did, in fact, spontaneously reduce seed rates. This showed that perhaps the researchers' faith in the strength of system interactions was well founded after all. One of the aims of the adoption survey described later in this chapter was to examine the issue of system interaction.

Research on soil fertility management

Soil fertility management was seen from the outset as an extremely important factor. Early diagnostic work had shown that farmers' maize yields were usually below 2 t/ha, despite applications of nitrogen fertilizer that often exceeded 150 kg/ha, as well as some use of manure. Leaf discoloration during early growth periods was fairly widespread. More mature plants often had spindly stalks and lodging was common.

The research team considered that the main soil fertility management issues were low efficiency of applied nitrogen and possible deficiencies of other nutrients. Agronomic knowledge, combined with contributions from the collaborating farmers, led to the identification of several possible causes of low nitrogen efficiency (*see* Figure 7.5). These included: high nitrogen application rates; late application (most farmers reported applying half of the nitrogen at about 30 days and the other half at about 60 days after planting); early season inter-plant competition (related to the plant stand management and plant protection issues described above); widespread use of unimproved germplasm (improved varieties are often more responsive to fertilizer); interactions between nutrients (the uptake of nitrogen might be constrained by deficiencies of phosphate or sulphur).

As research progressed, many of the above assertions were supported by evidence from surveys or trials. However, a causal link was rarely proven. For example, it was established that there was inter-plant competition but the extent to which this reduced nitrogen efficiency was never formally measured.[11]

Researcher-managed trials

The suspicion that farmers might be using excessively high levels of nitrogen was supported by the results from researcher-managed fertilizer trials. These trials evaluated levels of nitrogen and phosphate, and occasionally potash or sulphur, normally utilizing a full factorial arrangement of treatments. Non-experimental variables were set at different levels in different sets of trials, but typically included chemical shootfly control with reduced plant stand, as it was thought likely that farmer adoption of these practices would precede adoption of improved fertilizer management practices. The trial results, summarized in Table 7.6 (*overleaf*), suggested that farmers might be able to reduce nitrogen levels considerably with little or no yield loss. The results also confirmed that phosphate deficiencies limited maize grain yield. The application of phosphate consistently resulted in higher yields (*see* Table 7.7 *overleaf*). However, N x P interactions were not commonly found (data not shown).

Compared to the data on nitrogen and phosphate levels, other agronomic data on soil fertility management are fairly sparse. For example, the effect on maize yields of the timing of nitrogen application was studied in only one trial during one research cycle, with inconclusive results. Similarly, other issues, including variety/ fertilizer interactions, the effects of inter-plant competition on nitrogen efficiency, and the possibility that nutrients other than phosphate might be lacking, were never thoroughly assessed.[12]

Verification trials

The main reason why the above issues were not pursued with more vigour was that, early on, a soil fertility management package that appeared effective and profitable was put together. Joint researcher-farmer activities to verify this package largely replaced more technical agronomic research efforts. The results of the verification trials[13] with regard to soil fertility management are summarized in Table 7.8 (*overleaf*). The treatment descriptions in this table seem simple enough, but in fact they mask considerable variation between farmers, villages and locations. The 'farmer fertilizer management' treatment, for example, was exactly that — farmers

Figure 7.5 Hypotheses on the problems and causes associated with soil fertility management

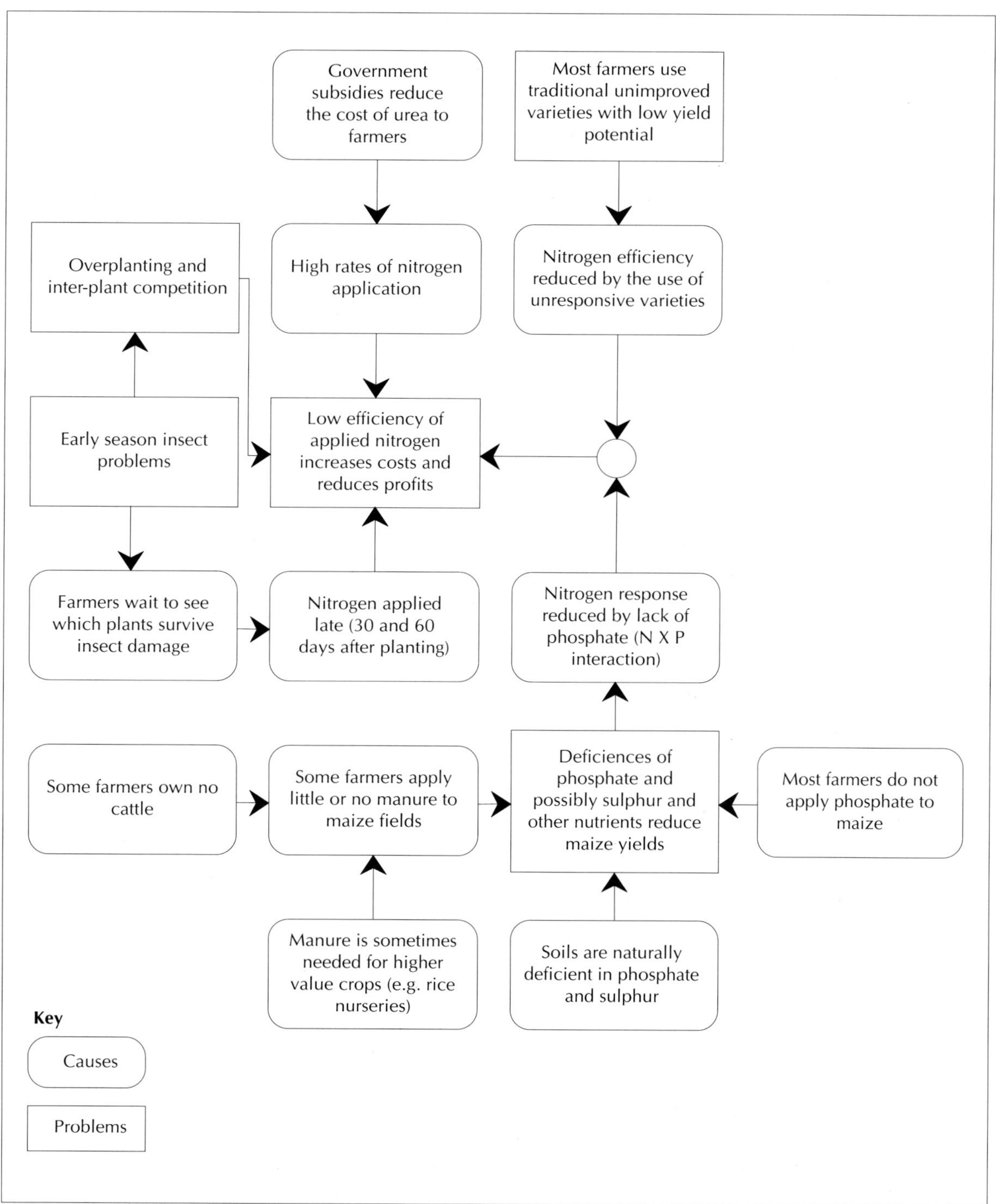

Table 7.6   Effect of nitrogen application on maize grain yield (several locations and research cycles)

Nitrogen level (kg/ha)	Cycle 4 Yield (t/ha)	Cycle 5 Yield (t/ha)	Cycle 6 Yield (t/ha)	Cycle 8 Yield (t/ha) [1]
69	—	—	$3.9\,b^2$	
92	3.8 a	4.4 a	4.3 a	n.s.[3]
138[4]	3.8 a	4.4 a	—	n.s.
Number of locations	2	1	16	3

Notes: 1. Data are not presented, but the ANOVA indicates that there is no significant yield response to the indicated increment in nitrogen levels.
2. Values followed by the same letters are not significantly different at 5% DMRT.
3. n.s. = not significant.
4. Farmers' practice.

Sources: MARIF, 1987, 1988; Dahlan et al., 1987

Table 7.7   Effect of phosphate application on maize grain yield (several locations and research cycles)

Phosphate level (kg/ha)	Cycle 4 Yield (t/ha)	Cycle 5 Yield (t/ha)	Cycle 6 Yield (t/ha)	Cycle 8 Yield (t/ha)
0	3.6	4.2 b	3.8 b	—
45	4.0	4.1 b	4.2 a	4.1 b[1]
90	3.9	5.1 a	4.2 a	4.5 a
Number of locations	2	1	16	3

Note: 1. Values followed by the same letters are not significantly different at 5% DMRT.

Sources: MARIF, 1987, 1988; Dahlan et al., 1987

were encouraged to select and use whatever practices they judged to be most suitable. As a result, the choice of level, method and timing of nutrient application differed from one farmer to another, as well as from location to location and from cycle to cycle.

Farmers often applied levels of nutrients, especially nitrogen, far above the levels used by researchers. The cost per ha of the farmers' practices was often significantly higher than the suggested alternative. This makes all the more striking the fact that the 'improved fertilizer management' treatment consistently led to higher yields. However, it is not known whether these higher yields resulted mainly from phosphate application or from changes in the timing of nitrogen application.

It should be noted that, in these trials, improved fertilizer management was not directly superimposed on the unmodified farmers' practice. Both the farmers' and the alternative fertilizer management practices were implemented on fields in which plant protection and complementary adjustments in plant population management had already been implemented. This reflected the researchers' belief that farmer adoption of shootfly control (and reductions in seed rates and plant density) would precede any adjustments in fertilizer

Table 7.8   **Summary of verification trials results on soil fertility management (several locations and research cycles)**

		Treatment		
Research cycle[3]	Farmer fertilizer management[1] (t/ha)	Improved fertilizer management[2] (t/ha)	Yield difference (t/ha)	Number of locations
3	3.2	4.4	1.2	11
4	4.4	5.4	1.0	1
5	3.4	4.1	0.7	4
6	4.6	5.1	0.5	7

Notes: 1. The farmers' practice, plus Arjuna seed, the improved plant population management practice and crop protection. The farmers' fertilizer management practice varied from location to location but typically included high nitrogen doses (in excess of 150 kg/ha) applied in two equal doses, the first of which coincided with the first weeding at about 3 weeks after seeding and the second about 6 weeks after seeding. This treatment rarely included a phosphate application.

2. 'Improved fertilizer management' normally included 46 kg/ha nitrogen and 92 kg/ha phosphate applied at seeding and an additional 90 kg/ha nitrogen applied at 30 days after seeding. Note that this is somewhat above the 92 kg/ha total dose that seems most profitable, as measured by the N x P factorial trials. In fact, the nitrogen level used in this treatment was reduced in later cycles of research. Other practices were as above.

3. In research cycles other than the ones shown, it was not possible to make a direct comparison of soil fertility management practices, as these were confounded with other factors.

Sources: Dahlan et al., 1987; MARIF, 1987

management practices. This, in turn, is a measure of the importance given by researchers to shootfly damage and early season inter-plant competition as causes of low nitrogen efficiency.

Economic analysis

A formal economic analysis of the fertilizer response data summarized above may seem beside the point. Recommended changes call for a reduction in nitrogen levels, earlier application of nitrogen and an increase in phosphate levels. For many farmers, the cost of the phosphate dose and any increase in labour costs associated with the change in nitrogen timing are likely to be entirely paid for by the reduced nitrogen cost. For some farmers, the combination of practices may actually reduce cost per ha as well as the cost per kg of maize grain produced. The extent to which this is true, of course, depends largely on the farmers' current nitrogen dose, which varies considerably.

Nevertheless, it is worth examining the very conservative case in which no savings are made through reducing the nitrogen dose, and all the costs of a phosphate application are recovered within one crop cycle.[14] Using yield response data from Table 7.7, a partial budget analysis reveals that even in this extremely conservative case, phosphate application produces acceptable rates of return (*see* Table 7.9 *overleaf*). In this analysis, 1988 farm-level maize prices from East Java are combined with 1988 world prices for a commonly used source of phosphate. As farm-level fertilizer prices tend to be lower than world prices because of government subsidies, the analysis underestimates phosphate profitability even further. The evidence suggests that phosphate use should be attractive and profitable to farmers.

Table 7.9   Partial budget analysis of phosphate levels

| | Phosphate level (kg/ha P_2O_5) | | |
Variable	0	45	90
Yield (t/ha)[1]	3.5	3.7	4.0
Gross benefits (000 Rp/ha)[2]	614	649	701
Phosphate cost (000 Rp/ha)[3]	0	11	22
Application cost (000 Rp/ha)[4]	0	2	2
Total costs that vary (000 Rp/ha)	0	13	24
Net benefits (000 Rp/ha)	614	636	677
Marginal rate of return		171%	361%

Notes: 1. Calculated from Table 7.7 (adjusted downwards by 10%).
2. Field price of maize estimated at Rp. 176/kg (MARIF, 1988); US$ 1 = Rp. 1730.
3. TSP (46% phosphate) price estimated at $155/t, or about Rp 248/kg (CIMMYT, 1990).
4. Assumes 1 person-day per ha of labour — more than is really required.

Research on variety

It will be recalled that the MARIF on-farm programme was launched initially to establish why so few farmers were using improved varieties. And yet, as the above discussion shows, varietal questions have not dominated the research agenda.

Research on variety has included surveys, researcher-managed trials and, for some research cycles, collaborative researcher-farmer verification trials. As shown in Table 7.3, diagnostic surveys conducted during the early research cycles confirmed that about 50% of the study area farmers used unimproved traditional varieties, such as Goter, Selli and Genjah Tongkol, while 25% used older improved varieties, such as Harapan, and 25% used more modern improved varieties, such as Arjuna, as well as hybrids developed by the private sector.

Widespread farmer use of unimproved varieties was considered to be a problem because improved materials were judged to have a higher yield potential. Four reasons were put forward to explain continued farmer use of unimproved materials: problems with the availability or price of improved seed; problems with germination rates of purchased seed; consumer preferences regarding such issues as taste and grain type; and low profitability of improved germplasm under farmers' management conditions. Most of the formal research activities focused on yield response and profitability of improved germplasm. Farmer assessment was relied on to identify any problems with non-yield characteristics of new varieties.

Researcher-managed trials were conducted in six seasons. A summary of the results of these trials is provided in Tables 7.10 and 7.11. Maize grain yield for different varieties and hybrids are reported by crop cycle, expressed in t/ha and as a percentage of the cycle-specific yield of Arjuna, the major improved open-pollinated variety (OPV). The implication of these results seems to be that yields from farmers' unimproved

Table 7.10   Maize grain yield for different varieties and hybrids, by research cycle (t/ha)

Variety	Cycle 1	Cycle 2	Cycle 4	Cycle 5	Cycle 7	Cycle 9
Farmers' varieties and old improved varieties:						
Genjah Tongkol	—	—	—	—	—	4.4
Harapan	5.0	3.7	—	—	—	—
Selli	—	—	—	—	—	3.4
Not identified	—	—	—	5.4	3.9	—
Improved varieties:						
Arjuna (MARIF)	4.8	4.1	4.0	5.1	4.1	4.1
Malang Composite	—	—	4.0	3.4	5.3	—
Muneng Synthetic	5.1	4.1	3.5	4.7	—	5.3
Suwan 1	5.0	4.1				
Hybrids:						
Hybrid CP-1	—	—	4.0	5.0	—	5.1
Hybrid C-1	—	4.2	4.3	4.7	—	4.6
Hybrid P-1	—	—	—	—	5.9	—

Sources: Dahlan et al., 1987; MARIF, 1988

Table 7.11   Maize grain yield for different varieties and hybrids, by research cycle (as a percentage of Arjuna yield for that cycle)

Variety	Cycle 1	Cycle 2	Cycle 4	Cycle 5	Cycle 7	Cycle 9
Farmers' varieties and old improved varieties:						
Genjah Tongkok	—	—	—	—	—	107
Harapan	104	90	—	—	—	—
Selli	—	—	—	—	—	83
Not identified	—	—	—	106	95	—
Improved varieties:						
Arjuna (MARIF)	100	100	100	100	100	100
Malang C9	—	—	—	—	—	—
Malang Composite	—	98	85	104	—	—
Muneng Synthetic	106	100	88	92	—	129
Suwan 1	104	100	—	—	—	—
Hybrids:						
Hybrid CP-1	—	—	100	98	—	124
Hybrid C-1	—	102	108	92	—	112
Hybrid P-1	—	—	—	—	—	144

Sources: Dahlan et al.,1987; MARIF, 1988

varieties are only slightly inferior to those of improved OPVs, and that hybrids have the potential to yield somewhat more than either of these. It should be noted, however, that available hybrids are of a longer maturity than Arjuna (and therefore can be difficult to fit into farmers' cropping systems) and have no downy mildew resistance.

The joint farmer-researcher verification trials described earlier also allowed direct comparisons to be made between the farmers' variety (chosen by the farmer) and an improved OPV (usually Arjuna). These comparisons provided little useful information, however, because of the variability in varieties chosen by the farmers. Many chose Arjuna and some chose a private sector hybrid; unimproved varieties were under-represented in these trials.

RESEARCH RESULTS AND TECHNOLOGY ADOPTION

During the course of several years of on-farm adaptive research, scientists at MARIF developed a number of practices — including shootfly control, reduced seed rates, limited thinning, reduced nitrogen application rates, the use of phosphate fertilizer and the possible use of improved varieties — which they considered might prove attractive to farmers. The scientists began by conducting diagnostic activities which identified productivity problems and then traced the causes of these problems through several different kinds of farming systems interactions. An understanding of the causes of productivity problems suggested, in turn, the appropriate research themes and activities to be followed. The practices seem immensely profitable, whether adopted singly or all together (*see* Table 7.12). The expectation of MARIF staff, however, was that adjustments in pest control and plant stand management practices were likely to precede adjustments in fertilizer management or variety selection.

During the latter part of the 1980s it appeared that the farmers in the study area were beginning to adopt some of the researchers' recommendations, and in 1989 it was decided to conduct an adoption survey to ascertain whether this was indeed the case. The survey fieldwork, comprising a brief questionnaire and two-stage stratified random sampling, was conducted during April 1990. It involved interviewing some 60 farmers living in 20 villages; six of these villages had had no previous exposure to MARIF's OFR programme. The preliminary results of the survey indicate that that farmers have been quick to adopt some of the recommended practices (*see* Table 7.13). Shootfly control has rapidly expanded in the study area and the farmers' former practice of overplanting and thinning in order to compensate for expected shootfly damage is now being dropped, as indicated in Table 7.14 (*overleaf*). Similarly, the use of improved varieties, especially Arjuna, has become widespread.

It should be noted that farmer adoption of recommended practices came during and after, not before, the implementation of the OFR programme, and that farmers collaborating in the trials tended to adopt improved practices before non-collaborators (*see* Figure 7.6 *overleaf*). This suggests a causal link between research effort and technology adoption (Traxler et al., 1990).

Apart from the consistency with which farmers use pest control practices, there now seems to be little difference in maize production technology between collaborating and non-collaborating farmers: as shown in Table 7.15 (*overleaf*) the two groups seem to have adopted pest control, reduced plant stand and improved germplasm about equally. Despite the time lag shown in Figure 7.6, non-collaborating farmers appear to have 'caught up' fairly quickly with the collaborating farmers. As indicated in Table 7.15, in most respects the

Table 7.12   Costs and benefits of farmers' practice vs. alternative practice[1]

Variable	Farmers' practice	Alternative practice
Yield (t/ha)	2.1	4.4
Adjusted yield (t/ha)	1.9	3.9
Gross benefits (Rp./ha)	Rp. 283 500	Rp. 594 000
Costs that vary (Rp./ha)[2]		
Seed	Rp. 6000	Rp. 37 750
Fertilizer	Rp. 68 250	Rp. 61 250
Insecticide		Rp. 27 500
Additional labour		Rp. 7500
Total: costs that vary	Rp. 74 250	Rp. 134 000
Net benefits (Rp./ha)	Rp. 209 250	Rp. 460 000
Marginal rate of return		420%

Notes: 1. Calculations based on subsidized prices for fertilizers, seed and insecticides prevailing in March 1989; US$ 1 = Rp. 1730.

2. Price of maize seed (local variety) about Rp. 150. Price of maize seed (Arjuna) about Rp. 1500. Farmers use about 40 kg/ha of seed whereas the alternative practice calls for 25 kg/ha.
Average fertilizer doses: farmer practice about 390 kg/ha urea; alternative practice about 150 kg/ha urea, 100 kg/ha ammonium sulphate and 100 kg/ha TSP (triple super phosphate). 1 kg fertilizer about Rp. 175, regardless of formula. Carbofuran 3G, 5 kg/ha at Rp. 1750/kg gives a cost of about Rp. 8750/ha. Ridomil 5 g/kg seed at Rp. 750 gives a cost of about Rp. 18 750 (included in experiments as protection against downy mildew, but had no effect on yield). Farm-level maize prices fluctuate over the season but are about Rp. 150 immediately after harvest.

Source: Van Santen and Dahlan, 1989

Table 7.13   Changes in farmer practices, 1984-90[1]

Practice	1984	1990
Plant unimproved varieties	48%	15%
Plant improved varieties	52%	70%
Plant hybrids	0%	15%
Seeds per hill at planting	3-5	2-3
Mean seed rate (kg/ha) [2]	42	28
Use pesticides for shootfly control	13%	78%
Thinning	83%	45%
Applied nitrogen this season	100%	100%
Applied phosphate this season	30%	38%

Note: 1. Sources include MARIF (1985), which reports on the 1984 baseline survey, and preliminary analysis of the 1990 adoption survey. Both surveys used two-stage random sampling, with sample sizes of about 60 farmers.

2. Seed rate is influenced by variety selection as well as by pest control. Lower seed rates are expected with improved varieties as well as in conjunction with shootfly control. The relative importance of each factor in the observed decline in seed rates since 1984 has not been estimated.

collaborating farmers are fairly representative of other farmers in the study area. Mean farm sizes are similar, although collaborating farmers are less likely than other farmers to specialize in maize and other field crops. Cropping patterns are similar but, of course, collaborating farmers received more attention from extension agents.

Many of the researchers' expectations regarding technology adoption were fulfilled. For example, shootfly control and modifications in plant stand management did, as expected, precede the adoption of other

Table 7.14   Farmer adoption of reduced plant stands

Variable	Estimate
Seed rate, 1984 (kg/ha)	42
Seed rate, 1980 (kg/ha)	28
Seeds per hill at planting, this season (median, farmer opinion)	3
Seeds per hill at planting 3 years ago (median, farmer opinion)	4
Plants per hill at 30 days (mean, direct field measurement)	2.32
Farmer opinion on changes in seed rate (this season vs. 3 years ago):	
Use the same amount of seed per ha	40%
Have reduced the seed rate	56%
Have increased the seed rate	3%

Sources: MARIF, 1985; preliminary analysis of the 1990 adoption survey

Figure 7.6   Cumulative adoption of pesticide use (Carbofuran) on maize

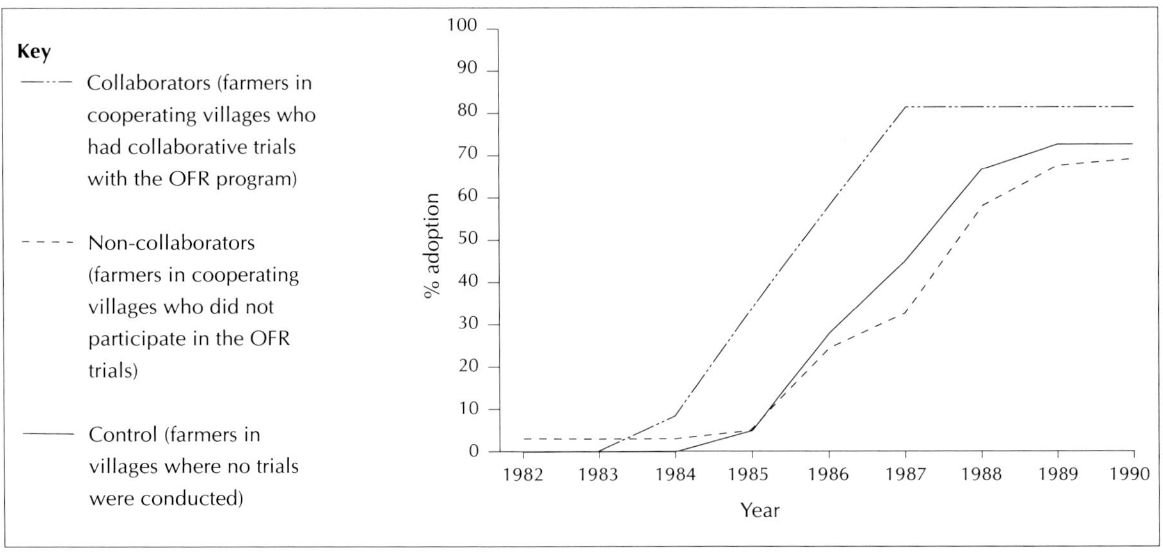

innovations. However, the researchers had also expected that farmers would find changes in fertilizer management practices more attractive than varietal changes, but instead farmer adoption of new varieties was widespread whereas there is little evidence that they have adopted the recommendations regarding phosphate and reduced nitrogen rates. Clearly, more work needs to be done in order to understand the farmers' viewpoint on these issues.

Table 7.15   Summary of results of the 1990 adoption survey, on-farm research collaborating farmers vs. other farmers

Variable	OFR collaborators	Others
Sample size	13	47
Farm size:		
mean *sawah* area (ha) [1]	0.72	0.19
mean *tegal* area (ha)	1.13	1.53
mean home garden (ha)	0.10	0.05
mean farm size (ha)	1.95	1.77
Grow maize on *sawah* land type	8%	6%
Grow maize on *tegal* land type	100%	100%
Cropping pattern:		
maize/maize	38%	25%
upland rice/maize	62%	75%
Frequency contact with extension: [2]		
never	0%	28%
once per season	0%	19%
twice or three times per season	92%	35%
whenever necessary	8%	18%
Use Carbofuran on other crops	69%	55%
Use Carbofuran on maize every season [3]	61%	21%
Use Carbofuran on maize this season	76%	78%
Seed rate (kg/ha)	28.7	29.3
Seeds per hill (this season)	2.7	2.8
Mean reduction, seeds per hill [4]	1.7	1.0
Thin maize (remove bad plants)	46%	44%
Variety selection:		
use hybrid	23%	12%
use Arjuna (improved OPV)	61%	72%

Notes: 1. Probability of t = .03 (one-tailed test).
2. Chi-square = 15.8, 4 df (probability of chance = .003).
3. Chi-square = 6.1, 1 df (probability of chance = .013).
4. Probability of t = .004 (one-tailed test).

RESEARCH-EXTENSION LINKS

There is increasing awareness among research managers that the impact of FSR or OFR on target farmers has been limited (Tripp et al., 1990). Some observers attribute this limited impact, at least partially, to weak research-extension links. Ewell (1989) argues that FSR or OFR should not be viewed as a substitute for extension, and describes five conditions useful for building effective links: a shared analysis of farmers' circumstances and problems; the existence of suitable technical alternatives; trained and committed professionals in both research and extension; a clear division of responsibilities between research and extension; and effective linkage mechanisms.

The official research-extension linkage mechanisms in the MARIF OFR programme are described by van Santen (1987) thus:

> ... formal lines of communication for passing relevant research results to extension services are not direct. New technologies ... are to be channeled via a central 'communication forum' consisting ... [of] AARD, the BIMAS program of the Directorate General of Food Crops (DGFC) and the Agency for Agricultural Education, Training and Extension (AAETE). This 'communication forum' refers the appropriate new technologies to the extension service, which is organized on a provincial basis.

Suggested recommendations, then, should in theory pass from researchers in Malang District, to their superiors in Jakarta, through the central 'communication forum', through the extension offices in Jakarta, back to provincial extension authorities, and finally to extension agents in Malang District. During the early cycles of the OFR programme, research-extension links were poor, mainly because the complexity of the research-extension linkage mechanisms inhibited the development of a common understanding among researchers and extension agents of farming practices and possible alternatives. Alternative practices had been developed by MARIF, but lack of extension involvement raised the prospect of slow rates of farmer adoption driven only by farmer-to-farmer contact.

However, it would be incorrect to say that there was no contact at all between MARIF researchers and extension agents. From the early cycles of the programme, researchers and extension agents collaborated in planning and implementing farmer field days, focused on verification trial field locations. In addition, village-level extension agents were normally involved in implementing trials located in their village. What was lacking, however, was active interaction regarding experimental themes and expansion of the programme efforts through a widespread implementation of demonstration trials in Malang District.

Informal contacts between MARIF researchers and district- and provincial-level extension agents eventually led to proposals for modifying the research-extension linkage mechanisms, and the interest of a senior provincial-level extension official led to improved contacts. However, the strengthened research-extension connection focused less on Malang District itself than on neighbouring districts with similar agroclimatic circumstances.

In 1988, plans were approved for joint research-extension diagnostic survey work and joint sets of verification trials. MARIF researchers and extension agents jointly conducted diagnostic exploratory surveys and verification trials in Mojokerto District to see if recommendations developed in Malang District could be transferred to similar neighbouring areas. In addition, a 'pre-production trial', covering 30 ha, was jointly implemented in Probolinggo District. It is hoped that these joint activities have laid the foundation for more fruitful research-extension liaison.

CONCLUSION

This chapter has described the activities of a team of commodity and disciplinary researchers with little previous 'systems' experience, how this team conducted diagnostic surveys and trials to better understand farmers' circumstances and problems, and how the survey results were used to design further trials. MARIF researchers have reason to be pleased that many of the recommendations developed as part of the OFR programme are being adopted by farmers. Part of this impact can be attributed to improved links with extension. Much work remains to be done, however, both in the study area reviewed here and in other study areas elsewhere in Indonesia.

Notes

1. The support provided by the ATA-272 project was not specifically aimed at strengthening the OFR programme but rather on strengthening MARIF as a whole.

2. This is in sharp contrast with the limestone area, in which cassava is far more important in farmers' cropping systems.

3. Although the OFR programme has also studied grain legumes and cassava, the early work, especially in the study area, focused on maize for reasons given later in the chapter, and this focus is reflected in this chapter,

4. There were a number of amusing misunderstandings over this practice until researchers realized that, to farmers, 'thinning' is a conscious operation whereby a certain proportion of plants are removed from the field at a particular stage of growth. In contrast, 'removing bad plants' is the continuous removal of plants damaged by downy mildew, insect attack or excessive inter-plant competition and thus is not considered to be a specific farming operation.

5. In the limestone areas of South Java, however, fodder scarcity is an extremely important issue. There, many farmers import fodder from other areas.

6. The chemicals tested included Carbofuran 3G (basal vs. whorl application), Promet 40 SD and Thiodicarb 75 WP (seed treatment). Carbofuran was tested at different doses: 0.15, 0.3 and 0.6 kg/ha a.i.

7. There were no significant differences between the chemical control treatments. These treatments yield, on average, 5 t/ha compared with 4.6 t/ha for the unprotected/high plant density treatment (farmers' practice) and 3.8 t/ha for the unprotected/low plant density treatment.

8. In the other cycles, the effect of plant protection was normally closely associated with the effects of fertilizer management and planting method.

9. Space does not permit a complete listing of verification trial treatments, as these tended to change over time.

10. Optimal plant and harvest densities for a wide range of maize varieties were fairly well known through on-station research. These optimal densities were well below those used by farmers, hence the assumption that adjustments would be towards lower seed rates.

11. This raises difficult questions of research resource allocation. Agronomic knowledge would endorse the inclusion of most of the listed factors as causes of low nitrogen efficiency. An insistence on formal quantification and proof of each causal link could result in excessive research resources being devoted to diagnostic activities, with little left over for assessing possible solutions to major problems.

12. This is not to say that these other factors were ignored. Sulphur, magnesium and other nutrients were occasionally studied, either as 'satellite' treatments in the N x P factorials or in 'double-pot' laboratory experiments. These studies were not pursued as intensively as the others, however.

13. These verification trials were the same ones used to assess plant protection and plant stand management, described earlier. For the research cycles and treatments chosen, it was possible to make direct comparisons between fertilizer management practices; here, treatment description is limited to these comparisons. For information on the full set of treatments for each cycle, see Dahlan et al. (1987), MARIF (1987) and MARIF (1988).

14. This is more conservative than it sounds, as benefits of phosphate application may be earned over a number of years. This approach, then, is likely to systematically underestimate benefits as well as overestimate costs.

The authors would like to acknowledge the work of the researchers and technicians whose efforts contributed to the success of the OFR programme described in this chapter. These include: Dr Sudaryono, agronomist and maize coordinator; Ir Chamdi Ismail, agronomist; Ir Muchdar Soedarjo, agronomist; Ir Herman Subagio, agronomist; Ir Fachrur Rozi, agroeconomist; Mrs Ulfah Dahlan, senior field assistant; Mr Cipto Prahoro, senior field assistant; and Ir M.J.P. van Staveren, formerly of the ATA-272 project. We also wish to acknowledge the support provided by Drs Soetarjo and Sumarno, former and current directors of MARIF, respectively, and by MARIF, the ATA-272 project and CIMMYT. The opinions expressed in this chapter, however, are not necessarily those of the institutions named. Lastly, we wish to express our gratitude to those extension agents and farmers who helped us understand the complexity of maize farming in Malang District.

References

CIMMYT. 1990. *1989-90 CIMMYT World Maize Facts and Trends: Realizing the Potential of Maize in Sub-Saharan Africa.* El Batan, Mexico: CIMMYT.

Dahlan, M. 1987. *Maize On-Farm Research in the District of Malang.* MARIF Monograph No. 3. Malang, Indonesia: MARIF.

Ewell, P. 1989. *Linkages between On-Farm Research and Extension in Nine Countries.* OFCOR Comparative Study No. 4. The Hague, Netherlands: ISNAR.

MARIF. 1985. *Report on the Maize Survey, December 1984.* MARIF Working Paper No. 8. Malang, Indonesia: MARIF.

MARIF. 1987. *Report on the Trial Results: Cycle 5 — Post-Rainy Season 1986, Cycle 6 — Rainy Season 1986/1987, Cycle 7 — Post-Rainy Season 1987.* MARIF Working Paper No. 19. Malang, Indonesia: MARIF.

MARIF. 1988. *Report on the Trial Results: Cycle 8 — Rainy Season 1987/1988, Cycle 9 — Post-Rainy Season 1988.* MARIF Working Paper No. 20. Malang, Indonesia: MARIF.

MARIF/RTI. 1989. *Project ATA-272: Strengthening of the Malang Research Institute for Food Crops: Annual Report 1988-89.* Malang, Indonesia: MARIF.

Supriatin. 1986. Control measures for shootfly on maize. Unpublished draft. Malang, Indonesia: MARIF.

Supriatin, Sri Wahuni, I. and Eveleens, K.G. 1985. *A Note on the Shootfly Issue.* MARIF Working Paper No. 10. Malang, Indonesia: MARIF.

Timmer, C.P. 1987. *The Corn Economy of Indonesia.* New York, USA: Cornell University Press.

Traxler, G., Renkow, M. and Harrington, L. 1990. Assessing the impact of new technology: Three levels of analysis. Paper presented at the 1990 Asian Farming Systems Research and Extension Symposium, Bangkok, Thailand, 19-22 November 1990.

Tripp, R., Anandajayasekeram, P., Byerlee, D. and Harrington, L. 1990. FSR: Achievements, deficiencies and challenges for the 1990's. Paper presented at the 1990 Asian Farming Systems Research and Extension Symposium, Bangkok, Thailand, 19-22 November 1990.

van Santen, C. 1987. Application of the results of the MARIF maize on-farm research program by farmers and the extension service. Unpublished manuscript. Malang, Indonesia: MARIF.

van Santen, C. and Dahlan, M. 1989. Five years on-farm research in the Malang Research Institute for Food Crops, 1984-88. Unpublished manuscript. Malang, Indonesia: MARIF.

Planned Change in Farming Systems
Edited by R. Tripp
© 1991 R. Tripp
A Wiley-Sayce Co-Publication

8

Revealing the Rationality of Farmers' Strategies: On-Farm Maize Research in the Swat Valley, Northern Pakistan

D. BYERLEE, K. KHAN and M. SALEEM

This chapter describes efforts to understand and improve the maize-based farming systems in the Swat Valley in the mountainous areas of Pakistan's North West Frontier Province (NWFP) (*see* Figure 8.1 *overleaf*). Particular attention is given to the diagnostic and verification stages of the on-farm research (OFR) process as these stages were critical in first understanding the system and then testing selected improved technology to fit the system. In many ways, the research process resembled the 'farmer-back-to-farmer' model described by Rhoades and Booth (1982). Several of the final conclusions of the research resembled what farmers already practised but were markedly different from the recommendations promoted prior to the OFR programme.

The analysis reported here is of special interest for two reasons. First, most farmers in the intensively cultivated Swat Valley farm on less than 1 ha. As on most small farms of South Asia, livestock play a major role in the farming system and are closely integrated with crop production. Second, farmers in the valley use fairly unconventional practices for producing maize, the major crop in the system. They all plant maize by the broadcast method, using a seed rate several times higher than is usually observed for maize. Moreover, they have consistently rejected the research and extension recommendations of line planting, early thinning, 'normal' (lower) seed rates and the use of insecticides.

In the 1970s a major research and extension programme operated in the Swat Valley to promote a package of improved practices, but only fertilizer and, to some extent, an improved maize variety were widely adopted.[1] The package had been developed through an extensive on-farm experimental programme conducted throughout NWFP during the 1970s and early 1980s. However, this programme had two flaws: it did not include an explicit diagnostic stage and, apart from distinguishing between irrigated and rainfed areas, it aimed to develop recommendations which would be applicable throughout NWFP.[2]

Figure 8.1   North West Frontier Province and the Swat Valley

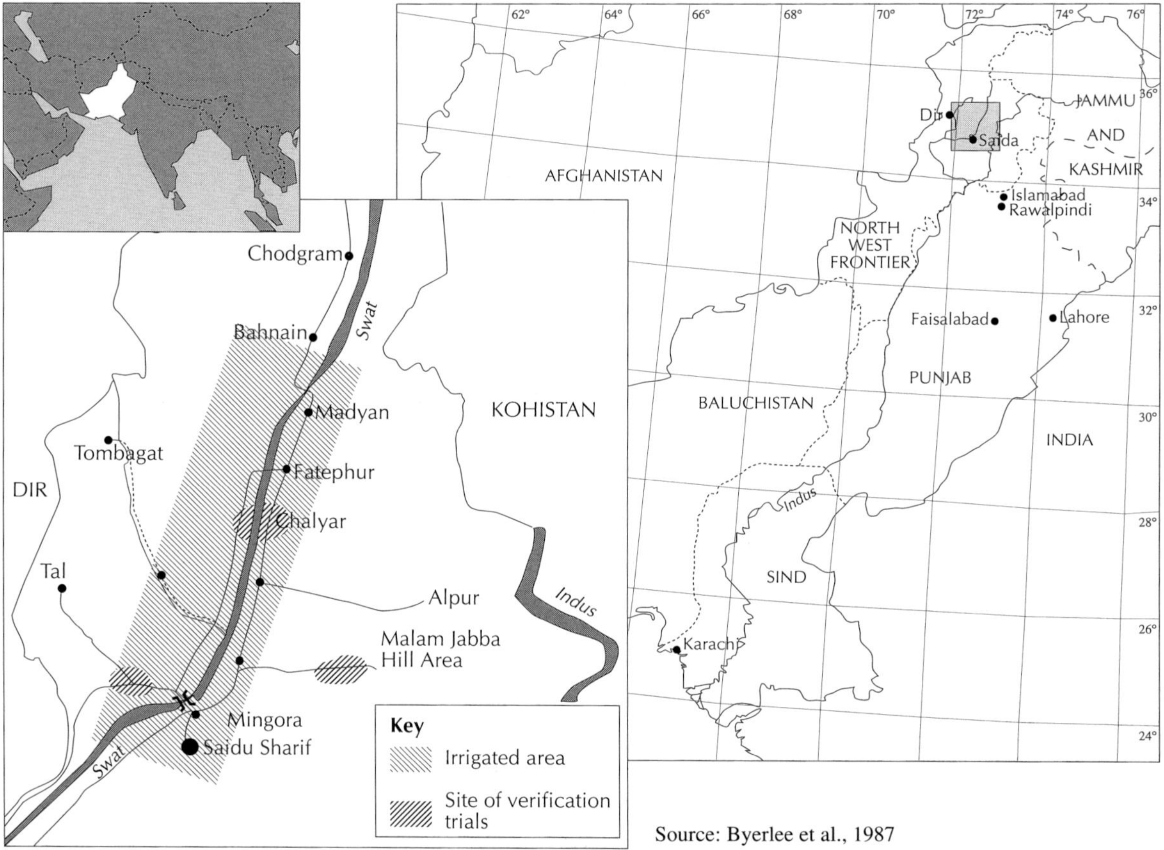

Source: Byerlee et al., 1987

Thus, despite several years of extensive research, the situation in the Swat Valley was unclear. The high potential for maize production was demonstrated in experiments both on the farm and on the research station, in which maize yields regularly averaged over 5 t/ha, whereas official statistics indicated that farmers' maize yields in the valley averaged less than 2 t/ha. Some researchers suggested that farmers used high plant densities and the broadcast planting method in order to produce fodder as well as grain and that these practices were responsible for low yields. As will become evident later in this analysis, maize requires different management practices depending on whether the objective is to optimize the production of grain or fodder.

Researchers had long debated the importance of maize as a fodder crop to farmers in the Swat Valley, but had made no effort to quantify its significance in terms of livestock nutrition. Moreover, even when fodder production was recognized as an important objective of farmers, no one agreed on what fodder production might imply for the development of maize technology. Indeed, at the beginning of the OFR programme, it was commonly accepted that the best way to produce both grain and fodder was in separate fields.

In 1983 it was agreed that a major effort should be made to resolve these issues so that researchers could formulate a strategy to increase maize production in NWFP. Maize research is conducted in NWFP by the Cereal Crops Research Institute (CCRI) in the irrigated plains. The CCRI has a number of substations, one of which is located in the Swat Valley. Until 1983 this substation was responsible for screening varieties on the station and conducting on-farm trials in the area; the on-farm work was then reoriented, with more emphasis being placed on diagnostic surveys to understand the system and on greater farmer participation in the experimental and technology verification programme. The OFR programme was led by the scientist in charge of the substation; as the only scientist at the station, he was also responsible for the on-station varietal screening. Agronomists and social scientists from the CCRI main station and the Pakistan Agricultural Research Council (PARC), a national research organization, played an important role in this work, especially in the diagnostic surveys and in farmers' assessment of the technology. The following discussion looks at how the results of key steps and findings of the research process led to a new set of recommendations. Preliminary evidence of the adoption of an improved variety identified through OFR is also given.

DIAGNOSTIC SURVEYS

A number of diagnostic surveys were conducted in the area (*see* Table 8.1 *overleaf*). Their structure and objectives can be summarized as follows:

- A short informal survey lasting about 1 week was carried out in 1984 to gain a better understanding of the issues in managing the farming system in view of the farmers' objectives. It was conducted by two maize agronomists and two social scientists.

- Maize harvest surveys were conducted in 1983, 1984 and 1985. They provided detailed information on maize crop management for specific fields and on grain and stover yields, harvest density, percentage of barren plants and other factors based on a crop cut at harvest; crop cuts consisted of three samples per field, each field being 2×4 m^2. The main survey was undertaken in 1984 but was repeated for a smaller sample in 1985 to confirm findings. In addition, data from a 1983 survey to estimate farmers' inputs and yields (again using the crop cut methodology) provided a 3-year sequence of information on some variables. In most cases, data from these surveys combined 3 years of information on a total of 214 fields (Byerlee et al., 1987).

- In 1984 and 1985, multiple visit surveys were conducted to quantify the importance of the green maize plants that farmers removed during the growing cycle to feed their livestock. Researchers observed a small sample of maize fields (15-20 each year) at 3-week intervals during the crop cycle and estimated plant density at each stage. They also visited the cultivator of each field and recorded information on the kind and amount of fodder fed to his/her animals on the previous day.

- A more detailed survey was carried out after the maize harvest in 1984. It focused on: system variables important for maize management, such as seasonal fodder supplies; farmers' knowledge and perceptions about improved practices; and maize utilization and marketing. For this survey, 15 villages were selected randomly and a total of 56 farmers were interviewed.

Together, these surveys provided an unusually rich data set on which to base investigations into farmers' decision criteria for managing their maize crops.

The surveys indicated that there was substantial variability in the agroclimatic conditions under which maize is grown in the area. Researchers decided to divide the maize area into several recommendation domains delineated by altitude and farmers' access to irrigation, as these two variables strongly influenced cropping

Table 8.1   Diagnostic surveys conducted in the irrigated area of the Swat Valley, 1983-85

Survey and year	Number of farmers
Informal survey	
1984	about 25
Surveys of maize production practices and yields	
1983	83
1984	77
1985	54
Multiple-visit surveys of density and fodder	
1984	19
1985	15
Farming systems survey	
1984	56

Source: Byerlee et al., 1987

patterns and production practices and problems. This chapter focuses only on the surveys and experiments in irrigated areas of the Swat Valley at altitudes between 900 and 1200 m; information on higher altitude and rainfed areas is provided in Khan et al. (1986) and Heisey et al. (in press).

In early 1985, after the initial survey results had been synthesized, a meeting was held to discuss the results and plan an experimental programme for the following cycle. This programme differed markedly from on-farm experiments previously conducted in the area. The strategy of the earlier work had been to conduct researcher-managed experiments using the recommended package of practices for the non-experimental variables. The new programme gave more emphasis to testing hypotheses emanating from the surveys through on-farm verification trials planted and managed by farmers. To measure grain and fodder production under different management strategies, complementary trials on the research station were also designed.

Survey findings on the farming system

The upper Swat Valley is surrounded by higher valleys and mountains rising to an altitude of 6000 m. Mild summer temperatures, readily available irrigation water and well-drained alluvial soils make the valley ideal for maize production; in fact, the highest experimental yields for maize in Pakistan are usually obtained in the valley. Swat District is also the country's largest maize-producing district, with some 100 000 ha under maize; about 25-30% of this district is in the Swat Valley. The winters are cold and some snow falls in the upper part of the valley. The main winter crops are wheat and a fodder crop, *shaftal* (*Trifolium persicum*). However, for several months in winter no green fodder is available and the value of maize is then at a premium as farmers in the valley and animal herders in the nearby mountains need the stover to feed their animals (*see* Figure 8.2).

Average farm size of the surveyed farmers was 1.1 ha, but well over half the farmers cultivated parcels under 1 ha. About a third of the farmers were tenants with quite variable tenancy arrangements. It is important to note

Figure 8.2   Farmers' use of animal fodder over the year in the Swat Valley

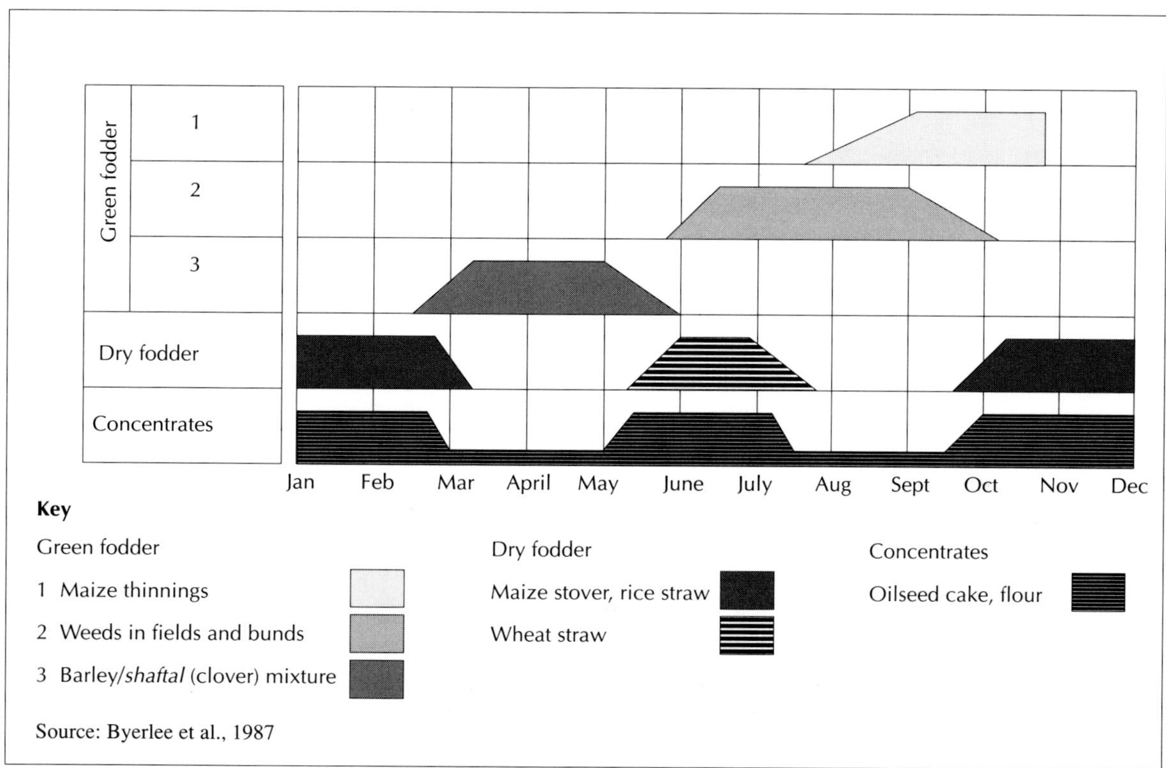

here that tenants, many of whom are specialist livestock farmers, often receive a higher share of crop by-products, such as green and dry fodder, than they do of grain, which may encourage management practices oriented toward fodder production.

Given the very small size of farms in the Swat Valley, supplementary sources of income are important. These include cash crops, such as vegetables, sale of livestock products and seasonal and long-term migration to other parts of Pakistan or abroad. Cash crops and other crops, especially rice, tend to be more important to larger farmers (*see* Table 8.2 *overleaf*). When there are conflicts in allocating resources (labour and cash) between subsistence crops, such as maize, and other crops, farmers favour the latter. This often leads to compromises in the management of subsistence crops. For example, planting maize conflicts with transplanting rice and tomatoes and is one of the reasons why farmers prefer the more rapid method of broadcast planting of maize to line planting.

Livestock are a major source of subsistence and cash income for the smallest farmers. The number of large animals per household (average of 3.3 animals) and per hectare of cultivated land (average of 6.3 cattle equivalents/ha) is substantially higher in Swat District than in any other maize-producing district of Pakistan (Byerlee and Hussain, 1986). The surveys showed that 60% of the farmers owned three or more cows or buffaloes, more than enough for their subsistence requirements, and 50% of the farmers ranked livestock above crops as their main source of cash income. However, only about half the farmers owned a bullock as tractors

Table 8.2   Variation in cropping pattern and animal ownership by farm size, Swat Valley, 1984

Farm size	No. farmers	% kharif[1] area under			No. cows and buffaloes	
		Maize	Rice	Vegetables	Per farm	Per ha maize
< 0.5 ha	20	87	10	3	2.6	9.9
0.5 - 1.0 ha	26	74	14	12	3.2	7.1
1.0 - 2.0 ha	21	81	15	4	2.8	3.9
> 2.0 ha	10	68	26	6	3.9	2.1
All farmers	77	75	18	7	3.3	6.3

Note: 1. Summer season, June-October.
Source: Byerlee et al., 1987

have rapidly replaced bullocks as the main source of draught power. Of particular importance is the very high ratio of animals to maize area for farmers with the least amount of land, as shown in Table 8.2. This statistic is especially significant given the role of crop by-products as a source of fodder. There is also evidence that the number of livestock has increased (Byerlee and Hussain, 1986). Farmers noted that there has been a marked shift from cattle, which depend in part on open grazing, to buffaloes, which are largely stall fed.

The importance of livestock, the small farm size and the relatively long winters combine to place a premium on fodder supplies in the area. Thus, fodder prices, such as the price of maize stover, tend to be higher in Swat than in adjoining lowland districts. In the maize growing season, thinnings from maize fields substitute for dry fodder and concentrates and are the main source of green fodder for animals.

Survey findings on maize crop management

A comprehensive description of farmers' management of their maize crop is given in Byerlee et al. (1987). A summary of this description is provided in Table 8.3. In the following discussion, emphasis is placed on how the management practices of farmers in the Swat Valley are influenced by crop/livestock interactions.

Variety

Most farmers grew a mid- to full-season white variety of maize, mixed to varying degrees with local and improved germplasm. Some farmers grew Sarhad White, the recommended variety, with seed obtained from the research substation. Several others stated that they grew previously recommended varieties such as Zia and Changez, although they had not replaced their maize seed for many years because NWFP lacked a formal system for producing and distributing seed. Hence these varieties were a mixture of local and improved types.

The main characteristic that farmers sought in improved varieties, besides grain yield, was earliness. Improved varieties such as Sarhad White matured significantly later than local varieties. Many farmers who were familiar with Sarhad White considered that it took too long to mature, as planting was often delayed

Table 8.3   Summary of maize production practices, Swat Valley, 1983-85

Production practice	
Land preparation:	
average number of ploughings	1.9
% prepare land with tractor	67
Planting:	
average seed rate (kg/ha)	96
% use improved variety	27
% broadcast seed	100
% maize planted late (in July)	31
Fertilizer:	
average dose of nitrogen (kg/ha)	72
average dose of phosphorus (kg/ha)	30
% apply phosphorus	55
Density management:	
% use *seel* practice	52
% thin maize	79

Source: Byerlee et al., 1987

because of irrigation water shortages or specialized cropping patterns involving late maturing vegetables. Other varietal characteristics that farmers considered important were: resistance to lodging; good stover yield and fodder quality; and resistance to diseases. In the 1984 survey, nearly half the fields had 50% or more lodged plants at harvest time. The lodging score was associated with the farmers' use of high plant densities.

Fertilizer

As shown in Table 8.3, farmers in the Swat Valley used fairly high levels of fertilizer. All nitrogen was usually applied at planting at an average rate of 60-80 kg/ha. Phosphorus was also applied by about half of the farmers — a much higher proportion than elsewhere in NWFP (Byerlee and Hussain, 1986). Farmyard manure was applied extensively to *shaftal* but generally not to maize or wheat. Data on fertilizer use in the various years of the surveys were reasonably consistent for phosphorus but the amount of nitrogen increased sharply, suggesting that farmers had still not reached the equilibrium stage of fertilizer adoption.

It had been hypothesized that some farmers might be applying phosphorus to the previous crop and relying on the nutrient carry-over for the maize crop. However, the survey results suggested that farmers either used phosphorus fairly regularly or not at all. For example, 70% of the farmers who applied no phosphorus to maize also had not applied phosphorus to the previous crop. It had also been hypothesized that farmers might substitute farmyard manure for phosphorus use but there was no evidence of this practice. The most important fertilization issue emerging from the diagnostic surveys was the appropriate dose of phosphorus, given that fertilizer expenditures accounted for a high proportion of the cash outlay of small farmers in the area.

Grain and fodder management

Although researchers and extension agents had consistently advocated and widely demonstrated line planting at a seed rate of 25-30 kg/ha, all farmers used the broadcast method, with a seed rate of 96 kg/ha. Over 80% of the farmers interviewed during the in-depth survey said they were familiar with line planting and some had used this practice in the past. They gave three major reasons for preferring the broadcast method:

- shortage of time and labour for maize planting, which conflicted with planting of cash crops;

- lack of bullocks required for the *kera* method of line planting (that is, dropping the seed in a furrow behind an animal-drawn local plough);

- the desire for a high plant density to provide fodder for animals.

Their reasons for using a high seed rate were closely related to their reasons for line planting; 86% favoured high seed rates so that they could obtain animal fodder. The use of high seed rates also insures against the germination and emergence problems that occur when there is too little moisture at planting or insects damage maize seedlings.

Farmers performed two operations after emergence which influenced final plant density. At about 3 weeks after planting, they cultivated the field with an animal-drawn local plough or a tractor-drawn cultivator (the *seel* practice). Because fields were not planted in lines, this operation uprooted many maize seedlings and reduced plant populations substantially. The farmers also claimed that the *seel* practice controlled weeds and reduced lodging. From about 45 days after planting and continuing up to harvest, farmers thinned the maize by hand, removing weak and barren plants and feeding them to livestock. The intensity of thinning varied according to the prevailing plant density and the need for livestock fodder.

These management practices (seed rate, the *seel* operation and thinning), along with other factors unrelated to management (such as insect attack at emergence), determined final plant density. The reduction in plant density over the cycle followed the curve shown in Figure 8.3, derived by observing 20 fields at 3-week intervals during the 1985 growing season and fitting a curve to represent a constant exponential rate of plant removal over the growing season (Byerlee et al., 1989). This curve indicated that about 100 000 plants/ha were removed by thinning for fodder, with an average rate of 1.1% of the remaining plants removed per day.

The variation in final plant densities between fields was analysed by the regression equation given in Table 8.4. Management practices that directly influenced plant density (the *seel* practice and thinning) had the expected significant negative effect on final density. Lower plant densities were also evident among farmers using improved varieties (but not significantly so); Sarhad White, the main improved variety in this sample, is a full-season leafy variety that is expected to respond better to lower density than the local varieties. Plant densities were much higher in late-planted fields (highly significant), probably reflecting the better germination percentage after the onset of the rains and the farmers' tendency to manage late-planted fields more for fodder than for grain production. As shown in Table 8.4, harvest density was also positively related to seed rate and the number of animals per hectare. Farmers with a larger number of animals would be less concerned about final plant density as they also placed a premium on dry stover production.[3]

Although farmers used a number of practices to influence plant density, the impact of these practices on density was fairly unpredictable. Average harvest densities varied from 69 000 plants/ha in 1983, when there was a severe insect attack after emergence, to 90 000 plants/ha in 1988, when heavy rains prevented many

Figure 8.3   Plant density counts plotted against days from planting, Swat Valley

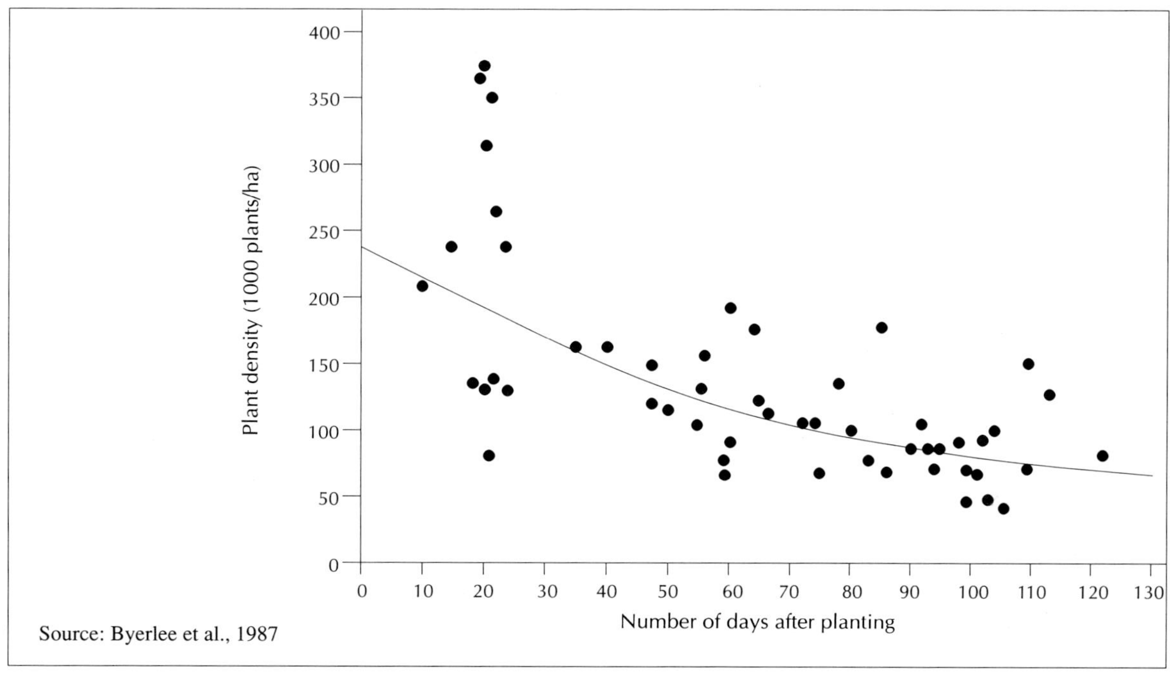

Table 8.4   Regression analysis of factors affecting final plant density, Swat Valley, 1984[1]

DENSITY   =   71.61 + 0.286 SEEDRATE - 11.68 SEEL -12.03 THIN
                          $(2.09)^2$ $(2.52)^2$ $(2.05)^2$

                   + 0.663 ANIMALHA + 15.39 LATEPLANT - 6.94 IMPVAR
                      $(2.16)^2$ $(3.15)^3$ (1.33)

                   N =73 $R^2 = 0.41$

where: DENSITY   =   harvest density in 1000 plants/ha
             SEED   =   seed rate in kg/ha
             SEEL   =   dummy variable = 1 if farmer did *seel* operation
             THIN   =   dummy variable = 1 if farmer did hand thinning
             ANIMALHA   =   number of large animals/ha
             LATEPLANT   =   dummy variable = 1 for planting after 1 July
             IMPVAR   =   dummy variable = 1 for improved variety

---

Notes: 1. t values in brackets.
             2. significance at the 5% level.
             3. significance at the 1% level.
Source: Byerlee et al., 1987

farmers from performing the *seel* operation and thinning. These results contrast sharply with the recommended density of 55 000 plants/ha to be obtained by thinning at 3 weeks after planting. In 1985, farmers were asked whether they thought the plant density at harvest in one of their own fields was too high, too low or about right. Only about half the farmers said they had achieved their desired density. Moreover, they seemed to be well aware that high plant densities would lead to a reduction in grain yields. Two-thirds of farmers who had fields exceeding 90 000 plants/ha felt that the final density was too high.

It is well known that grain yields are optimized at densities of 50 000-80 000 plants/ha and fodder yields are optimal at densities over 100 000 plants/ha, but little research has been done to establish the optimal plant density for producing grain and fodder in the same field. A second critical question in managing plant density is whether there are ways of reducing the uncertainty that farmers face in achieving their desired density.

Survey findings on maize productivity

Maize grain and stover yields

Table 8.5 summarizes the harvest data from 214 maize fields sampled from 1983 to 1985. Grain yields averaged almost 4 t/ha, with little variability from year to year. This figure was considerably higher than the official estimates of maize yields for the Swat Valley of about 1.5 t/ha. Stover yields were similar to grain yields with a harvest index (ratio of grain yield : total dry matter) of 45%. A significant percentage of plants were barren, especially in 1984 when harvest density was high. Barrenness is positively correlated with density and is the

Table 8.5   Overview of harvest survey data, Swat Valley, 1983-85

	1983	1984	1985	Average
		Average statistics		
Number of fields	83	77	54	
Grain yield (t/ha)[1]	3.79	3.88	4.10	3.90
Stover yield (t/ha)[2]	4.00	3.92	4.23	4.00
Harvest density (1000 plants/ha)	69	90	75	79
% barren plants	12	26	14	18
% grain moisture	n.a.[3]	43.5	42.7	43.3
Days from planting to harvest	n.a.	104	108	106
		Coefficient of variation (%)		
Grain yield	32	40	31	35
Stover yield	31	35	28	32
Harvest density	21	27	27	28

Notes: 1. 15% moisture.
          2. 10% moisture.
          3. n.a. = not available.
Source: Byerlee et al., 1987

reason that grain yields decline beyond a certain density. Perhaps the most striking finding was that the farmers harvested at a very high grain moisture content of 45%, before the grain had reached physiological maturity.[4] This practice is still imperfectly understood but it probably relates to farmers' desire to preserve the quality of the dry fodder, as well as the difficulty of drying out grain and stover in the field in October. This again demonstrates farmers' willingness to trade off grain and fodder yields.

Yields in the sample ranged from less than 1 t/ha to over 7 t/ha. As all the samples were taken from a relatively small and homogeneous environment, a multiple regression analysis was used to relate differences in yields between fields to the practices employed. Several specifications of the model were employed (Byerlee et al., 1987); only one equation is presented in Table 8.6.

The management variable with the largest single effect on yields is variety. The use of an improved variety is estimated to increase yields by about 800 kg/ha and this effect is highly significant. Plant density has the expected effect, with yields rising to a maximum and then declining, as demonstrated by the significant negative coefficient for the quadratic term, DENSQ. The interaction term between improved variety and density (IMPVAR*DEN) is also negative and is significant at the 5% level, indicating that the optimum density is lower for the improved variety than for the local variety. This finding is discussed later in the chapter.

The level of nitrogen applied to maize was also highly significant in explaining yield differences. A quadratic term for NIT was also included to estimate physical and economic optima. The maximum yield occurred at 156 kg N/ha, and, at the price ratios prevailing in 1985, the economic optimum for a marginal return on fertilizer investment of 50% was estimated to be 100 kg/ha. In practice, farmers used an average nitrogen

Table 8.6   **Regression analysis of yields and management practices, Swat Valley, 1983-85**

Variable	Definition	Coefficient and t value	
IMPVAR	Dummy variable = 1 if used improved variety	2383	(2.94)[1]
NIT	Nitrogen level (kg/ha)	14.5	(2.57)[1]
NITSQ	$(NIT)^2$	-.440	(1.53)
PUSER	Dummy variable = 1 if applied P_2O_5	130.3	(.67)
DEN	Harvest density (plants/ha)	75.6	(3.09)[1]
DENSQ	$(DEN)^2$	-.451	(3.54)[1]
VARDEN	VAR*DEN	-21.1	(1.99)[2]
LOC	Dummy variable = 1 if upper Swat	377	(1.88)[3]
Constant		-334	
N		180	
R^2		0.29	

Note: 1. Denotes significance at the 1% level.
        2. Denotes significance at the 5 % level.
        3. Denotes significance at the 10% level.
Source: Byerlee et al., 1987

dose of 72 kg/ha, which implies a marginal return on fertilizer investment of 120% — a reasonable figure for farmers with little land and capital (CIMMYT, 1988). The response to phosphorus was far less conclusive. Average response was 162 kg/ha for a phosphorus application of 55 kg/ha and was not significant. Given this response, it is not profitable to apply phosphorus.

The same regression analysis was also run for stover yields. Use of the improved variety, which is relatively leafy and favourably rated by farmers for its fodder characteristics, was found to have a significant and positive effect on stover yields. As hypothesized above, stover yields are less sensitive to density than grain yields, and the variables DEN and DENSQ had small and insignificant effects on stover yields. Thus the penalty for farmers overshooting the desired density may not be very high where stover has a relatively high value. In fact, the penalty for low densities (for example, 40 000 plants/ha) is probably higher as the farmer loses both grain and fodder. Farmers who are averse to risk will therefore prefer management strategies that favour high densities.

The optimum density calculated from the yield determination equation in Table 8.6 was 60 000 plants/ha for the improved variety and 83 000 plants/ha for the local variety. However, the yield/density curves shown in Figure 8.4 indicate that although the improved variety was less tolerant of high densities, it dominated the local variety except at very high densities over 100 000 plants/ha. Further analysis showed that the improved variety had a higher number of barren plants for a given density.[5] These results suggested that maize breeding work, which had been selecting varieties planted at densities of 55 000 plants/ha, might be able to provide further yield gains for farmers if varieties were screened for tolerance to higher densities.

Figure 8.4   Estimated relationship between yield and plant density for local and improved varieties (derived from maize production survey, Swat Valley)

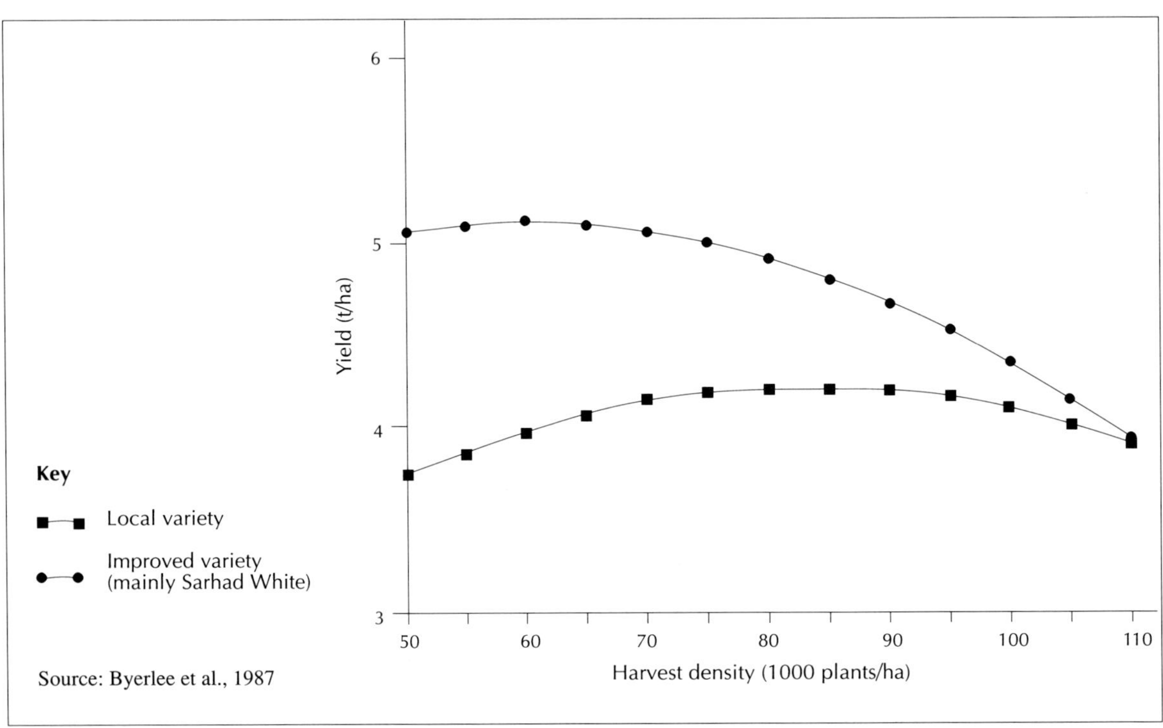

Source: Byerlee et al., 1987

Relative value of grain and fodder production

Data from the surveys were used to estimate the relative value of maize grain and fodder yields in the Swat Valley maize fields. The quantity of dry fodder was measured and valued by taking crop cuts at harvest time. The value of green fodder was estimated indirectly by the savings in concentrates and dry fodder fed to animals in the maize season (Byerlee et al., 1989). The results indicated that the total value of green and dry fodder was about equal to the value of grain production (*see* Table 8.7).

Given these relative values, it is not surprising that farmers have rejected the recommended technology package, which produces no green fodder. Although this package increases grain yields by nearly 40%, the cost increase (more than double) and loss of green fodder make it unprofitable compared to the farmers' technology.

Table 8.7 **Comparison of profitability of farmers' technology and the recommended package of technology (Rs/ha)[1]**

	Farmer technology	Recommended technology
Total variable costs	995	2387
Gross revenue:		
Grain [2]	5320	7315
Fodder[3]	5000	3000
Gross margin	9325	7928

Notes: 1. The official rate of exchange was approximately US$ 1 = Rs 17.
      2. Assumes yield of 4 t/ha for farmer technology and 5.5 t/ha for the recommended technology (based on results of on-farm experiments).
      3. Assumes stover yields equal for each technology, as concluded by Khan (1986), and green fodder yields in the farmer technology as estimated by Byerlee et al. (1989); the green fodder yield in the recommended technology assumed to be zero.
Source: Byerlee et al., 1989

ON-FARM EXPERIMENTS

Planning the experimental programme

In early 1985, when the diagnostic surveys had been completed and analysed, researchers held a 3-day planning workshop to synthesize the available information and plan a set of on-farm experiments for 1985. The survey findings had provided valuable new insights into farmers' management objectives and practices, which led to considerable changes in the experimental programme, both in terms of the technological components selected for testing and in the management of the experiments. It was decided that the programme would comprise the following components:

- verification of an improved variety under farmers' management practices;

- verification of response to phosphorus, in the verification trial and in a researcher-managed fertilizer trial;

- measurement of grain and fodder production under recommended management practices and under simulated farmer management practices in on-station trials;

- continuation of trials to evaluate promising varieties.

The results of only the first three components of the programme are reported here.

The verification trials

Design

It was recognized that use of an improved variety in the high potential irrigated conditions of the Swat Valley represented a priority opportunity for increasing productivity. Because the diagnostic surveys had indicated that the recommended variety Sarhad White did not perform well under the high plant densities characteristic of farmers' fields, the variety Azam was selected for inclusion in the verification trial. Developed from CIMMYT Population 30 x Zia, Azam had performed well in on-farm trials in the Swat Valley in 1984. In addition, farmers had expressed interest in a medium maturity variety such as Azam, similar in maturity to the local variety and earlier than Sarhad White.

With regard to fertilizer, the surveys had indicated an economic optimum dose of about 100 kg/ha of nitrogen and in fact most farmers generally used only slightly less than this amount. They had also shown that about half the farmers were using phosphorus on maize and that phosphorus users did not obtain significantly higher yields. It was therefore decided to include phosphorus as a treatment in the verification trial in order to determine response over a larger sample of fields.

After considerable discussion the researchers decided to include a treatment in the verification trial to compare the effect of the recommended practice of early thinning (3 weeks after emergence) aimed at reducing plant density to 60 000-70 000 plants/ha with the farmers' practice of continuous thinning throughout the production cycle. Some researchers felt this treatment would show farmers the benefits of early thinning. Others saw it as a way of comparing the increased yield from early thinning with the value of fodder production lost.

The final design chosen was a simple four-treatment trial with step-by-step changes in technology:

Treatment 1   Farmers' practice with farmers' variety;

Treatment 2   Improved variety (Azam) under farmer management;

Treatment 3   Improved variety (Azam) with phosphorus (50 kg/ha) under farmer management;

Treatment 4   Improved variety (Azam) with phosphorus (50 kg/ha) and early thinning at 3 weeks after emergence to reduce density to 65 000 plants/ha.

Selection of sites and farms

Two villages which were representative of the recommendation domain, Kabal (elevation 875 m) and Chalyar (elevation 1085 m), were selected. Five verification trials were laid out in Kabal and seven in Chalyar. Of these

12 sites, only 10 gave useable results; one site was flooded when a canal broke its banks and a second was harvested for fodder.

Experimental methods

The plot size averaged about 500 m^2 but was varied in order to suit the size and shape of the farmers' fields. Inputs were weighed in the field after determining plot size, and applied by the farmer. A basal dose of 100 kg N/ha was applied to the whole experiment.

The farmers planted the trial their own way and managed the plots throughout the season. They were also responsible for the early thinning that was part of Treatment 4, although in practice they were often reluctant to use this treatment and sometimes delayed thinning well beyond the recommended time. Researchers visited the fields at 10- to 14-day intervals and recorded their observations on plant densities, insects, weed problems, drought and other factors.

Results

The yields from the verification experiments were high, averaging 5.5 t/ha (*see* Table 8.8) and the final harvest density averaged 65 000 plants/ha. The only significant treatment difference was that Azam yielded about 10% more than the local variety. No consistent response to phosphorus or early thinning was found.

These results, although they do not show many statistically significant responses, were important in two respects. First, yields under farmer management, especially with broadcast planting at high densities, were shown to be quite high. They approached the yields which were generally observed in researcher-managed experiments conducted in farmers' fields that had been line planted using recommended seed rates and other

Table 8.8   Summary of yield responses in farmer-managed trial, Swat Valley, 1985

Treatment	Grain yield (t/ha)	CV yield (%)
1. Farmer practice (FP)	5.05	23
2. FP + improved variety (IV)	5.63	13
3. FP + IV + phosphorus (P)	5.44	10
4. FP + IV + P + early thinning	5.67	9
Mean	5.45	15
LSD (5%)	.59	
F-ratio: Treatment	2.14	
Sites	3.31[1]	

Note: 1. Denotes significance at the 5% level.
Source: Khan et al., 1986

practices. Thus it appeared that line planting and low seed rates would have little impact on yields and, in addition, would incur significant costs in terms of the labour needed for planting and the reduced supply of green fodder. Second, Azam performed well under farmer management, as indicated in Table 8.9. Stover yield from this variety was somewhat higher than the yield from the farmers' variety, an important criterion for farmers' selection of variety in this area of Pakistan. Azam was harvested at slightly higher moisture content for grain, although stover moisture content was less than for the farmers' variety. (Farmers may have selected their variety to have some 'stay green' quality for fodder.) Despite similar harvest densities, Azam had a higher percentage of fertile plants than the farmers' variety. Lodging in Azam was only one third of that seen in the farmers' variety.

Table 8.9   Comparison of harvest results of farmers' variety and Azam, Swat Valley, 1985

	Farmers' variety	Azam variety	Significance level
Grain yield (t/ha)	5.05	5.58	.06
Stover yield (t/ha)	5.86	6.65	.11
% fertile plants	89.7	94.8	.11
Moisture % at harvest:			
grain	43.4	45.8	.05
stover	75.1	73.4	.08
Shelling %	82.9	84.3	.02
% plants lodged	30	10	.00

Source: Khan et al., 1986

Despite the farmers' reluctance over early thinning, densities after thinning were close to the target density of 65 000 plants/ha, compared to over 100 000 plants/ha under traditional management (subsequently thinned to 65 000 plants/ha at harvest). In other treatments, variability in early densities correlated closely with seed rates (r = .61 and .46 between seed rate and densities at knee high and silking, respectively) but not with harvest density. As the season progressed, farmers' management of the maize crop (plant removal) was more important than seed rate in determining plant density.

Early thinning had no effect on grain and stover yields, but it did reduce the green fodder yield. Overall, plant density did not seem to be a major factor determining yields under the management conditions of this sample of farmers. Both the degree of lodging (r = .29) and barrenness of plants (r = .56) were significantly and positively correlated with harvest density, but there was no effect on the yield of either grain or stover. These results are consistent with the results of the on-station experiment described below.

Multiple regression analysis of the verification trial results confirmed the significant effect of the improved variety on maize yields under farmer management. In addition, rotation after *shaftal*, grain moisture at harvest and lodging all had significant effects on grain yields in the yield estimation model (Khan et al., 1986).

Farmers' assessment of treatments

An important part of OFR is to solicit farmers' active involvement in order to benefit from their experiences and opinions. In addition to interacting from day to day with the farmers during the maize season, researchers interviewed the farmers 3 months after harvest to assess their reactions to the technological components being tested; this would have given them enough time to evaluate Azam's cooking and storage qualities. The interviews, using a short questionnaire, were conducted by a rural sociologist who was skilled in farmer interview techniques but had not been part of the research team that laid out the trials. It was hoped that this would give some objectivity to farmers' responses.

The interviews produced several valuable insights, especially with respect to Azam. Farmers slightly preferred Azam for its higher grain yields, which is consistent with the relatively modest yield advantage of Azam observed in the trials (*see* Table 8.9). However, farmers also said they preferred their own variety for its higher stover yields, but experimental results showed Azam stover yields to be higher. This misjudgement probably reflects the fact that Azam is shorter and less leafy than the local variety.[6] For other characteristics, such as earliness and stover and *chapati* quality, farmers ranked Azam quite favourably and said they intended to plant 60% of their maize area to it in the following year, using seed retained from the verification trials.

Overall, the trials gave the researchers enough confidence to recommend Azam for the Swat Valley. They also confirmed that the farmers' system of joint production of maize for grain and fodder is very productive.

Experiments on interactions between variety and plant density

The verification trials were complemented by a number of experiments to evaluate how alternative systems of plant density management affected grain and fodder production and interacted with use of an improved variety. Most of these experiments were carried out on the research station.

The survey data strongly suggested that the previously recommended variety, Sarhad White, was less tolerant of higher plant densities. A review of earlier on-station trials supported this finding. An experiment in 1982 had evaluated the local variety and Sarhad White at two levels of fertilizer application and at two densities. The results showed a highly significant variety x density x fertilizer interaction (*see* Table 8.10). At the farmers'

Table 8.10   Interaction effects between variety, fertilizer and density, Swat Valley[1]

Fertilizer (kg/ha NPK)	Plant density (p/ha)	Grain yield (t/ha)	
		Local variety	Improved variety[2]
68:34:34	50 000	4.75	6.83
68:34:34	100 000	5.48	5.28
136:68:34	50 000	6.40	7.65
136:68:34	100 000	6.73	7.73

Notes: 1. Calculated from a 2^4 experiment of variety, fertilizer, density and weed control. The three-way interaction reported in this table was significant at the 1% level.
2. Improved variety = Sarhad White.

fertilizer level (75 kg/ha of N), Sarhad White produced significantly higher yields than the local variety at the low density (50 000 plants/ha) but lower yields than the local variety at the high density (100 000 plants/ha).

In 1985 a more comprehensive attempt was made to evaluate grain and fodder production of local and improved varieties under different systems of management (Khan, 1986). The recommended practice of line planting at a target density of 60 000 plants/ha was compared with a simulation of farmers' practice in which seed was planted with uniform spacing at an initial density of 300 000 plants/ha and then thinned by 150 000 plants/ha at 20 days, 37 500 plants/ha at 35 days, and another 37 500 plants/ha at 50 days after planting. (This rate of thinning was in fact somewhat accelerated compared to the farmers' practice of thinning plants continuously until harvest.) Green fodder yields were measured at each thinning. Four improved varieties and the local variety were evaluated under these management systems.

A surprising finding of this study was that grain yields under simulated farmer management were as high as yields obtained following the recommended practice (that is, production of green fodder up to 50 days after planting had no negative effects on grain yields). The uniform spacing of plants in the simulated broadcast trial may account for the apparent lack of competition early in the growing season at these high densities (Fischer and Javed, 1986). The improved varieties also performed well under simulated farmer management although, as observed in the surveys, Sarhad White had a significantly higher number of barren plants compared to both the local variety and Azam.

In 1987, six on-farm experiments were conducted to compare the effect of conducting the *seel* operation at a spacing of only 15 cm with the farmers' practice of using a 30-cm spacing. The closer spacing was aimed at achieving a lower plant density early in the season. No yield advantage was observed in the plots with close spacing, again confirming the benefits of farmers' practice over alternative practices aimed at reducing plant density early in the season.

Experiments on fertilizer response

Expenditure on fertilizer at Rs 500/ha accounted for over 50% of the variable cost of maize production and over 80% of the cash costs. Hence, for the small farmers in the Swat Valley, ways of improving the efficiency of fertilizer use, especially the appropriate balance between nitrogen and phosphorus, represented a potential research opportunity.

Before and during 1984, when the first in-depth diagnostic survey was undertaken, several fertilizer experiments had been planted in farmers' fields to assess the response to both nitrogen and phosphorus. An economic analysis of five experiments conducted from 1982 to 1984 showed, as expected, a highly significant response to nitrogen and indicated that the economic optimum for nitrogen was 100 kg/ha (Byerlee and Khan, 1990). This result confirmed the survey finding (*see* Table 8.11). In addition, farmers applied an average of 72 kg N/ha, and this application rate increased over the 3 years of the survey. Researchers concluded that there was little point in conducting further research on optimum nitrogen levels; farmers already approached the optimum and insufficient capital was the main factor limiting nitrogen use.

The situation for phosphorus was quite different. Of the five experiments conducted in 1982-84, a response to phosphorus was observed in only two, and then only at the 10% level of significance. In none of the five experiments was it economic to apply phosphorus at a minimum acceptable rate of return of 100% (Byerlee and Khan, 1990). However, about 55% of the farmers were applying phosphorus to maize, so it was decided to focus on response to phosphorus in future fertilizer experiments.

Table 8.11 Summary of statistical and economic analysis of fertilizer response, Swat Valley, 1982-86

Type of experiment	No. of years	No. of sites	Management	No. of sites where response significant		Economic optimum[1] (kg nutrient/ha) m = 0.5	m = 1.0
N x P	2	5	Researcher	N - 4	N	102	94
				P - 2[2,3]	P	22	0
P	1	3	Researcher	1	P	45	33
P	1	10	Farmer	1[2]	P	38	23

Notes: 1. m = minimum acceptable rate of return on capital (CIMMYT, 1988).
       2. Significant at the 10% level only.
       3. A significant N x P interaction was observed in one experiment.
Source: Byerlee and Khan, 1990

In 1985, three researcher-managed trials were conducted in farmers' fields to estimate the response to various levels of phosphorus. A significant response was observed at only one site and there was no apparent relationship between phosphorus response, previous phosphorus use and use of farmyard manure. As shown in Table 8.11, the overall optimum derived from the response curve was 33 kg/ha (at a marginal rate of return of 100%). In 1986, 10 farmer-managed experiments on phosphorus response were conducted. Again, at only one site was a significant response to phosphorus observed when statistical analysis was conducted individually for each site. However, the response over all the sites was significant at the 5% level and indicated an economic optimum of 23 kg/ha, similar to the results obtained in the 1985 trials.

These results suggested that the recommended phosphorus level of 57 kg/ha was almost certainly too high. A general dose of perhaps 25 kg/ha seems more appropriate. However, even more efficiency could be achieved if research could determine with greater precision the conditions under which a statistically significant and economically profitable response to phosphorus is obtained. To do so, a set of experiments to analyse phosphorus response in the maize-wheat cropping pattern, with and without use of farmyard manure, would be required. Such research is expensive, and the benefits may not justify the cost. Alternatively, phosphorus response could be calibrated to soil tests, but it is unlikely that farmers in the Swat Valley will have access to reliable soil testing facilities for many years.

SEED MULTIPLICATION, EXTENSION AND ADOPTION

On the basis of the results of the verification trials and the farmers' assessment of the technological components, researchers felt confident that Azam should be promoted more widely to farmers in the Swat Valley.[7] As there was no maize seed industry in NWFP, researchers initiated a pilot project to produce and distribute seed in the area. In 1986, certified seed of Azam was distributed to farmers in the two villages where the verification trials had been conducted. These farmers were asked to form groups so they could distribute the seed they produced to each farmer in the group on an exchange basis. Farmers with neighbouring fields were selected to produce the seed, which protected the seed crop from out-pollination with local materials and helped ensure seed purity. Overall, it is estimated that some 54 t of seed were produced — enough to plant over 500 ha of improved maize if all the grain produced was used for seed.

In 1988 a survey was undertaken to assess the adoption of Azam in the two pilot villages as well as in nearby villages within a radius of 5 km. The survey involved 60 farmers, half of them from the pilot villages and the other half from the neighbouring villages. The farmers were asked which maize varieties they grew and what they would plant in the following season. The results, summarized in Table 8.12, indicated that Azam had been adopted by almost all farmers in the pilot villages and by the majority of farmers in nearby villages. As most of the farmers (78%) grew only one variety, they had therefore switched completely from their previous variety to Azam. This, together with the farmers' earlier favourable rating of Azam, indicated that the great majority of farmers preferred Azam for home use, both for grain and for fodder.[8]

Table 8.12   Adoption of Azam variety, Swat Valley, 1988

	Two villages with pilot seed programme	Neighbouring villages
Number of farmers surveyed	30	30
% farmers adopting Azam	90	63
% area planted to Azam	83	50

The farmers planned to continue to plant Azam although many noted that the Azam they now grew had become mixed with the local variety. This response highlights the problem of introducing improved open-pollinated varieties to small farmers in a single infusion of new seed. Research is now being undertaken to estimate the rate of seed deterioration and loss of vigour of improved cultivars under farmers' seed management practices.

CONCLUSION

Although researchers had long recognized the Swat Valley as an area of high potential for maize production, it was widely believed that farmers' 'poor management' of their maize crop constrained the realization of this potential. Farmers' use of high seed rates, broadcast planting and high plant densities seemed particularly at variance with practices aimed at maximizing grain yields. The official statistics, which grossly underestimated maize yields in the valley, seemed to confirm those beliefs.

The analysis of the farmers' management practices and the productivity of the farming system strongly confirms the rationality of the farmers' strategy of jointly producing maize grain and fodder. Fodder accounts for approximately 50% of the total value of production from maize fields. Moreover, the farmers' system is highly productive. The production of fodder has a surprisingly low cost in terms of grain yields, which averaged

4 t/ha in surveys and over 5 t/ha in farmer-managed verification trials. The results from the OFR programme explain the farmers' rejection of a number of recommendations, especially those related to plant density management, that have been extensively demonstrated to farmers in northern Pakistan since 1970. The farmers' system is simply more profitable than the recommended system when fodder is given an appropriate economic value.

In a high potential area such as the Swat Valley, it is to be expected that use of an improved variety will be important in increasing system productivity. Improved varieties will be acceptable to farmers in the area if they yield well at high densities and possess such fodder production characteristics as leafiness, the ability to 'stay green' and good storability and palatability of stover. Farmers also prefer a medium maturity variety that fits into their cropping patterns and provides greater flexibility in planting date. Fortunately, with the information on varietal needs gathered from the surveys, researchers were able to identify a newly released variety, Azam, with the required characteristics. Azam has now been widely adopted in the villages where the OFR programme and subsequent seed multiplication programmes were conducted. Unfortunately, wider use of this variety is constrained by the lack of a maize seed industry in NWFP. Even in the villages where this research was undertaken, farmers' seed must be renewed continuously with certified seed.

The high productivity of the farmers' system does not mean that there is no room for improvement. While farmers' management of plant density is, *on average*, appropriate for the joint production of grain and fodder, there is a significant risk that farmers will undershoot or overshoot their target density. Hence, it may be useful to continue investigating ways to provide greater control over plant density and reduce inter-plant competition. For example, it may be worth investigating planting maize in closely spaced rows, with some rows removed for fodder while others are cultivated for grain. In addition, the farmers' high expenditure on fertilizer in relation to their limited cash resources suggests that research to increase the efficiency of fertilizer use should continue. The recommended phosphorus dosage should be reduced by 50% and efforts made to manage phosphorus use within the whole cropping system — that is, by analysing phosphorus use on both wheat and maize and in relation to the use of farmyard manure.

The experience of the OFR programme in the Swat Valley demonstrates the complexity of farming systems, especially those of small farmers, and the importance of conducting thorough diagnostic surveys. The use of farmer-managed trials with full farmer participation was also important in demonstrating to researchers that reliable responses can be generated from such trials. In fact, a strong feature of the research described here was the consistency of the results obtained on the effects of plant density, improved variety and fertilizer, whether from surveys, researcher-managed trials or farmer-managed trials. As farmer-managed trials are much cheaper to execute, it is recommended that future OFR programmes place more emphasis on this type of approach.

Notes

1. By 1983, almost all the farmers had adopted nitrogenous fertilizer and most were using varieties which were a mixture of local and improved materials. As there is no well-developed seed supply system, the new varieties spread by the extension programme rapidly became mixed with the local materials.

2. The population of NWFP is about 15 million. The province produces about 500 000 ha of maize in a wide range of environments (Byerlee and Hussain, 1986).

3.   Seed rate is also significantly and positively correlated with number of animals per hectare ($r = .26$), although the difficulties of obtaining accurate measures of seed rate probably cause this correlation to be underestimated.

4.   Plant physiologists estimate that harvesting maize at moisture levels above 40% will almost certainly lead to yield losses. Many farmers in the sample harvested maize at moisture levels above 50%.

5.   This difference was confirmed by a regression of percentage of barren plants on density for each variety. Differences between regressions were significant at the 10% level, as indicated by the Chow test.

6.   A similar inability to judge yields of by-products, especially for a shorter improved variety, was observed by Husain (forthcoming).

7.   However, a parallel research effort in the nearby hills (above 1500 m) found that Azam was unsuitable because of its susceptibility to smut (*Ustilago maydis*).

8.   Some farmers (26%) who adopted Azam, however, continued to produce their previous variety in addition to Azam.

The synthesis of research results presented in this chapter draws heavily on several published reports of the OFR programme in the Swat Valley, including Khan (1986), Khan et al. (1986), Byerlee et al. (1987) and Byerlee et al. (1989). Among the many other researchers who made valuable contributions were H.J. Iqbal, A.D. Sheikh, M. Ahmed, M. Aslam and E.J. Stevens. The work undertaken by field assistants M. Nasser and F. Qayum was critical to the success of the programme.

References

Byerlee, D. and Hussain, S.S. 1986. *Maize in North West Frontier Province: A Review of Technological Issues in Relation to Farmer Circumstances*. PARC/CIMMYT Paper No. 86-1. Islamabad, Pakistan: PARC.

Byerlee, D., Sheikh, A., Khan, K. and Ahmed, M. 1987. *Diagnosing Research and Extension Priorities for Small Farmers: Maize in the Swat Valley*. PARC/CIMMYT Paper No. 87-19. Islamabad, Pakistan: PARC.

Byerlee, D., Iqbal, M. and Fischer, K.S. 1989. Quantifying and valuing the joint production of grain and fodder in maize in northern Pakistan. *Experimental Agriculture* 25: 435-45.

Byerlee, D. and Khan, K. 1990. *Fertilizer Response in Irrigated Maize in the Swat Valley: A Synthesis of Five-Years' Research*. Draft paper, PARC/CIMMYT Collaborative Research Program. Islamabad, Pakistan: PARC.

CIMMYT. 1988. *From Agronomic Data to Farmer Recommendations: An Economics Training Manual*. (revd edn). El Batan, Mexico: CIMMYT.

Fischer, K. and Javed, H.I. 1986. *Production of Maize Grain and Fodder in North West Frontier Province*. PARC/CIMMYT Paper No. 86-3. Islamabad, Pakistan: PARC.

Heisey, P., Ahmad, M., Stevens, E.J. and Khan, K. (in press). Crop intensification in maize-based mountain agriculture: The Swat Mountains. In Byerlee, D. and Hussain, T. (eds) *Farming Systems of Pakistan. Diagnosing Priorities for Agricultural Research*. Lahore, Pakistan: Vanguard Books.

Hussain, T. (forthcoming). Resource interactions and innovation in the wheat-livestock system of Gilgit. In Byerlee, D. and Hussain, T. (eds) *Farming Systems of Pakistan. Diagnosing Priorities for Agricultural Research*. Lahore, Pakistan: Vanguard Publishing.

Khan, K., Byerlee, D., Ahmad, M., Saleem, M. and Stevens, E.J. 1986. *Farmer Managed Verification of Improved Maize Technology: Results from Swat, 1985*. PARC/CIMMYT Paper No. 86-12. Islamabad, Pakistan: PARC.

Khan, K. 1986. A study of different varieties of maize under traditional and modified management practices in Swat, NWFP. MS thesis. Peshawar, Pakistan: Department of Plant Breeding and Genetics, NWFP Agricultural University.

Rhoades, R.E. and Booth, R.H. 1982. Farmer-back-to-farmer: A model for generating acceptable agricultural technology. *Agricultural Administration* 11: 127-37.

9

On-Farm Research in Support of Varietal Diffusion: Bean Production in Cajamarca, Peru

W. JANSSEN, N. RUIZ DE LONDONO, J.A. BELTRAN and J. WOOLLEY

Small-scale farming systems have a reputation for being difficult to change. One common explanation is the traditional nature of small farming, but this stands in sharp contrast to the equally popular theory of farmer rationality. In fact, rather than explaining lack of change, tradition is simply another way of describing the delicate balance in which small-scale farming systems have to operate. Through indigenous experimentation, small farmers have developed intricate and apparently stable interactions with what, in most cases, is a complex climatic, institutional and economic environment (Byerlee et al., 1982; Richards, 1985). That stability is not necessarily achieved at high levels of productivity, however. The technological means that are available within the farming community are restricted and are not always effective solutions to existing production problems.

These observations suggest two situations in which on-farm research (OFR) would be particularly productive. First, it could try to provide technological solutions that are beyond the reach of indigenous knowledge and yet sufficiently feasible to be adopted in the farming systems. Improved bean varieties are one such solution. Farmers do not have access to the same germplasm or plant breeding techniques as researchers. Second, OFR could try to identify elements of farming systems which are subject to change. Such changes may be caused by exogenous factors, such as population growth or market development, or by the availability of new technology. In either case, OFR can help farmers adjust to, and take advantage of, these changes.

This chapter analyses the effectiveness of an OFR effort initiated after a farming system had been altered by the introduction of a technological innovation from outside the system. In the Department of Cajamarca in Peru, an improved bean variety, Gloriabamba, had been released in 1985 after extensive on-station and on-farm testing from 1981 to 1984. It was released initially for its resistance to anthracnose (*Colletotrichum lindemuthianum*), a major disease affecting traditional varieties in the region (Ruiz de Londoño and Janssen, 1990). Because of the interest stimulated by the on-farm tests and the imminent varietal release, the Department of Cajamarca was chosen as the base for a three-phase course designed to strengthen OFR in Peru on cropping systems which included beans. The course was coordinated by the national agricultural research institute in

Peru, Instituto Nacional de Investigación Pecuaria e Agrícola (INIPA), and by the International Center for Tropical Agriculture (CIAT). It was funded by the Ford Foundation and by INIPA. Seven of the 25 course participants were from the Cajamarca highlands, from the provinces of Chota, Cajabamba, Cutervo, Celendín, San Marcos and Contumazá. Diagnostic data and experimental results, although not always complete, were collected in these six provinces. Wide coverage was intended and Chota province was chosen as the site of course fieldwork shared by all the participants.

The course included diagnoses of the problems in farmers' bean cropping systems, followed by the design of on-farm trials. There were some doubts, arising from experience with the earlier on-farm trials, about how widely Gloriabamba was adapted to the various cropping systems and temperature and rainfall variations in the Cajamarca highlands. The course coordinators also considered the possibility that other technological innovations might be suitable as well as, or even instead of, Gloriabamba. The diagnostic work identified the diffusion of Gloriabamba and the adaptation of cultural practices to the new variety as the most appropriate features for OFR. Although the identification and release of Gloriabamba cannot be considered the product of the OFR programme, increasing the impact of this improved variety was one of its main objectives.

The following analysis examines the effectiveness of OFR for reinforcing the benefits of such technological innovations. To assess this impact, it was considered necessary to first measure the benefits. This was a complex exercise, with few examples from the literature to provide guidance. Thus, the issue of measuring the impact of OFR programmes receives considerable attention in this analysis. The case selected for study is unusual in that several different geographical areas in a diverse region were examined using similar methodologies, trials and technologies. Indeed, one of the questions facing OFR is its adaptability and cost in very heterogeneous areas; this case shows the feasibility of a flexible approach to the management of OFR in such circumstances.

RESEARCH SITE AND RESEARCH PROGRAMME

Research site

The Department of Cajamarca is located in northern Peru, some 550km from Lima (*see* Figure 9.1). It stretches from the edge of the coastal desert to the fringes of the Amazon forest, but most of it falls within in the Andean cordillera. The department covers an area of 35 417 km², of which 4716 km² are dedicated to agriculture. Pastures, cereal grains, potatoes and food legumes account for 51%, 24%, 14% and 7% of total land use, respectively. Of the 33 000 ha used for food legume production, 13 000 ha are under common beans (INIPA/CIPA, n.d.). However, as fields of beans intercropped with maize are often counted in agricultural censuses as maize, the total area under beans is probably greater (Woolley et al., 1990).

This study focuses on the highland areas of Cajamarca, which range from 1800 m to 2800 m above sea level. Between the small agricultural areas there are deep valleys and high mountains. With the exception of the road to the coast from the departmental capital, roads within the department and the few between the department and the population centres on the coast are in poor condition, with travel often limited to 30 km/hour. The highlands have a cool climate with a rainfall of about 900 mm per year. However, rainfall varies considerably between the different provinces included in this study and also from year-to-year in the same location. Soils are shallow and stony in some areas (for example, in parts of Chota province) and therefore store little water between showers. There are two rainy seasons, from September to November and from April to May, and temporal distribution is often poor. The periods from January to March and from June to August are very dry.

Figure 9.1   Department of Cajamarca, Peru showing provinces involved in OFR programme

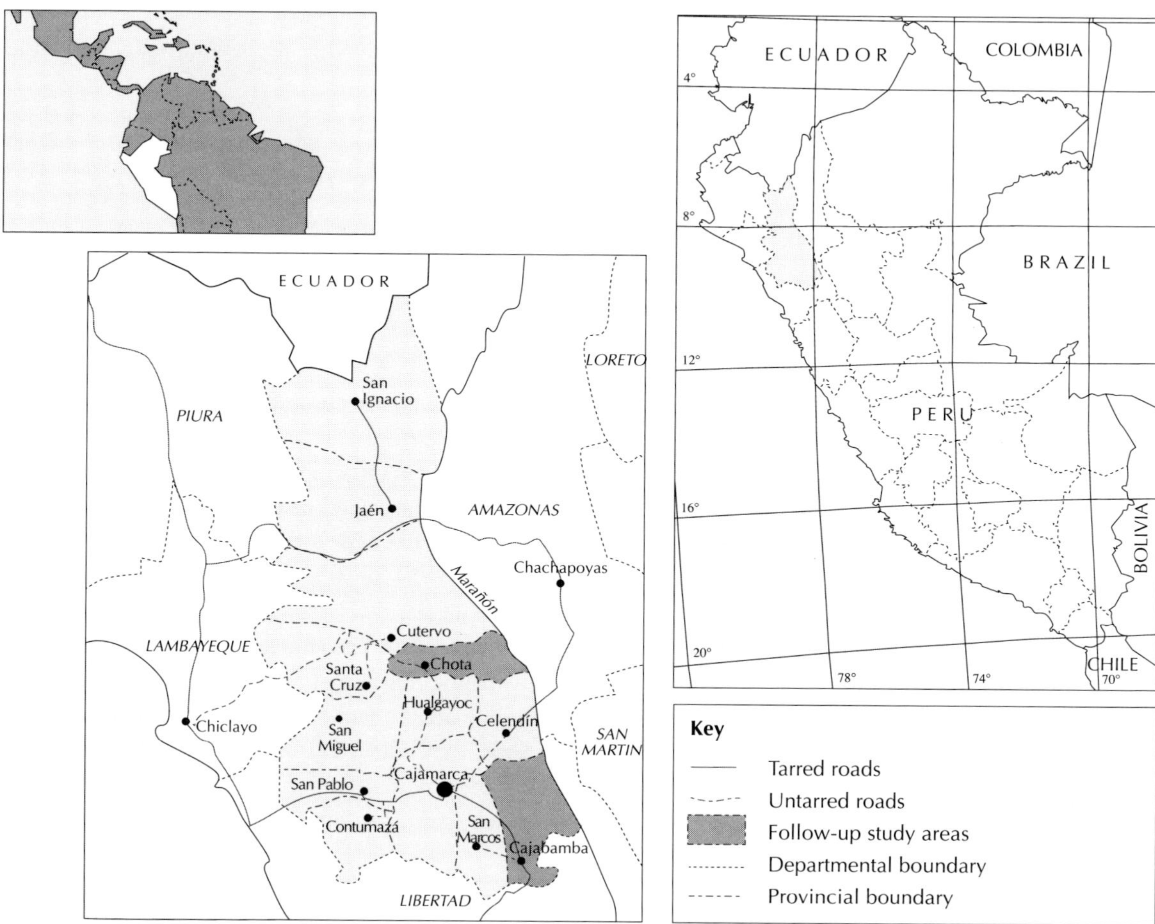

The average size of farms in the highlands is 4 ha, and most farms have a poorly developed infrastructure, with little in the way of farm equipment or irrigation facilities. Soil fertility is low, especially in organic matter and phosphorus (Ruiz de Londoño and De la Cruz, 1986). Although these limitations are typical of the bean-growing areas of Cajamarca Department, it should be pointed out that around the departmental capital are some of the more favourable areas of the Peruvian highlands in terms of soils and rainfall. The physical and socioeconomic conditions of the department mean that most farms are subsistence oriented and have low cash incomes. Instead of improving their farms, many people migrate in search of employment in large-scale rice production or in the industrial and service sectors of the coastal urban centres.

Although about 20% of the bean area in Peru is found in Cajamarca, the department accounts for only 13% of national bean production (Ministerio de Agricultura, 1981). The most important bean-producing provinces in the department are Cutervo, Celendín, Hualgayoc, Chota, Cajamarca and Cajabamba, where the areas under bean cultivation in 1985 were 2834 ha, 1576 ha, 1381 ha, 1027 ha, 980 ha and 492 ha, respectively. As noted earlier, these are probably underestimates. Apart from the release of Gloriabamba, there were other reasons which favoured the selection of Cajamarca as a pilot region for OFR:

- the average bean yield in the department was low (260 kg/ha), while the average maize yield was acceptable (1700 kg/ha);

- INIPA's experiment stations in Cajamarca and Cajabamba could provide adequate support to the OFR programme and had some technologies that appeared appropriate for on-farm testing;

- the bean/maize intercrop system in the region is characteristic of many parts of Latin America.

The on-farm research process

The main features of the OFR programme on bean production are summarized, in chronological order, in Table 9.1. A brief description of this process is given here.

In 1984 the first researcher from Cajamarca Department was trained in OFR methods at CIAT. On his return to the department he initiated the first diagnostic survey. Earlier, INIPA's legume programme had identified the CIAT germplasm accession G 2829 from Mexico as promising for the region and had conducted extensive on-station and on-farm testing, leading to the release in 1985 of G 2829, renamed Gloriabamba (De la Cruz, 1985).

In 1985 the formal OFR programme began with a three-phase course. The strategy of the course was that the key stages of OFR were conducted in Chota province jointly by the course participants and instructors. Each participant then conducted that part of the process in his/her home area (other parts of Cajamarca Department or elsewhere in Peru) and reported the results to the next phase of the course. This had the effect of improving quality of the work planned and establishing a collegial group to give mutual support after the end of the course. This strategy is common in multiple-phase OFR courses conducted by CIAT and other institutions (Tripp and Woolley, 1990).

Diagnostic studies

Before the first phase of the course, a questionnaire was prepared by the course instructors on the basis of their knowledge of bean production practices and problems in the different ecological zones of Peru. After an introduction to OFR and survey techniques, the participants conducted a 1-day informal survey in parts of Chota province, and modified a few of the questions which were found to be inappropriate. They then carried out a 2-day formal survey, which involved conducting 140 interviews with farmers and visiting most of the farmers' fields to identify apparent production problems. The survey results were analysed (2 days) and plans for trials and further diagnostic work were drawn up (2.5 days). The planning methodology used was a rudimentary version of the methodology later published by CIAT and the International Maize and Wheat Improvement Center (CIMMYT) (Tripp and Woolley, 1989).

In the first interphase of the course, the participants from Cajamarca Department carried out a further 223 interviews with field observations in selected study areas of five other provinces (*see* Table 9.2 *overleaf*). Cropping systems and practices for beans varied in the areas surveyed, particularly with regard to varietal use

Table 9.1   Chronology of the on-farm research programme and preceding activities in Cajamarca Department

Date	Activity
1978	G 2829 (later Gloriabamba) first grown on Cajabamba experiment station as an introduction from CIAT (originally from Mexico)
October 1982	Gloriabamba planted in 10 on-farm trials in Cajabamba; seed multiplication commenced
October 1983	Gloriabamba planted in 30 demonstration plots in Cajabamba
October 1984	Gloriabamba planted in demonstrations in Cajabamba and in on-farm variety trials in 7 regions
February 1985	First phase of OFR course (2 weeks); group diagnosis and planning in Chota (140 short questionnaires)
March-June 1985	223 questionnaires applied in Cajabamba, San Marcos, Contumazá, Cutervo and Celendín as course practical
June 1985	Gloriabamba officially released by INIPA
October 1985	Second phase of OFR course (1 week)
October 1985	On-farm trials planted in Chota, Cajabamba, San Marcos, Contumazá and Cutervo, as course practical; 33 trials successfully harvested
August 1986	Final phase of OFR course (2 weeks)
October 1986	Trials planted in all five areas; 40 trials successfully harvested
October 1987	Course follow-up meeting to plan 1987-88 trials (4 days)
April 1989	Adoption study of Gloriabamba

and planting arrangement. Maize and bean associations were predominant in all areas except Contumazá, where bush bean sole crops were more common. In all study areas maize was planted in hills, with several grains per hill. In some areas beans were planted in the maize hills; in others they were planted between the hills in the same row and or broadcast between the maize. In Chota, farmers planted climbing beans in the maize hills and broadcast bush beans. Overall, 49% of the farmers broadcast the bean seed and 51% planted in rows.

A very wide range of climbing and bush bean varieties, 41 in all, was mentioned by farmers. Most farmers planted two varieties. Blanco Caballero, a large round white-seeded, aggressive, late-maturing climbing bean, susceptible to most diseases, was planted in 33% of the fields surveyed, but was important in only three of the eight areas surveyed. Other common varieties were Poroto (a general name for bush beans), Nuñas (a pop-bean), Panamito and Colorado. Maize and beans were planted mainly in October or November. In many of the provinces, especially Chota, males emigrated from the area after planting to seek seasonal work in rice production in the Amazon basin, and did not return until after the harvest.

Table 9.2   Diagnostic information on beans, Cajamarca Department

Characteristics of study area	Chota	Cajabamba	San Marcos	Cuervo	Contumazá	Celendín
No. of interviews	140	56	39	49	36	43
Mean altitude (m a.s.l.)	2290	2620	2400	2450	2700	2450
Cropping system for beans	M+B[1] S	M+B S	M+B S	S[2] (61%) M+B	M+B	M+B
Predominant planting system	Broadcast (67%)	Holes (75%)	Broadcast (83%)	Holes (64%)	Broadcast (100%)	Broadcast (90%)
Main varieties (and % of farmers using them)	Criollo Bco[3] (90%) Caballero (46%) Pintado (24%)	Nuñas (70%) Bco Caballero (46%) Poroto (36%)	Panamito (74%) Pilón Bco (64%)	Bco Caballero (70%) Pintados (77%)	Bayo Chimú (39%) Canario (33%) Bco Caballero (19%)	Pintado (47%) Bco Caballero (23%)
Fertilizer use (%)	0	0	0	0	58	0
Pesticide use (%)	7	0	0	2	47	50

Notes: 1. M + B = maize + climbing and/or bush beans
       2. S = sole crop
       3. Bco = Blanco
Sources: Astupilco, 1985; Deza, 1985; Terrones, 1985; Vargas, 1985

Common planting densities were 3.2 plants/m^2 for maize and 2.6 plants/m^2 for beans. The low densities should be interpreted in the light of the extreme vigour of the traditional bean varieties (leading to lodging in the supporting maize plants) and the low soil fertility. Most farmers used their own maize and bean seed, and mentioned lack of cash as a limitation in obtaining seed. The implication was that bean densities were also often limited by lack of seed. The use of fertilizer, pesticides and credit was negligible for maize and beans, but common for potatoes; the majority of farmers, however, used hired labour. Potatoes were produced mainly for sale, whereas 80-90% of the bean, maize and other crops were for home consumption. Within the farm system, therefore, potato was the most important crop and received most inputs and attention. Maize and beans were grown on hillsides instead of in the valley bottoms.

Despite the differences in practices, the main technical problems identified by researchers and farmers were similar in all six study areas: foliar disease, low bean density, low soil fertility and drought (*see* Table 9.3). Although not explicit in survey reports, it seems likely that low yield potential of existing varieties (because of their inefficient plant habit) was implicit in many of the diagnoses. Detailed results of the surveys in Cajamarca Department and other areas of Peru are given in Ruiz de Londoño and De la Cruz (1986); the survey results specifically for Chota province are given in Arbulu et al. (1986).

Table 9.3   Problems, proposed solutions and type of trials for bean cropping systems in Cajamarca Department

Problem	Proposed solutions	Type of trials
Foliar diseases (anthracnose, ascochyta, oidium and virus)	Tolerant varieties Selection of clean seed Chemical control	A. Varieties B. Seed selection x seed treatment + fertilization
Low plant density	Row planting Increased plant density Seed selection Seed treatment	C. Variety x cropping systems or variety x population density + fertilization[1] B. Seed selection x seed treatment + fertilization
Low soil fertility	Chemical fertilization Use of organic fertilizers Incorporation of green manure Rotation with potato Inoculation with *Rhizobium*	C. Variety x population density + fertilization[1] B. Seed selection x seed treatment +fertilization[1] E. Inoculation trial (commenced 1986)
Drought	Early varieties Tolerant varieties Vegetative cover Use of manure	A. Varieties
Seed quality (Contumazá)	Seed selection Seed treatment Production of clean seed	B. Seed selection x seed treatment + fertilization
Late and aggressive varieties (Contumazá)	Early varieties	A. Varieties
Pests and stem borers (Contumazá, Cutervo)	Chemical control Increased bean population	C. Variety x population density + fertilization[1]
Root rot (Cajabamba)	Tolerant varieties Seed treatment	A. Varieties B. Seed selection x seed treatment + fertilization
Reduced diffusion of Gloriabamba	Seed multiplication	D. Seed multiplication on farm[2]

Notes: 1. Semi-commercial plots in 1986.
         2. Only in 1986.

In the second phase of the course, participants practised site selection and planting (2 days) and then presented and discussed the diagnosis and the trials planned for their own areas (2.5 days). They also learnt how to collect cost-of-production data from farmers and the bases of economic analysis of trial results (1.5 days). The potential technological innovations idenitified by participants in the first phase had been selected according to chance of technical success, the need for additional farm resources, the need for special institutional support, the compatibility with the farming system and the potential benefits. As shown in Table 9.3, eight possible solutions had been identified for experimentation in different parts of Cajamarca Department. In the second interphase, the participants conducted the trials, and each trial was visited by one of the course instructors.

A summary of the trials conducted in Cajamarca Department each year is provided in Table 9.4. The trials designed for Chota province were more complex and ambitious than those for the other areas, probably because their planning was done by all participants and instructors. Selected results are presented and discussed here to illustrate the similarities and differences between the work areas. Most of the results are from 1985, the year in which most of the activities were initiated.

Table 9.4   Movement of technologies in on-farm trials (with number of trials harvested)

Proposed solutions	Pre-1985	1985	1986
Gloriabamba	Variety trials Demonstrations	14 variety trials 17 verification trials[1]	13 variety trials 24 verification trials 11 semi-commercial
Other disease-tolerant climbing beans	Variety trials	14 variety trials	14 variety trials
Disease-tolerant bush beans		14 variety trials	14 variety trials
Row planting instead of broadcasting		Added to 4 verification trials (Chota only)	8 verification trials (Chota only)
Higher density in row planting + chemical fertilizer	Included in a few of the demonstrations	13 verification trials (other areas)	7 verification trials (other areas)
Seed selection		2 exploratory trials (Chota only)	2 exploratory trials (Chota only)
Seed + soil treatment (fungicide + insecticide)		2 exploratory trials (Chota only)	2 exploratory trials (Chota only)
Nutrient addition		N, P, K Micronutrients separately in Chota exploratory trials	N-P-K in 3 inoculation trials
Rhizobium inoculation			3 inoculation trials (Chota and Cutervo)

Note: 1. Also used as demonstrations where appropriate.

1985 trials

For the second interphase 51 experiments were planned, including 19 variety trials and 26 verification trials. In Chota province only, there were also three exploratory trials and three determinative trials. Farmers participated in implementing and evaluating the verification trials and, in most study areas, the variety trials.

Many variety trials were planned because varieties apparently offered a solution to a number of the problems facing farmers, as indicated in Table 9.3, and because a number of elite climbing and bush bean lines were the most readily available and adoptable technology in the region. The variety trials usually had 16 entries, two replications per site and three to four sites per study area. In all provinces except Contumazá, where bush beans had been found to grow very poorly, four bush bean lines were included in the trial. In Chota and Cutervo provinces a mixture of climbing beans was included as the local check, whereas in other areas pure bush and/ or climbing bean cultivars were considered an appropriate check. In Chota, a treatment with a farmers' mixture of bush and climbing beans was also included. Data on bean and maize yield, earliness, disease resistance and suitability for intercropping with maize were collected. In terms of mean performance in each work area, Gloriabamba was close in yield to the local checks in Contumazá and Cutervo provinces, but superior in San Marcos, Cajabamba and Chota province (*see* Table 9.5). However, these results mask some within-area variation. Gloriabamba, when planted at farmers' densities in Chota, outyielded the local check at an infertile site, but was inferior to it at a high-yielding site where its vigour was insufficient to exploit the high fertility. This would be overcome by higher planting density.

Table 9.5   Yields of climbing bean varieties and associated maize in five work areas, Cajamarca Department, 1985-86

Variety	Bean yield (kg/ha)							Mean maize yield (t/ha)
	San Marcos	Cajabamba	Cutervo	Contumazá	Chota fertile	Chota infertile	Mean	
Gloriabamba	1028	917	347	443	347	920	731	2.6
Puebla 444	1112	588	322	434	486	463	624	2.8
Guatemala 1076	674	528	489	n.t.[1]	1146	339	608	2.5
G 2333	1383	420	241	n.t.	347	344	596	2.5
Local check	897	290	377	445	625	178	491	2.2
ZAV 8398	810	470	151	n.t.	625	265	469	2.8
Cajamarca 64-1	672	548	176	n.t.	340	70	400	2.5
G 10889	791	258	305	n.t.	264	84	382	2.7
PG 149 x PI 311915	737	376	169	274	208	223	374	2.4
Mean	900	488	286	373[2]	488	321	519	2.6
LSD (5%)	239	181	45	101	373	196	—	—

Notes: 1. n.t. = not tested.
            2. Estimated mean if all nine lines had been tested.
Source: de la Cruz and Rojas, 1987

As some on-farm trials had already been conducted in Cajamarca and new technologies were urgently needed for promotion, emphasis was also placed on verification trials. Twenty-six simple verification trials were implemented, with four treatments and two replications per site and four sites per work area. In the course, it had been suggested that six to eight sites per work area would be more suitable, but all participants lacked the necessary manpower and financial resources. The verification trial in Chota province sought to compare Gloriabamba with the farmers' local mixture of hill-planted climbing beans and broadcast bush beans. To obtain a comparison between hill-planted beans and broadcast beans, the large plots of the verification trial proved suitable, giving a 2 varieties x 2 practices trial. In Chota, hill planting was not strictly under verification at this stage. All the other participants modified this trial to one which tested the interaction of two varieties (Gloriabamba and a pure farmers' variety) with two practices (increased bean density plus chemical fertilization and a typical farmers' hill-planting).

The results of the verification trials are given in Table 9.6. The choice of technology for Gloriabamba depended on the relative importance to the farmer of maize and beans. In Chota province, row planting gave

Table 9.6   Bean and maize yield results of verification trials 1985-86 (kg/ha)

Variety	Technology	San Marcos (4 trials)		Cajabamba (4 trials)		Cutervo (1 trial)		Contumazá (4 trials)		Chota (4 trials)	
		B[1]	M	B	M	B	M	B	M	B	M
Farmers'	Farmers'	263	1380	339	3420	_550[2]_	_1580_	995	2240	584	1690
Farmers'	Researchers'	579	3000	420	3320	758	1450	817	2750	516	1870
Gloriabamba	Farmers'	432	1540	_628_	_3980_	225	1590	1290	1760	309	1860
Gloriabamba	Researchers'	_818_	_2940_	863	3880	318	1850	_888_	_3410_	_401_	_2010_
Mean		523	2220	563	3650	463	1550	997	2540	453	1810
DMS (5%)		309	—	211	—	n.a.	n.a.	394	—	126	—
Farmers' variety		Mixture climbers + bush		Poroto (bush cultivar)		Mixture climbers				Mixture climbers + bush	
Bean spacing in farmers' technology (m)		0.8 x variable		0.8 x variable		0.8 x variable		0.6 x 0.6 Alternate rows		Broadcast	
Bean spacing in researchers' technology (m)		0.8 x 0.6 x 3p[3]		0.8 x 0.6 x 3p		0.8 x 0.6 x 3p		0.6 x 0.6 x 3p All rows		0.8 x 0.8 x 1.5p	
Fertilizer (40 N - 60 P_2O_5 - 40 K_2O)		Yes		Yes		Yes		Yes		No	

Notes: 1. B = beans; M = maize.
          2. Treatment of highest net benefit underlined.
          3. p = number of seeds per hill.

similar yields to broadcast planting of beans at the same densities, confirming that the testing of beans in rows in small plot trials should give accurate yield estimates for farmers who still prefer to broadcast beans. Gloriabamba yielded relatively poorly, as in the variety trial in Chota, but impressed many farmers by its adaptation to the maize crop and its earliness.

The objective of the three exploratory trials in Chota was to determine the causes of low bean population. The factors included in these trials were seed selection (Gloriabamba from the station, local seed cleaned by physical selection, and unselected local seed), fungicide treatment of seed and insecticide/nematicide (Carbofuran) treatment placed in the soil with the seed. A simple exploratory test of the four factors of N, P, K and combined micronutrients was also included. The three determinative trials in Chota included different spatial arrangements and densities within row planting and broadcasting systems. However, these trials were planted last and all three were lost as a result of frost before the harvest (De la Cruz and Beltran, 1987).

1986 trials

In the third phase of the course, the participants met to present and analyse the trial results (1 week) and to prepare and discuss their plans for the 1986-87 trials (1 week). In Cajamarca Department, the number of experiments planned and planted in the second year increased to 53 (13 variety trials, 24 verification trials, three determinative trials to study *Rhizobium* inoculation, two exploratory trials to study seed selection and soil treatment, and 11 semi-commercial trials). Seed multiplication of Gloriabamba was emphasized because seed availability was identified as a constraint to more rapid diffusion, and 13 seed multiplication plots were planted.

The results of the variety trials were similar to those obtained in 1985. In the semi-commercial trials, two large plots of 500 m^2 to 1000 m^2 were planted; they were managed entirely by the farmer, including decisions on planting date and time of all the operations (*see* Table 9.7). Because of the positive results obtained in

Table 9.7   **Results of semi-commercial trials, Cajamarca Department, 1986-87**

| | Yield (kg/ha) | | | |
| | With Gloriabamba and with increased planting density | | With traditional bean variety and with traditional density | |
Location	Beans	Maize	Beans	Maize
1	255	n.a.[1]	144	n.a.
2	485	n.a.	228	n.a.
3	1135[2]	—	634	—
4	1518[2]	—	564	—
5	768	n.a.	384	n.a.
6	326	1870	55	1471
7	645	2950	232	2960
8	985	3700	522	4135
Average	765	2840	345	2855

Notes: 1. Data not available.
         2. Monoculture.

verification trials in 1985, the emphasis in these semi-commercial trials was on testing Gloriabamba at high densities in row planting alongside each farmer's current practice.

The objective of the inoculation trials conducted in Chota province was to find low-cost solutions to the problem of low soil fertility. They made use of scientific advice and inoculum available from the University of Cajamarca and were conducted with both Gloriabamba and traditional varieties. The first year's results were very encouraging. There was a mean yield increase of 59 kg/ha for beans and 287 kg/ha for maize. As inoculation is an inexpensive practice, these results represent a very high marginal rate of return. For beans, the yield increase was the same as that obtained by using chemical fertilizer; for maize it was slightly lower (*see* Table 9.8). A further yield increase was obtained for both crops when phosphorus and potassium were applied.

Table 9.8   **Inoculation x fertilizer trial, Chota province 1986-87 (with yield of beans and maize in kg/ha)**

Rhizobium inoculation	Fertilizer $(N\text{-}P_2O_5\text{-}K_2O)$	Site 1 Beans	Site 1 Maize	Site 2 Beans	Site 2 Maize	Site 3 Beans	Site 3 Maize	Mean Beans	Mean Maize
No	0-0-0	192	1950	391	3260	206	1980	263	2400
No	40-60-40	213	1920	412	2940	341	3770	322	2880
Yes	0-0-0	270	2240	434	3450	263	2360	322	2680
Yes	0-60-40	263	2240	448	3640	377	4350	362	3410
Average		235	2080	421	3320	297	3120	317	2840
DMS (10%)		146	796	195	2580	56	254	70	756

FOLLOW-UP ON THE ON-FARM RESEARCH COURSE

Most of the data for the follow-up on the course were derived from three studies. The first was the 1985 diagnostic study described earlier which was undertaken during the course. This study provided the baseline data that allowed evaluation of the changes that occurred later (Astupilco, 1985; Arbulu et al., 1986).

In 1986 a study was carried out to measure the acceptability of Gloriabamba in the provinces of Chota and Cajabamba. Interviews were conducted with 90 farmers who had had access to Gloriabamba. This study tried to assess how the new variety performed under farmers' management and to record farmers' perceptions of its characteristics. Detailed information was obtained on bean production management and the relative importance of different varieties (Ruiz de Londoño, 1986).

The third study, undertaken in 1988 and 1989, assessed the extent to which Gloriabamba had been adopted in Chota and Cajabamba. Interviews were conducted with 141 randomly selected farmers; the group included farmers who had received Gloriabamba from the OFR programme, those who had received it from other farmers and those who had never planted it. Again, the performance of the variety was measured, but attention focused on how adoption of the new variety changed farmers' production systems and marketing strategies. Issues covered in the study included origin of the seed, the role of the interviewed farmer as a diffusion agent, and other diffusion mechanisms. Farmers who did not plant Gloriabamba were asked which varieties they did plant and why they had not planted the new variety (Ruiz de Londoño and Janssen, 1990).

From these studies, an analysis can be made of the changes that occurred between 1985 and 1989 and the role played in this process by the OFR programme.

EVALUATION OF THE ON-FARM RESEARCH PROGRAMME IN CAJAMARCA

The final test for any agricultural research programme should be how much it has contributed to improved consumer and producer welfare. However, those parameters are not easy to measure and it is necessary, therefore, to use more accessible proxy variables, such as the production increases in the region caused by the introduction of the improved technology. Nevertheless, in the case of OFR programmes, even this type of measurement is not necessarily very clear.

Many OFR programmes introduce various technological improvements at the same time. Not all these improvements will be adopted; in addition, improvements that are not related to the research programme might occur. The difficult question to answer is which aspects of the yield increases can be accounted for by the programme and which are the result of autonomous improvements. The intricate interactions that often occur between technological components further complicate the analysis. To address this question, a complete inventory of the technological innovations in the region should be made. After this, the extent to which these innovations arose from the OFR programme or from autonomous efforts should be determined. Then a method to identify what part of the overall yield increase is caused by each technological innovation has to be developed. Lastly, those parts attributable to technological innovations induced by the OFR programme should be aggregated in order to obtain a fair assessment of the programme's significance.

The evaluation procedure outlined above does not start with a predetermined set of technological innovations, as in the methodology described by Hardaker et al. (1984), but with the technological innovations that were observed. By attributing those innovations to the OFR programme or to other sources, an assessment can be made of the incremental effect of the programme, instead of merely measuring the productivity difference before and after the programme. This allows an evaluation of the speed of technological progress with and without the programme.

In the following discussion we present a method for making an inventory of technological innovations and for attributing them to the OFR programme. We then describe the technological innovations that occurred in Cajamarca Department and examine whether or not they were attributable to the OFR programme. In the final part of the discussion we try to identify the principal ways in which the programme contributed to increased bean productivity in the region.

Evaluation method

Although the release of Gloriabamba took place before the on-farm programme started, its initial diffusion was limited. Hence, the innovations described here are those which took place between 1985, when the programme started, and 1989. The evaluation method we used involved the following steps: making an inventory of all technological innovations in bean cultivation; assessing the extent of diffusion; identifying the reasons for change, as expressed by the farmers, and any OFR experiments and demonstrations that could have contributed to this change; establishing whether there is a causal link between experiment and change; and evaluating the relationships between different technologies.

To investigate causality we tried to find both direct and circumstantial evidence. In some cases it was clear that farmers had adopted technologies that had been tested and promoted by the OFR programme. In others, OFR was responsible for suggesting opportunities for technological change to farmers, but the farmers had developed their own innovations; such instances warrant special attention, as they may indicate inadequacies in the way in which research programme priorities are set. Finally, there was technological change that could not be traced to the OFR programme. The fact that such change is common during the conduct of OFR suggests that the exhaustive diagnostic studies and priority setting which are characteristic of much OFR need to be carefully balanced against more intensive monitoring and frequent readjustment of goals and experiments.

With regard to the final step of evaluating the relationships between different technologies, in the present case the main issue addressed was whether farmers changed their practices in response to the adoption of the new variety. This provided insight into the question asked in the introduction to this chapter concerning the ability of OFR to enhance other technological innovations.

The steps described above were followed for each of the identified technological innovations. The matrix presented in Table 9.9 was constructed to provide a concise overview of the role that OFR played in the technological evolution of bean production in the region. The evaluation method helped us greatly to understand the contribution of OFR to the observed changes. However, in a number of instances the innovations were difficult to relate to the research process and the causal interpretation of the events was somewhat ambiguous.

Technological changes in bean production in Cajamarca

Nine technological changes in bean production were identified in Cajamarca Department (*see* Table 9.9). These changes can be grouped into four categories:

A   Varietal changes — the adoption of Gloriabamba (1);

B   Changes in the planting system — the shift from broadcasting to row planting (2); the increase from 20 000 plants/ha to 27 000 plants/ha (3); the reduction of planting distances between beans and maize (5); more shallow planting depth (7);

C   Changes in use of inputs — the use of fertilizers (4); the use of fungicides (6);

D   Changes in farmer orientation — the trend towards marketing products, especially green shelled beans (8); the increase in the area devoted to beans (9).

Six of these changes were related to the adoption of Gloriabamba. The changes in planting densities were linked with the less vigorous nature of Gloriabamba compared with the traditional climbers. The shallow planting was adopted to avoid the problems that Gloriabamba might have with root rots, probably because of its less vigorous nature. The changes in farmer orientation were caused, indirectly, by the improved yields, which provided the incentive to plant more, and by the reduced marketability of the dry grain (the grain is smaller than the usual bean varieties for sale and slightly different in colour), which inspired the sales of green

Table 9.9   An evaluation matrix for technical change and on-farm research activities in Cajamarca Department, Peru

Changes in production systems	% of farmers who changed and reasons for change	Experiments that could have contributed to change[1]	Influence of the experimentation[2]	Direct evidence for this influence	Indirect/ circumstantial evidence	Relationship with adoption of Gloriabamba
1. Adoption of Gloriabamba	65; yields and earliness	A, C, D	Significant	% farmers stating they knew of variety through technicans	—	n.a.[3]
2. Change from broadcasting to row planting	43; easy weeding; better maize support to beans	C	Significant	Farmers stated that they saw it in experiments	Change only where experiments were located	Partly; change also took place in traditional varieties
3. Increased plant density	59; new variety has less vigour; does not cause maize to lodge; traditional varieties react positively to increased density	C	Significant	Farmers appreciated yield increases at higher densities	Experimental densities consistently higher than regional average	Partly; change also took place in traditional varieties
4. Increased fertilizer use	12; enhanced yields	C, B	Very reduced	Farmers used formula different from researchers	Slight relationship between adoption and experiment site	Yes
5. Reduced distance between beans and maize	32; extra support for new variety; easy weeding	—	—	—	—	Yes
6. Increased fungicide use	9; new variety responds well	—	—	—	—	Yes
7. Change in planting depth	16; to avoid root rot problems	—	—	—	—	Yes
8. Increased market orientation, esp. for green beans	27; surplus production limited marketability of new variety as dry grain	—	—	—	—	Yes
9. Area increases in beans	16; profitability increased for beans; profitability decreased for potatoes	—	—	—	—	Yes

Notes: 1. For explanation of experiment types, *see* Table 9.3.
       2. Authors' judgement of influence based on direct as well as circumstantial evidence.
       3. n.a. = not applicable.

shelled beans. Some of the other changes, such as the shift to row planting, were only partially linked with the adoption of Gloriabamba, as they also took place in the traditional varieties.

Were the observed changes related to the OFR programme? The programme carried out experiments on improved varieties, row planting, plant densities, fertilization and seed selection, as well as planting multiplication plots of Gloriabamba on farmers fields. In principle then, those experiments might have triggered off the observed innovations. However, before we accept or reject that causality, we should review the available evidence from the study region.

With respect to the diffusion of Gloriabamba, 40% of the farmers interviewed said that they had learnt of the variety from local technicians and researchers. They could have observed Gloriabamba in varietal trials, which might have produced a demonstration effect, or they could have obtained seed from the multiplication plots. The influence of the programme can be further validated by comparing the performance of the variety in the experiments with the reasons for adoption given by the farmers: in the experiments, the variety yielded 50% more than the local varieties and matured two months earlier; farmers mentioned yield advantages and earliness as principal motives for adoption. All this suggests that the programme has contributed significantly to the diffusion of the variety.

In the case of the shift from broadcasting to row planting, the influence of the experiments appears strong. Before the programme was initiated, row planting was common in Cajabamba but not in Chota. After some years of experimentation with row planting in Chota, this practice is now common in Chota as well. The percentage of adopting farmers was high and in the experiments the bean yields increased by 30% without harming the maize yields. But there was little convergence between the objectives of the OFR programme and the motives of the adopting farmers: while the aim of the experiments was to better protect the planted population, farmers gave other reasons for adopting row planting — ease of weeding, improved support by the maize for the beans, and a reduction in root rot problems. The effects of row planting on weeding and support for beans, although not primary objectives, were known to the researchers. The effect of row planting on root rots, however, was new.

What is interesting here is that neither weed control nor lack of support for climbing beans was identified as a production problem in the diagnostic study. This suggests that there are factors that the farmer does not consider as problematic but that he is willing to improve when he has the opportunity. This becomes clear only during the research process, and shows that it is impossible to formulate a complete diagnostic study at the beginning of the OFR process, and certainly not on the basis of farmers' views only. Rather, one should maintain communication throughout the research process to adjust the focus of the experimentation.

The density increases were directly related to the research experiments as well as to the less vigorous nature of Gloriabamba. The consistent yield increases of 30%, obtained in the experimentation at the higher densities, motivated farmers to adopt this innovation, as they confirmed in the interviews. Farmers also followed this rationale in case of the traditional varieties, although the evidence of this in the survey is weak.

The increase in the use of fertilizers by some farmers could have been caused by the research experimentation, as fertilizers were applied in conjunction with the increased densities. However, its diffusion was very limited and we doubt whether it was related to the experiments. The fertilizer effects were confounded with the density effects and any demonstration influence appears unlikely. The fact that the farmers adopted a fertilizer formula other than the one used in the experiments reinforces the assessment that this innovation was not attributable to the OFR programme.

The reduction of the planting distance between beans and maize appears to be an innovation that was not inspired by any experimentation. In the on-farm experiments the distance between maize and beans was traditional, and there were no specific experiments on planting distance. It is possible, however, that farmers were inspired by the different spatial arrangements of the experiments, but no such evidence emerged in the interviews. It would have been difficult in any case to foresee such a change and the only way to support this innovation would have been by very close monitoring of farmers and their opinions during the time of the experiments.

Four other changes were clearly not directly related to the OFR programme. The first was the increased use of fungicides by some farmers, mainly with Gloriabamba; farmers stated that the yield from Gloriabamba was worth the investment in fungicide. This shows that excluding a chemical solution from experimentation does not prevent its adoption, even in situations where the farmers are short of cash, as in Cajamarca Department. The second change concerned planting depth. There were no experiments that could have triggered off this innovation and no evidence that it was inspired by any experiment was found. Apparently, the shallow planting overcomes the root rot problems associated with Gloriabamba. The two final changes, increased market orientation and increased area under beans, were indirectly related to the enhanced profitability and the increased marketable surplus of beans, which, in turn, stemmed from both the varietal release and the OFR programme. Home consumption was apparently satisfied at the levels of productivity of the old varieties and most of the additional production from Gloriabamba could be sold. There is now a very high market demand for Gloriabamba, especially in the form of green shelled beans.

In summary, the on-farm experimentation made a sigificant contribution, directly or indirectly, to some of the observed changes in farmer practices. Between 45% and 65% of the farmers interviewed had changed their bean production practices, which had had a considerable impact on yield levels. However, although the innovations, particularly the adoption of the improved variety, solved certain production problems, they brought in their wake a number of new problems. Three unresolved problems appear to be linked with the diffusion of Gloriabamba. The first is the increased presence of bean weevils, possibly because of the variety's earliness or because of its intrinsic susceptibility. The second problem is root rots, which the farmers have tried to reduce by shallow planting. The third problem is the increased use of fungicides and insecticides by those farmers selling green beans, to ensure that an immaculate product reaches the market, without spots or holes. Solutions to these problems are essential if the impact of the innovations on bean production in the region is to be maintained.

Impact of on-farm reseach on bean production in Cajamarca

Technological change in agriculture is often divided into varietal improvements and non-genetic or management improvements. This distinction is useful for the organization of agricultural research, but does not provide a clear analytical framework for OFR evaluation. Many people stress the crop management potential of OFR, but it can also contribute to the participatory testing of improved lines or by enhancing the diffusion of released varieties (Lightfoot, 1988).

Another difficulty, as mentioned earlier, is whether change has been induced by the OFR process, or has taken place autonomously. Here we shall try to make a clear distinction between these two factors as well as between varietal impact (among adopters of Gloriabamba) and management impact (for both adopters and non-adopters). This analysis will provide greater insight into the relative influence of the OFR programme.

A third difficulty is the dynamic nature of technological change (Feder et al., 1985). Research often advances the speed of change, rather than creating changes that would not have occurred otherwise. For example, planting density might have been successfully increased because of an OFR programme, but without the programme a similar change might have taken place some years later. In that situation, the impact that should be attributed to the programme is the production increase from the moment of the induced adoption to the moment of the expected autonomous adoption.

The analysis here is based on the observed changes in Chota province, focusing first on management and varietal impact and then on the impact of OFR and autonomous changes.

Management versus varietal impact

As shown in Table 9.10, 32% of the bean area in Chota is planted to Gloriabamba and the rest to traditional varieties. We assumed that in 1986, a season with comparable rainfall to 1989, the area planted to Gloriabamba was zero. Thus, any yield increase in the area still planted with traditional varieties would have stemmed from improved management, while in the area planted to Gloriabamba both improved management and the varietal change would have have affected yields.

Among the traditional varieties, yields increased by 44%, from 236 kg/ha to 340 kg/ha; the difference of 104 kg/ha is attributed entirely to management. For Gloriabamba, yields increased by 358 kg/ha as result of both varietal change and improvements in management. It is difficult to calculate the effect of either change alone, because this depends on the sequence of adoption and the magnitude of the interactions between variety and management. We estimated the effects of two scenarios (improved variety or improved management adopted

Table 9. 10   Yield improvement mechanisms in Chota province, Cajamarca Department (kg/ha)

Mechanism	Total diffusion	Autonomous diffusion	OFR-induced diffusion
Introduction of management improvements in traditional cultivars (68% of area)			
management effect	104	19	85
Introduction of Gloriabamba (32% area)			
varietal effect	214	128	86
management effect	144	75	69
total effect	358	203	155
Total average yield improvements (100% of area)			
varietal effect	69	41	28
management effect	117	37	80
total effect	186	78	108

first) and took the average. The management improvement equalled a figure of 144 kg/ha and the varietal change 214 kg/ha. The varietal impact assumed here is comparable to that found in other adoption studies (Pachico and Borbón, 1987; Janssen et al, 1990).

Research impact versus autonomous changes

The first change that we analysed in distinguishing between the impact of OFR and that of autonomous changes was the diffusion of Gloriabamba. Some 40% of farmers first received Gloriabamba through regional research or extension staff. Between 1986 to 1989 most of these staff were heavily involved in the OFR programme. We therefore assumed that 40% of the varietal diffusion was induced by the programme. Of the total varietal impact of 214 kg/ha, 86kg/ha could be attributed to the OFR effort and 128 kg/ha to autonomous diffusion.

For the area planted to Gloriabamba, we separated research impact from autonomous impact by studying the effect of improved planting practices (Ruiz de Londoño and Janssen, 1990). The average yield increase resulting from improved planting practices was estimated to be 119 kg/ha. We did not distinguish the individual components, such as row planting, planting distance, planting depth and maize-bean positioning, that made up the improvements, but looked only at whether farmers had incorporated any changes from the traditional low density broadcasting practice to a more formal planting practice. If the total yield increase resulting from improved management of Gloriabamba was 144 kg/ha, the remaining 25 kg/ha (17% of the total) can be attributed to increased use of fertilizer and fungicides. However, not all of the 119 kg/ha increase resulting from improved planting methods can be attributed to the OFR programme, as some of these methods had been autonomously introduced; 58% of the newly adopted methods were OFR induced.

Among the traditional varieties, the yield increase attributed to improved management was 104 kg/ha. This increase resulted from changes in planting (row planting and increased density) as well as some increase in fertilizer and fungicide use. We decided to attribute 17% of the yield increase to changes in input use (as in the case of Gloriabamba) and the rest to changes in planting method. Thus row planting and density increases for local varieties accounted for 85 kg/ha (83% of 104 kg/ha). Both these changes were associated with the OFR programme. The remaining 19 kg/ha was attributed to autonomous change.

A similar impact analysis to that carried out in Chota was conducted in Cajabamba. Management practices in Cajabamba were more advanced than in Chota, and thus management improvements had less impact and were hardly associated with the OFR effort. The main impact was through varietal diffusion, but this was more limited than in Chota and most of it took place autonomously. Table 9.11 (*overleaf*) shows the yield improvement mechanisms in the two provinces. Average yields increased by 151 kg/ha, of which 82 kg/ha was attributed to the OFR programme. The OFR programme had more impact on management improvements than on varietal diffusion. This suggests that varietal diffusion was a rather autonomous process but that the adaptation of management practices to the new variety was an important task for the programme. Thus, the OFR effort has been very instrumental in accelerating adjustment to new production opportunities arising from the introduction of the new variety.

Although Gloriabamba cannot be considered as a product of OFR, it was the product of an institutional research effort. For measuring the total institutional effect, we added the total varietal impact of Gloriabamba (both autonomous and OFR induced) and the management impact resulting from OFR. This gave a total institutional impact of 121 kg/ha (80% of the productivity increase). Of this, 82 kg/ha can be attributed to OFR and 39 kg/ha to genetic improvement research and the autonomous diffusion of the results of this research.

Table 9.11   Yield improvement mechanisms in Chota and Cajabamba provinces, Cajamarca Department (kg/ha)

Mechanism	Total diffusion	Autonomous diffusion	OFR-induced diffusion
Introduction of management improvements in traditional cultivars:			
management effect	77	17	60
Introduction of Gloriabamba:			
varietal effect	207	128	79
management effect	114	60	54
total effect	321	188	133
Total average yield improvements: (100% of area)			
varietal effect	63	39	24
management effect	88	30	58
total effect	151	69	82

Economic impact of the on-farm research programme

Apart from Chota and Cajabamba, the area of influence of the OFR programme included four other provinces, accounting for an additional 5000 ha of beans. If we assume that the yield increase in these provinces was the same as the average yield increase for Chota and Cajabamba, then in 1988 the OFR programme would have resulted in additional bean production of some 529 t, worth about US\$ 265 000 (the figures given here should not be considered as net benefits as they have not been corrected for possible incremental costs of production or for the cost of the OFR itself). Over 70% of the production increase would have been attributable to improved management. Under a more conservative assumption that the yield increases in the other provinces were the same as the Cajabamba figures, the programme would have increased bean production by 256 t, worth about US\$ 130 000. In this case the relative importance of genetic improvement would increase.

The Net Present Value (NPV) of increased bean production directly related to the OFR programme is a final important evaluation criterion. To estimate the NPV we needed to know the diffusion rates of genetic improvements and management improvements with and without the programme.

For the genetic improvements, the data on the actual diffusion of Gloriabamba were available from the questionnaire. We assumed that, without the programme, the 1988 diffusion would have been equal to the autonomous diffusion (60% of total diffusion in Chota and 68% in Cajabamba), and that the time taken to complete the diffusion process would have been 7 years instead of 5 years; studies in Mexico and Nicaragua had estimated the length of the diffusion period to be 7 years (Borbón and Janssen, 1989; Janssen et al., 1990). We also assumed that final diffusion stayed 5% lower, because the technology provided by the programme widened the range of adoption of Gloriabamba, and that the economic lifetime of the variety is 15 years.

For the management improvements we assumed a time lag of 5 years between present diffusion and diffusion without the programme. This assumption is somewhat arbitrary, but does not particularly favour the expected benefits calculations as the evidence that these changes would have occurred without the programme is inconclusive. We used a 7% discount factor.

Table 9.12 presents the outcomes of these estimations. Under the favourable assumptions, the NPV of the programme was over US$ 1.9 million, 75% of which was attributable to management improvements. Under the less favourable assumptions, the NPV was US$ 0.9 million, 62% of which was attributable to management improvements. Because of the many assumptions we have had to make, the interpretation of these figures should be treated with caution. Nevertheless, it appears safe to say that the minimum impact of the OFR programme on production was US$ 0.9 million and the maximum impact US$ 1.9 million and that 60-75% of this impact was attributable to the development of improved management methods.

Table 9.12   The impact of the on-farm research project in Cajamarca Department

	Favourable assumptions[1]		Less favourable assumptions[2]	
	Genetic improvements	**Management improvements**	**Genetic improvements**	**Management improvements**
Production increases, 1988:				
estimated bean production increases attributed to OFR, 1988 (t)	150	379	115	141
estimated monetary value (US$ 000)	75	190	58	71
Net Present Value of programme:				
estimated bean production increase 1985-2000 (t)	883	2872	678	1082
estimated monetary value (US$ 000[3])	438	1495	335	559

Notes: 1. Impact of the programme in the region of influence is equal to average impact in Chota and Cajabamba provinces.
2. Impact of the programme in all other areas of influence is equal to impact in Cajabamba; in Chota, impact is as estimated with questionnaire data.
3. NPV of increased production in US$ in 1990. An interest rate of 7% was used.

CONCLUSION

The OFR programme in the Department of Cajamarca has had significant influence on regional bean production. That influence was partly because the programme constituted a successful follow-up to the release of Gloriabamba. The fact that Gloriabamba was not a product of the programme provided the opportunity to review how OFR can interact with technological innovations that include a strong on-station component. The analysis shows that the release of Gloriabamba without a complementary on-farm programme would have had a significant, but much lower, impact on production. Similarly, if an OFR programme had operated without the previous release of Gloriabamba, it would have had only a modest impact on the management of local bean varieties.

In the introduction to this chapter it was suggested that OFR is appropriate for situations where the traditional balance of the farming system is disrupted by a technological innovation from outside the system. The data from the Cajamarca case tend to support this suggestion in that they show that OFR and on-station research are not mutually exclusive but, rather, enhance each other's potential. This observation has some important implications. First, the exclusion of one of these types of research from national research programmes to the advantage of the other one appears to be counter-productive. Second, on-farm and on-station researchers should jointly determine which regional priorities to address. Third, in order to captialize on new developments, the timing of on-farm trials should depend to some extent on progress made in on-station research.

A number of other observations also deserve mention. The first one concerns the measurability of innovations arising from OFR. Before we started evaluating the Cajamarca programme we were rather sceptical about its impact. We tried to apply a rigorous analysis and this produced a rather favourable evaluation, with considerable yield increases attributable to the programme. Why had we been so pessimistic initially? We consider that one of the main problems concerning OFR is that its results are often not very tangible. This problem has been exacerbated by the lack of attention in OFR methodology to the systematic follow-up of results. The actual research procedures are well developed and described but both short- and long-term feedback mechanisms have not received adequate attention. In this respect, we hope that the evaluation method described in this chapter proves useful and stimulates researchers to develop alternative methods.

It is important that evaluation methods make a clear distinction between the technological innovations induced by OFR and those that would have occurred anyway. A subjective analysis might easily find OFR responsible for more changes than was actually the case. An evaluation exercise should also include mapping out the chronology of the OFR; by linking this with the chronology of technological innovations, causal effects can be more easily identified. In practical terms, this implies that on-farm researchers who intend documenting the impact of a programme should make an inventory of technological changes taking place in the research area at the start of the programme. Apart from providing a clearer picture of the relative impact of the programme and of autonomous changes, such an inventory might suggest activities which would support or complement the technological changes already taking place.

The second observation concerns farmer rationality and the complexity of small farming. In the Cajamarca case, a diverse range of technological changes, resulting from the OFR programme and from the autonomous efforts, was identified in a poor and isolated region of Latin America. As a consequence, many small farmers were able for the first time to obtain cash income from beans. This supports the premise that farmers continuously search for ways to improve their farming systems, but, more importantly, it indicates the potential of OFR for enabling farmers to increase production and obtain additional income in difficult conditions.

The final observation concerns OFR procedures. We found that many of the innovations that were tested and eventually adopted were viewed somewhat differently by researchers and farmers. This would suggest that the time and effort involved in the diagnostic studies should be carefully weighed against better monitoring of the experimentation to allow for the identification of potentially successful technologies and farmers' motives for technical change. Such a modification would change the traditional character of OFR. The follow-up studies would increase the degree and length of interaction between farmer and researcher, requiring the use of more participatory research methods. In addition, a more continuous monitoring of agricultural change may make less demands on data collection and at the same time increase the ability to fine tune research orientation.

Notes

The credit for the success of this OFR project lies mainly with the Peruvian researchers involved in the project. Special mention should be made of Hipolito de la Cruz, who selected the Gloriabamba variety from an international varietal nursery and tested it extensively with many farmers, and Elmer Rojas, Segundo Terrones, Ramon Astupilco, Alberto Sanchez and Hector Vargas, who designed, planted and analysed most of the experiments. Ramon Astupilco also collected much of the diagnostic data.

References

Arbulu, P., Ruiz de Londoño, N. and Pachico, D. 1986. *Diagnóstico de la Producción de Frijol en la Provincia de Chota, Departamento de Cajamarca, Perú, 1985.* Working Paper No. 12. Cali, Colombia: CIAT.

Astupilco, R. 1985. *Diagnstico de la Producción de Frijol en la Provincia de Cajabamba, Departamento de Cajamarca, Perú.* Mimeograph prepared for first intensive OFR course in Peru, 1985. Lima, Peru: INIPA.

Borbón, E. and Janssen, W.G. 1989. Estudio de adopción de variedades mejoradas de frijol en las regiones I y IV de Nicaragua, cosecha veranera 1987. Paper presented at the 35th PCCMCA, April 1989, San Pedro Sula, Honduras.

Byerlee, D., Harrington, L. and Winkelmann, D.L. 1982. Farming systems research: Issues in research strategy and technology design. *American Journal of Agricultural Economics* 64(5): 897-904.

Deza, J. 1985. *Implicaciones para la Investigación en Finca sobre Frijol para la Provincia de Celendín, Departamento de Cajamarca, Perú.* Mimeograph prepared for first intensive OFR course in Peru, 1985. Lima, Peru: INIPA.

De la Cruz, J.H. 1985. Frijol Gloriabamba. *Extension Bulletin.* Cajamarca, Peru: INIPA/CIPA.

De la Cruz, J.H. and Beltran, J.A. 1987. *La Investigación de Frijol a Nivel de Chacra. Primer Curso Intensivo Postgrado, Perú 1985-1986, Resultados del Primer Año de Ensayos.* Mimeograph. Cali, Colombia: CIAT.

De la Cruz, J.H. and Rojas, E. 1987. Investigación en frijol en chacras de agricultores del departamento de Cajamarca, Perú. In Woolley, J.N. (ed). *La Investigación de Frijol en Campos de Agricultores de America Latina, Memorias de un Taller.* Working Paper No. 27. Cali, Colombia: CIAT.

Feder, G., Just, R.E. and Zilberman, D. 1985. Adoption of agricultural innovations in developing countries: A survey. *Economic Development and Cultural Change* 33(2): 255-99.

Hardaker, J.B., Anderson, J.R. and Dillon, J.C. 1984. Perspectives on assessing the impacts of improved technologies in developing countries. *Australian Journal of Agricultural Economics* 28: 87-108.

INIPA/CIPA. (n.d.) *Estadisticos Varios.* Mimeograph. Cajamarca, Peru: Programa de Leguminosas de Grano, INIPA.

Janssen, W., Luna, C.A. and Lopez, E. 1990. *La Adopción de la Variedad Negro Huasteco 1981 en las Huastecas de México.* Mexico: Economía Agricola, Universidad Autonoma de Chapingo.

Lightfoot, C. 1988. On-farm trials: A survey of methods. *Agricultural Administration and Extension* 30: 15-23.

Ministerio de Agricultura. 1981. *Anuario Estadstico.* Lima, Peru: Ministerio de Agricultura

Pachico, D. and Borbón, E. 1987. Technical change in traditional small farm agriculture: The case of beans in Costa Rica. *Agricultural Administration and Extension* 26: 65-74.

Richards, P.F. 1985. *Indigenous Agricultural Revolution.* London, UK: Unwin Hyman.

Ruiz de Londoño, N. and De la Cruz, J.H. 1986. *Diagnostico de la Producción de Frijol, Perú. Informe Preliminar.* Mimeograph. Cali, Colombia: CIAT.

Ruiz de Londoño, N. 1986. Informe de viaje a Perú, junio 1986. Internal Document. Cali, Colombia: Bean Program, CIAT.

Ruiz de Londoño, N. and Janssen, W. 1990. *La Adopción de Tecnologia: El Caso de la Variedad Gloriabamba en Perú.* Working Paper No. 61. Cali, Colombia: CIAT.

Terrones, S. 1985. *Diagnóstico del Cultivo de Frijol en la Provincia de Contumazá, Departamento de Cajamarca, Perú.* Mimeograph prepared for first intensive OFR course in Peru, 1985. Lima, Peru: INIPA.

Tripp, R. and Woolley, J. 1989. *The Planning Stage of On-Farm Research: Identifying Factors for Experimentation.* El Batan, Mexico/Cali, Colombia: CIMMYT/CIAT.

Tripp, R. and Woolley, J. 1990. Training in farming systems research: Review and prospects. *Experimental Agriculture* 26: 247-62.

Vargas, H. 1985. *Diagnóstico del Cultivo de Frijol Realizada en los Distritos de Cutervo y Socot, Departamento de Cajamarca, Perú.* Mimeograph prepared for first intensive OFR course in Peru, 1985. Lima, Peru: INIPA.

Woolley, J.N., Lepiz, R., Portes e Castro, T. de A. and Voss, J. 1990. Bean cropping systems in the tropics and their determinants. In van Schoonhoven, A. and Voysest, O. (eds) *Common Beans: Research for Crop Improvement.* Cali, Colombia: CIAT.

10

On-Farm Research in Caisán, Panama

J.C. MARTINEZ, G. SAIN and J.R. ARAUZ

During the 1970s the Government of Panama initiated a number of steps to increase the domestic production of basic grains and reverse the growing trend towards imports. A programme of guaranteed prices for basic grains was established and specific attention was given to the organization of agricultural research, extension and credit. Until that time, agricultural research in Panama had been in the hands of the Ministry of Agricultural Development (MIDA), the University of Panama, and various other public and private institutions. In 1975 the national agricultural research institute, Instituto de Investigaciones Agropecuarias de Panamá (IDIAP), was created, drawing principally on scientists who had been employed by MIDA.

One of IDIAP's first tasks was to develop a strategy that would incorporate the identification of research priorities and the development of technologies for well-defined groups of farmers. As part of this effort, IDIAP began discussions with the International Maize and Wheat Improvement Center (CIMMYT) and formulated a plan to test the feasibility of the on-farm research (OFR) methods that CIMMYT was developing and promoting. The idea was to select one area of Panama where OFR could be carried out, using only the resources and personnel available to IDIAP. If the results were promising, then the methods and strategy could be extended to other areas of the country. This chapter reviews the first 3 years of research in the area chosen for the initial work (Arauz and Martínez, 1983; Martínez and Arauz, 1983; Martínez and Arauz, 1984).

THE RESEARCH AREA

The area selected for the first OFR effort in Panama was the *corregimiento* of Caisán, in Renacimiento District, Chiriquí Province. It lies in the western part of the country, near the Costa Rican border (*see* Figure 10.1 *overleaf*). This area was chosen because its generally small- and medium-sized farms produced commodities that were priorities in the National Agricultural Development Plan and seemed to offer potential for technological development.

At the time of the research effort there were about 300 farm families in Caisán, distributed among nine small communities. The area was fairly well served by agricultural institutions, including an office of MIDA's

Figure 10.1   The District of Renacimiento, Panama, showing the Caisán study area

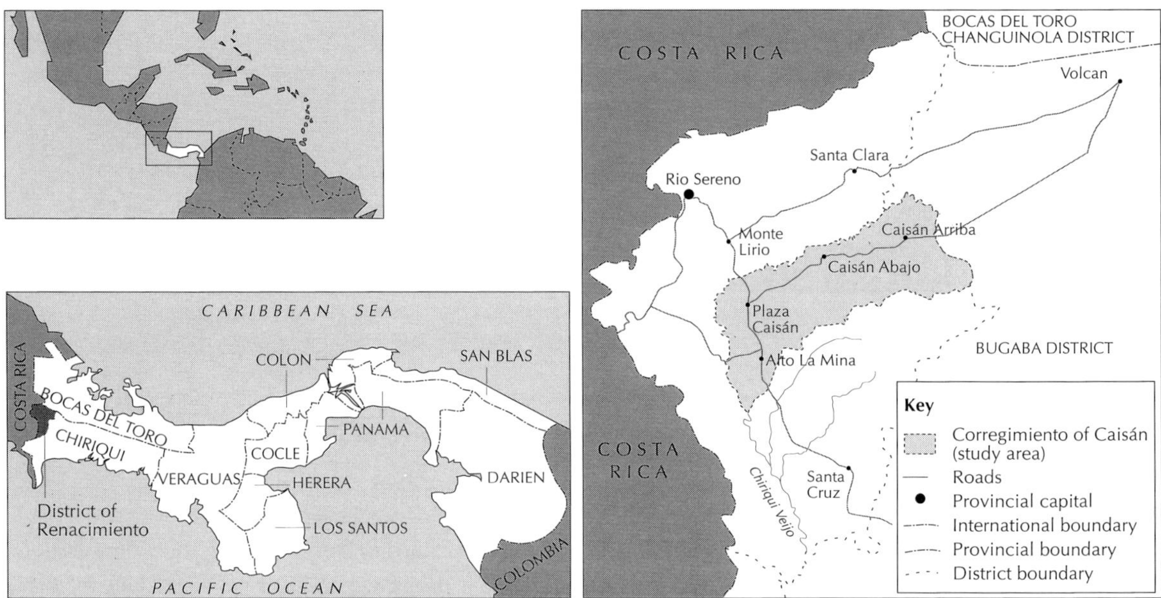

extension service, an office of the agricultural credit bank, Banco de Desarrollo Agropecuaría (BDA), in the town of Volcan, about 30 minutes' drive from Caisán, and representatives of the government agricultural mechanization service, Empresa Nacional de Maquinaria Agrícola (ENDEMA). Most farmers in the area belonged to one of three *juntas agrarias*, farmer associations established by the government during the early 1970s to deal with the purchase of inputs and to represent farmers' interests in government institutions. Caisán had 23 km of all-weather roads, but access to many of the farming areas was severely limited during the rainy season.

The western part of Caisán is a small valley formed by the Chiriquí Viejo and Caisán Rivers and characterized by flat or gently rolling terrain, where most of the annual crops of the area are grown. The climate is tropical, with an average annual rainfall of 4000 mm, about 90% of which falls between May and November (*see* Table 10.1). Average temperatures range from 18°C in the dry season to 22°C in the rainy season. The soils in the area are of volcanic origin, deep, well drained and fertile, with a high organic matter content.

The rainfall pattern in Caisán permits two cropping seasons per year. The predominant cropping pattern, accounting for about 44% of the cropped land in the area, is a maize/bean rotation in which maize is planted in March or April, harvested in September and then immediately followed by a crop of beans. This pattern became the focus of the OFR. Some maize and beans are also planted outside of this rotation. In addition, about one-third of the farmers in the area keep cattle, mostly for commercial milk production. The average size of maize planting in the first season is 5.3 ha; 57% of the maize fields are under 4 ha.

In 1978, IDIAP assigned an agronomist (a former CIMMYT trainee) to work full-time in the area and later provided a field assistant as well as the financial resources required to meet the operational costs of the programme. CIMMYT regional staff made periodic visits to Caisán and helped organize surveys and on-farm experiments.

Table 10.1   Average monthly rainfall (mm) at the meteorological station, Caisán Arriba, 1973-78

Jan	Feb	Mar	Apr	May	Jun	Jul	Aug	Sept	Oct	Nov	Dec	Total
52	44	126	206	679	464	313	313	554	612	446	89	4097

Source: Estadística Panameña, Metereología, 1973-78

DIAGNOSTIC ACTIVITIES

After IDIAP and CIMMYT had agreed that the OFR would focus on Caisán, available secondary data from the area were reviewed to help plan diagnostic activities. Researchers decided to concentrate on the first cropping season, when maize was the major crop, and in August 1978 an informal survey was conducted. The survey consisted of visits to farmers' fields and informal conversations with farmers (Byerlee et al., 1984).

In December 1978 a formal survey, based on a questionnaire, was carried out by IDIAP personnel among a sample group of farmers in Caisán and the neighbouring area of Bajo Chiriquí. The areas turned out to be two distinct recommendation domains. Whereas Caisán was characterized by a fairly high use of purchased inputs and tractor hire services, Bajo Chiriquí was very isolated and farming practices reflected a lack of access to inputs and machinery. This reaffirmed the decision to concentrate the initial OFR on the recommendation domain of Caisán. The formal survey results reported here summarize the practices of 35 randomly selected farmers in Caisán. Information from the national census, updated during the informal survey, was used to select the survey sample. The questionnaire itself was quite short (it took about 30 minutes to complete with each farmer) and it addressed issues that needed to be confirmed or further explored after the informal survey, focusing in particular on practices in the first season maize crop and their relation to the following crop.

MAIZE PRODUCTION PRACTICES

Farmers in Caisán attempt to plant their maize as early as possible in the first season. Part of this urgency can be explained by the need to harvest the maize and establish a bean crop in the same field before the rains end. Another reason for early planting is the prevalence of strong winds in June or July which can cause serious lodging in the maize crop. Land preparation thus begins in the dry season. The formal survey showed that most farmers (74%) used rented tractors to prepare their fields. Credit was available for tractor rental, as part of a loan package provided by the BDA to maize farmers. A small percentage of farmers did not use tractors, mainly because of the slope of their fields. In these cases, land was prepared manually, by cutting the weed growth with a machete. Evidence from the informal survey, however, showed that much of the incipient soil erosion observed in the area was the result of tractor tillage.

Maize is planted with the first rains, most farmers completing their planting before April. Almost all the farmers surveyed planted their maize manually, and of these all but one planted at random, rather than in rows, using a stick to make holes for the seed. The average distance between holes was about 1 m, and farmers placed four or more seeds in each hole. The large distance between holes meant that farmers were not taking maximum advantage of their land, and the high number of seeds per hole led to excessive competiton among plants.

Virtually all the farmers planted a local maize variety called Caisán and saved their maize seed from one harvest to plant in the following season. Panama has not been an important area for maize production and has been subject to the introduction of maize varieties from Central America, the Caribbean and South America, both through trade and agricultural development programmes. Thus the exact origins of the Caisán variety are not known. Although this variety yielded well, it suffered from several disadvantages. The most important was its height, which exceeded 3 m, making the plant very susceptible to lodging. Most of the farmers interviewed (77%) mentioned that one of their main production problems was yield loss resulting from lodging. A shorter variety had been made available by the research and extension services, but farmers found it unacceptable because its poor husk cover made it susceptible to ear rots.

When farmers were asked which factors limited their maize production, the most common answer was weeds. This response confirmed a conclusion reached during the informal survey. Farmers listed several important broadleaf and grassy weeds as contributing to the problem, and the range of weed control practices observed in the survey reflected the farmers' attempts to find an adequate solution (*see* Table 10.2). Most of them used some form of chemical weed control, but nearly 50% had found it necessary to complement this practice with manual weed control, using a machete. The fact that farmers did not plant their maize in rows made chemical weed control less efficient. In the year of the survey, farmers were testing six different chemicals, using a wide range of application rates and timings. As the cost of labour in the area was rising rapidly, and heavy rains during the growing season made tractor cultivation impracticable, they were very concerned to find a viable practice for controlling weeds.

Table 10.2   Weed control methods in maize, Caisán, 1978

Methods	No. of farmers	%	Hectares	%
Manual	9	27.3	54.5	31.6
Chemical	13	39.4	62.5	36.2
Manual and chemical	11	33.3	55.5	32.3
Total	33	100.0	172.5	100.0

Source: Arauz and Martínez, 1983

Fertilizer was made available to maize farmers through the BDA; farmers who used the bank's credit facilities had to accept fertilizer as part of the package. The credit programme recommended an application of 4 quintals of compound fertilizer per hectare, equivalent to about 20 kg N/ha, 50 kg P_2O_5/ha and 20 kg K_2O/ha. More than half the farmers applied the fertilizer to their maize but at levels below the recommendation (*see* Table 10.3), while a number of farmers used the fertilizer on other crops or sold it. The relatively low use of fertilizer, despite farmers' familiarity with it, and observations on the soils of the area, led researchers to question the BDA's fertilizer recommendation. The recommendation was not based on local research, nor did it meet the requirements of the maize crop in Caisán. However, it was the only fertilizer available.

Field observations during the informal survey did not indicate serious problems with insects in maize. Only about 25% of the farmers were using insecticides on maize, but nearly all of them used insecticides on beans.

The maize crop is harvested in early September. The period of maize harvest, followed by field preparation and planting the bean crop, represents the peak demand for hired labour during the year. On average, about 25% of the maize harvest was kept for home consumption and the rest marketed.

Table 10.3   Fertilizer use in maize, Caisán, 1978

Fertilizer type	No. of farmers	%	Hectares	%
None	14	42.4	70.0	38.8
10-30-10 N-P-K	14	42.4	72.5	40.2
12-24-12 N-P-K	3	9.1	28.0	15.5
Foliar	2	6.1	10.0	5.5
Total	33	100.0	180.5	100.0

Source: Arauz and Martínez, 1983

FIRST CYCLE OF EXPERIMENTS

Information from the diagnostic studies was reviewed in order to plan a series of on-farm experiments to be planted in the first season of 1979. Emphasis was placed on distinguishing short-term from more long-term research strategies and on defining a work programme that would parallel the sequence farmers would be likely to follow in changing their practices. The research priorities included weed control, spatial arrangement, fertilizer use, and varieties more resistant to lodging. The selection of weed control and lodging corresponded with the farmers' expressed priorities but it was the researchers who recognized the inadequacy of farmers' plant spacing (and its relation to weed control) and the need to investigate current fertilizer recommendations.

Some of these research topics could be dealt with immediately, while others required a longer time horizon. Experimentation to determine adequate weed control and plant spacing, for instance, offered the possibility of developing recommendations fairly quickly. The issue of fertilizer requirements, on the other hand, would need several years of research and also implied some interaction with the policy makers responsible for designing and administering the credit programme for maize. The development of a more suitable, shorter maize variety for the area would also involve a number of years of research. Other research themes, including soil erosion problems and the labour bottleneck at the time of maize harvest and bean planting, were discussed, but their investigation was postponed pending the results of the first cycle of experiments.

The research strategy for the first cycle is outlined in Table 10.4 (*overleaf*). The four components that were to be included in on-farm experiments were weed control, plant density and spatial distribution, nitrogen, and phosphorus. Two basic types of experiment were outlined. One was an exploratory experiment to examine the relationships between the four factors; each factor would be represented in the experiment by two levels, that of the farmer and an alternative. The other type of experiment would examine the effects of various levels of particular components. Three of these 'levels' experiments were designed: herbicide type, herbicide timing, and nitrogen and phosphorus levels.

Because the herbicide experiments looked at technological alternatives that were potentially immediately available to farmers, they were planted with non-experimental variables at the level of usual farmer practice.

Table 10.4   Experimental strategy and trial management, first year of experiments in Caisán

Components	Nature of problem	Timing of research	Exploratory trials (2^4)			Level trials	
			Components	Range of experimental variables	Non-experimental variables	Components included	Non-experimental variables
A. Weed control B. Plant density and spatial distribution	Production	Short term	A B	Farmers' practice and alternative	Farmers' practice	A	Farmers' practice
C. Nitrogen requirements D. Phosphorus requirements	Production/ agricultural policy	Medium term	C D			C D	Farmers' practice and improved weed control and plant density
E. Lodging	Breeding	Medium term	Program to reduce plant height of local variety				

Source: Martínez and Arauz, 1983

On the other hand, the fertilizer experiments were planted using 'best bet' weed control practices, planting density and spatial distribution — that is, the alternative practices that researchers hoped farmers would employ by the time that the fertilizer research yielded recommendations.

The experiments were planted in farmers' fields that were judged to be representative of the area. The types of experiments and number of sites are shown in Table 10.5. All the experiments in this first cycle were planted in fields with less than 5% slope, because this type of field predominated in the maize/bean rotation. Researchers expected that the results of the experiments would be applicable to fields on steeper slopes, particularly in terms of the interaction between slope and fertilizer response. The farmers prepared their land as usual, and the local maize variety was used in all the experiments. The experiments were planted by the IDIAP agronomist and field assistant, with help from hired labour and local farmers.

The results of the exploratory experiment were particularly interesting (*see* Table 10.6, and Tables 10.7 and 10.8 *overleaf*). There were significant responses to improved weed control and plant density and distribution (and a significant interaction between these elements), but not to nitrogen and phosphorus. This tended to confirm researchers' hypotheses that better weed control and planting arrangements could improve maize production and placed further doubt on the adequacy of current fertilizer recommendations. The interaction between plant density and weed control (*see* Figure 10.2 *overleaf*) indicated that a change in farmers' planting arrangements would be more effective if combined with better weed control. A partial budget analysis of these components showed that changes in planting practices and weed control would be very profitable (*see* Table 10.9 *overleaf*).

Table 10.5   Types and number of experiments, 1979

Type	Components	Sites planted	Sites harvested
Exploratory	Weed control, density and spatial distribution, nitrogen, phosphorus (2 levels each, 16 treatments)	6	5
Levels	Types and combinations of herbicides (10 treatments)	2	2
Levels	Timing of herbicide application (12 treatments)	2	0
Levels	Nitrogen and phosphorus (16 treatments)	2	2

Source: Arauz and Martínez, 1983

Table 10.6   Treatments in exploratory experiment, Caisán, 1979

A. Weed control

W_0 = 2,4-D, 1 l/ha of commercial product applied 30 days after planting

W_1 = Gesaprim 80, 2.5 kg/ha of commercial product applied at pre-emergence

B. Density and spatial distribution

D_0 = 40 000 plants/ha, randomly planted at 1 m distance, no thinning

D_1 = 50 000 plants/ha, 2 plants per hill (after thinning), planted in rows with 50 cm between hills, 80 cm between rows

C. Nitrogen

N_0 = 0 kg/ha

N_1 = 90 kg/ha

D. Phosphorus

P_0 = 25 kg/ha

P_1 = 75 kg/ha

Source: Arauz and Martínez, 1983

Table 10.7   Results of exploratory experiment, Caisán, 1979

Treatment	Weed control	Density	Nitrogen	Phosphorus	Average yield, 5 sites (t/ha)
1	0	0	0	0	2.9
2	1	0	0	0	4.1
3	0	1	0	0	4.7
4	1	1	0	0	5.2
5	0	0	1	0	3.8
6	1	0	1	0	4.2
7	0	1	1	0	3.6
8	1	1	1	0	5.5
9	0	0	0	1	4.1
10	1	0	0	1	4.2
11	0	1	0	1	4.2
12	1	1	0	1	5.3
13	0	0	1	1	3.5
14	1	0	1	1	4.2
15	0	1	1	1	4.2
16	1	1	1	1	6.1

Note: 0 = farmers' practice
         1 = improved practice

Source: Arauz and Martínez, 1983

Table 10.8   Principal effects, exploratory experiment, Caisán, 1979

Treatment	Average yield 5 sites (t/ha)
W_0	3.9
W_1	4.8
D_0	3.9
D_1	4.8
N_0	4.3
N_1	4.4
P_0	4.3
P_1	4.5

Source: Arauz and Martínez, 1983

Figure 10.2   The interaction between weed control and plant density

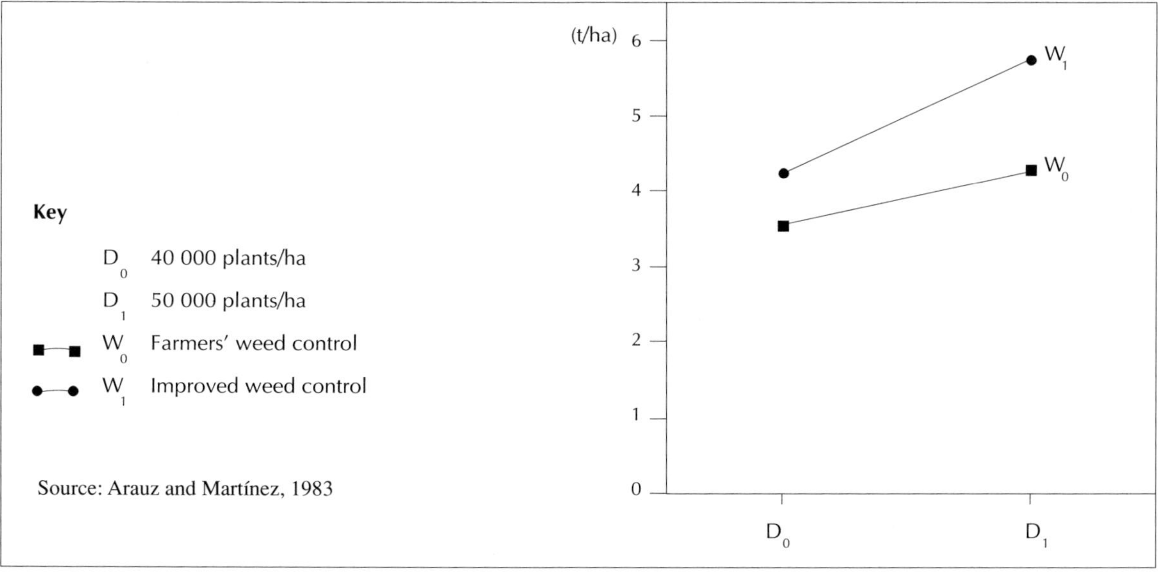

Table 10.9   Economic analysis of weed control and plant spacing factors

| | Treatment | | | |
	$W_0 D_0$	$W_0 D_1$	$W_1 D_0$	$W_1 D_1$
Yield(t/ha)	3.60	4.20	4.20	5.50
Adjusted yield (10%) (t/ha)	3.24	3.78	3.78	4.95
Gross field benefit (US$/ha)	369.36	430.92	430.92	564.30
Herbicide:				
2,4-D (US$/ha)	1.63	1.63	—	—
Gesaprim (US$/ha)	—	—	17.97	17.97
Planting:				
seed (US$/ha)	2.86	3.52	2.86	3.52
labour (US$/ha)	10.74	17.90	10.74	17.90
Total costs that vary (US$/ha)	15.23	23.05	31.57	39.39
Net benefits (US$/ha)	354.13	407.87	399.35	524.91

Change	Marginal rate of return
$W_0 D_0 \rightarrow W_0 D_1$	687%
$W_0 D_1 \rightarrow W_1 D_1$	716%

Source: Arauz and Martínez, 1983

The herbicide experiments tested various products (and combinations) against the most common local practice of applying 1 l/ha of 2,4-D at about 30 days after planting. The results of these experiments showed a clear yield superiority of almost every other treatment over farmers' practice. The economic analysis of the experiments showed that the use of Gesaprim 80 at 2 kg/ha was the most attractive weed control treatment, adding credence to the results of the exploratory experiment.

Unfortunately, the experiments on the timing of herbicide application were lost because of wind damage. The fertilizer experiments were harvested but suffered from a combination of animal and soil insect damage. The analysis of the partial results available, however, showed no clear response to either nitrogen or phosphorus, confirming the results of the exploratory experiment.

SECOND CYCLE OF EXPERIMENTS

The results of the first cycle of experiments provided clear guidelines for designing the 1980 experiments (*see* Table 10.10). The identification of weed control and planting methods that were clearly superior to those used by farmers allowed the researchers to develop and refine possible recommendations. Because the pre-emergence herbicide Gesaprim was found to be particularly attractive, it was incorporated as a non-experimental variable in a new exploratory experiment. Attention turned to the possibility of combining this chemical weed control with zero tillage in order to begin to address the erosion problem that had been identified, and thus a tillage variable was substituted for the weed control variable which had been used in the 1979 exploratory experiment. In addition, experiments were designed to further explore the interaction between weed control and planting density, to provide more information on fertilizer response and to provide data on the residual effects on the bean crop of the application of atrazine (the active ingredient of Gesaprim) to the maize crop.

The results of this exploratory experiment were again very instructive. The treatments are shown in Table 10.11 and a summary of the results is presented in Table 10.12. There were significant responses to planting practice and to phosphorus, and no interactions were observed between any of the factors. A review of each factor in the exploratory experiment, combined with the results of other experiments, provided important information for developing recommendations.

Table 10.10   Types and number of experiments, Caisán, 1980

Types	Components	Sites planted	Sites harvested
Exploratory	Tillage, density and spatial distribution, nitrogen, phosphorus (16 treatments)	4	3
Levels	Herbicide by plant density (8 treatments)	3	0
Levels	Nitrogen, phosphorus, potassium (10 treatments)	3	3
Observation	Residual effect of atrazine	1	1

Source: Arauz and Martínez, 1983

Table 10.11   Treatments in exploratory experiment, Caisán, 1980[1]

A. Tillage

T_0 = Conventional preparation with tractor, 3 harrowings

T_1 = Zero tillage; hand chopping of weeds and crop residues, Gramoxone 1 l/ha commercial product

B. Density and spatial distribution

D_0 = 40 000 plants/ha, randomly planted at 1 m distance, no thinning

D_1 = 50 000 plants/ha, 2 plants per hill (after thinning), planted in rows with 50 cm between hills, 80 cm between rows

C. Nitrogen

N_0 = 0 kg/ha

N_1 = 90 kg/ha

D. Phosphorus

P_0 = 25 kg/ha

P_1 = 75 kg/ha

Note: 1. Weed control in all treatments was with Gesaprim 80, 2.5 kg/ha of commercial product.

Table 10.12   Main effects of factors in exploratory experiment, Caisán, 1980

Treatment	Average yield from 3 sites (t/ha)	Economic analysis
T_0	4.7	Zero tillage gives equivalent yield at
T_1	4.8	40% lower cost
D_0	3.8	Very high rate of return to improved
D_1	5.6	density and spacing
N_0	4.8	(No response)
N_1	4.8	
P_0	4.6	Low rate of return to additional
P_1	4.9	P_2O_5 (8%)

Source: Arauz and Martínez (1983)

Tillage

There was no difference in yield between conventional tractor tillage and zero tillage. The economic analysis, however, showed a clear cost saving where zero tillage was used. This economic advantage, combined with the increased urgency of addressing the erosion problem (a growing minority of maize/bean rotations were being planted on slopes) gave these results added importance. In addition, the adoption of zero tillage would enable farmers to plant on time and would contribute to more effective weed control. An experiment to examine the residual effects of atrazine showed that under the heavy rainfall of Caisán, a bean crop could be planted within 90 days of atrazine application. Given that the period between maize planting (when atrazine would be applied) and bean planting was about 6 months, the possibility of using atrazine in the maize cycle of the maize/bean rotation appeared to pose no problem.

Planting practices

The response to improved plant density and spatial arrangement was as strong in the second year's experiments as it had been in 1979 and provided a clear recommendation for farmers. Unfortunately, the experiments that were designed to examine the interaction between density and weed control were lost because of high winds.

Fertilizer use

The lack of response to nitrogen in the exploratory experiment repeated the previous year's experience. In the second year, however, a small response to phosphorus was noted, but an economic analysis of these results showed that the investment in additional phosphorus gave very low returns. The experiment that examined response to nitrogen, phosphorus and potassium gave similar results: small yield increases in some sites for each of the nutrients but no evidence of economic advantage. The equivocal nature of these results encouraged researchers to further question the current fertilizer recommendation but also to gather more evidence of the long-term effects of various fertilizer management practices.

THIRD CYCLE OF EXPERIMENTS

The 1981 experiments placed more emphasis on demonstrating and extending some of the information gained in the previous two years, while at the same time addressing research topics that had proved particularly difficult (fertility management) and taking on some new problems (soil insects) (*see* Table 10.13).

The results of all the experiments in 1981 were severely affected by an unusual rainfall pattern. The rainy season began late, which delayed planting, and the mid-season drought (*canicula*) that often occurs in this region was exceptionally severe. Another factor, which was almost certainly related to the delayed planting and the drought, was that the maize crop in Caisán suffered a severe attack of leaf blight (caused by *Helminthosporium* spp.) and many fields were damaged by white grubs (*Phyllophaga* spp.). As a result, the experimental yields were 50-60% lower, on average, than in the previous two years. Nevertheless, most experiments were harvested and the results confirmed some previous observations as well as providing new information.

Table 10.13   Types and number of experiments, Caisán, 1981

Type	Components	Sites planted	Sites harvested
Exploratory	Tillage, density and spatial distribution, soil insect control, phosphorus (16 treatments)	5	3
Levels	Herbicide x plant density (12 treatments)	4	4
Levels	Nitrogen, phosphorus, potassium (16 treatments)	5	2
Verification	Farmers' practice Zero tillage, improved density Zero tillage, improved density, fertilizer	3	3
Demonstrations	Farmer-managed zero tillage	3	3

Source: Arauz and Martínez, 1983

The experiments which included weed control treatments showed that Gesaprim and/or Gramoxone were the most attractive options. The fertilizer trials, which were planted on the same sites used in 1980, began to show some response to nitrogen, although three of the five experiments could not be harvested. The experiments on the interaction between herbicide and plant density showed a response to densities above 50 000 plants/ha, although under admittedly unusual conditions. The soil insect control factor in the exploratory experiment showed a positive response. The economic analysis showed the treatment to be profitable, but it was uncertain whether the treatment would be profitable in years when the white grub problems were less severe.

The verification trials were planted on large plots, with increased farmer participation. The results indicated that zero tillage (manual cutting and an application of Gramoxone, followed by an application of Gesaprim), line planting, increased density and no fertilizer application gave higher yields, at lower cost, than the farmers' practice. In the zero tillage demonstration trials, on fields of 1-2 ha, the zero tillage practices varied somewhat between sites, depending on previous land management. In one site it was necessary to clear weeds and stubble before herbicide application, while in the others animals had grazed the field after the previous bean harvest. The costs of the herbicide application were recorded and on average were lower than those estimated in the economic analysis of the 1980 researcher-managed experiments.

Farmer field days were held at experimental and demonstration sites. The *juntas agrarias* proved to be an effective means of organizing field days. In addition, the farmers themselves played an active role in the process of technology generation. For instance, they learned how to fashion a shield for the sprayer nozzle out of an empty Gramoxone container so that the herbicide could be directed more accurately. It was difficult for some farmers to find the labour required for the manual chopping of old stands before applying Gramoxone. In consultation with the research team, they adapted a light harrowing method instead of hand chopping to cut back the weeds and crop residues, and thus developed their own minimum tillage system.

Researchers initiated discussions with the BDA on the fertilizer and credit package, with the result that the bank continued to provide credit to farmers in the area but no longer made fertilizer use a necessary condition for receiving credit. The farmers responded by reducing their fertilizer use.

TECHNOLOGY ADOPTION

Progress in the first 3 years of on-farm experimentation was rapid enough, and farmer interest and participation so high, that by 1982 it was clear that the combination of experiments and demonstrations had led to significant changes in maize practices in Caisán. IDIAP and CIMMYT considered it important to carry out a small adoption survey in the area to measure progress. Conducted in July 1982 among a sample group of 52 farmers, the survey sought to measure change in chemical weed control, spatial arrangement and density, zero tillage and fertilization. Table 10.14 gives the adoption criteria for these practices, and Table 10.15 gives the rates of adoption.

With respect to appropriate weed control, the survey found that 61.4% of the farmers were now using Gramoxone and/or Gesaprim, in adequate doses and at appropriate application times. In 1978, only 6% had used similar weed control practices. Nearly all the farmers surveyed knew of the new recommendations, and the most common reason given by those not following them was difficulty in obtaining the chemicals.

Over 70% of the farmers were now line planting maize at a density of at least 45 000 plants/ha. Among those farmers who had not adopted this practice, most said the reason was that it required more labour than traditional random planting. Those who had adopted the practice tended to have smaller fields than those who had not. Fertilizer use in maize had dropped from 58% of the farmers to 21.5%, partly as a result of the removal of this

Table 10.14 **Discriminant variables and acceptance criteria for each technological alternative in the adoption study, Caisán, 1982**

Technological alternative	Discriminant variables	Acceptance criteria
Chemical weed control	1. Chemical weed control	1. If the farmer uses chemical control
	2. Type of product	2. If the control is with Gesaprim or Gramoxone
	3. Application time	3. (i) Gesaprim: 0-15 days after planting (ii) Gramoxone: 0-35 days after planting
	4. Doses	4. (i) Gesaprim: 1-3 kg/ha (ii) Gramoxone: 1-3 l/ha
Spatial arrangement and density	1. Planting arrangement	1. If planting is in rows
	2. Density	2. 45 000-60 000 plants/ha
Zero tillage	1. Tillage system	1. If the farmer does not use mechanical tillage
	2. Application of herbicides	2. If the farmer applies herbicides prior to planting
Fertilization	1. Application of fertilizers	1. If the farmer does not apply fertilizers

Source: Martínez and Sain, 1983

Table 10.15 Adoption survey results on levels of adoption of recommended technologies

Technological alternatives	Farmers (%)	Maize area (%)
Appropriate weed control	61.4	60.9
Planting in rows/higher density	70.5	62.7
No fertilizer used	79.5	79.5
Zero or minimum tillage	43.5	23.0

Source: Martínez and Sain, 1983

requirement from the BDA credit package. The use of some form of minimum or zero tillage had risen from 0% to 43.5% of the farmers. The farmers who had not adopted these practices gave a variety of reasons for their decision, including lack of information, lack of money (reduced tillage was still not included in the credit package offered to farmers), the feeling that mechanical tillage was more effective, and machinery ownership. As indicated in Table 10.15, reduced tillage tended to be adopted by farmers with smaller maize fields.

The data from the adoption survey were later used to do a benefit-cost study of the Caisán OFR programme and to calculate the returns to the research resources invested (Martínez and Sain, 1983). Such an analysis depends on a number of factors and assumptions involving the flow of research costs, adoption rates and price changes, but under the most reasonable scenario the annual average rate of return on IDIAP's investment in the Caisán programme was estimated to be 60% (Martínez and Sain, 1986), which compares favourably with the rates of return to successful plant breeding programmes.

CONCLUSION

The OFR experience in Caisán offers several lessons for consideration. The importance of ordering research topics both in terms of their potential impact and adoption sequence should be obvious. A number of possible research themes were identified during the diagnosis and as a result of early experimentation. These themes were carefully arranged in an experimental programme. The need to improve weed control was apparent to researchers as well as farmers. There were also interactions between weed control and plant spacing and density (row planting would provide better weed control, and better weed control would give greater response to improved spacing and density). This meant that it was best to test these factors together, and recommendations for farmers were available within 2 years. As experience was gained with improved weed control it became clear that several options were available for reduced tillage, which was included in demonstrations by the third year. Information on fertilizer response was more problematic, and research on the long-term effects of fertilizer management continued beyond the third year. The recommendations of the credit bank were recognized as inadequate, however, and were changed. Other research themes, such as the control of soil insects, were identified in the early years of experimentation and provided a focus for future work.

The interaction with farmers during the Caisán programme should also be emphasized. Farmers were actively experimenting with various methods of weed control, for example, before the programme began and the research helped to focus farmers' interest and tie the weed control problem to other factors, such as high tillage costs and inadequate plant populations. The result was a much more rapid and effective identification of a series of new practices than farmers could have developed on their own. During the experimentation, farmers played an important role in adjusting the new technologies to their own conditions, and thus the practices that were adopted in Caisán owe much to farmer innovation.

The Caisán programme was not an isolated effort. For example, communication with the BDA was an important part of the change in fertilizer recommendations. In addition, the OFR helped identify maize breeding priorities for the research station, and work was initiated on developing shorter varieties, based on crosses between the local variety and other materials, which would be suitable for Caisán and similar areas of Panama.

The programme also had a profound effect on the organization of IDIAP. While the programme was being implemented, IDIAP went through a systematic planning effort, which resulted in a reorganization of institutional activities into national level programmes and commodity research projects complemented by regional research programmes in which the basic operational unit was the area-specific OFR project. By 1983, OFR in agriculture was being carried out in five priority areas, involving 24 national researchers; a similar effort in livestock involved 21 researchers in three areas.

IDIAP's OFR continued to expand, and in 1985-86 CIMMYT organized a course for researchers and extension agents from IDIAP and MIDA who had been assigned to one of the seven areas in Panama where OFR was being carried out. The course took place in the San Andrés area, where IDIAP and MIDA staff had recently been assigned to begin OFR (Sain and de Gracia, 1990). The course consisted of a series of seven 1-week or 2-week meetings over two agricultural seasons. The OFR process — surveys, planning, experimental management and data analysis — was taught using the San Andrés area as a 'hands-on' research site. Participants provided reports of progress from their own research areas during the course. A recent survey has shown that several of the technologies developed through resesearch carried out during the course have been adopted by farmers in San Andrés, providing further evidence of the effectiveness of the OFR approach.

References

Arauz, J.R. and Martínez, J.C. 1983. *Desarrollando Tecnología Apropiada Para el Agricultor*. IDIAP Special Studies Series No. 1. Panama: IDIAP.

Byerlee, D., Collinson, M. et al. 1984. *Planning Technologies Appropriate to Farmers: Concepts and Procedures*. (2nd edn). El Batan, Mexico: CIMMYT.

Martínez, J.C. and Arauz, J.R.. 1983. *Institutional Innovations in National Agricultural Research: On-Farm Research within IDIAP, Panama*. Economics Program Working Paper 02/83. El Batan, Mexico: CIMMYT.

Martínez, J.C. and Arauz, J.R. 1984. Developing appropriate technologies through on-farm research: The lessons from Caisán, Panama. *Agricultural Administration* 17: 93-114.

Martínez, J.C. and Sain, G. 1983. *The Economic Returns to Institutional Innovations in National Agricultural Research: On-Farm Research in IDIAP, Panama*. Economics Program Working Paper 04/83. El Batan, Mexico: CIMMYT.

Martínez, J.C. and Sain, G. 1986. On the cost efficiency of on-farm research: Social costs and benefits in Caisán, Panama. Economics Program Draft Paper. El Batan, Mexico: CIMMYT.

Sain, G. and de Gracia, R. 1990. Generación y transferencia de tecnología agrícola a través de metodología de investigación en fincas de agricultores: El modelo de Caisán — San Andrés, Panama. Paper presented at the Workshop on Estrategias de Transferencia de Tecnología en América Central, March 1990, IICA, San José, Costa Rica.

Planned Change in Farming Systems
Edited by R. Tripp
© 1991 CIMMYT
A Wiley-Sayce Co-Publication

11

Post-Harvest Technology Development with Farmers: The Diffused Light Storage Case

R. RHOADES, R. BOOTH, E. SCHMIDT and O. CUYUBAMBA

National, international and private development organizations attach the highest priority to maintaining and improving the nutritional status of an ever increasing world population. One of the more successful efforts in this regard has been the development of high yielding varieties which are resistant to pests, diseases and abiotic stresses and respond well to agrochemical inputs. However, acceptance of these varieties by farmers in the developing world usually depends on farmers' ability to store at least some of their production and keep seeds from harvest for planting in the following season.

Many new varieties promoted by plant breeding programmes differ from traditional landraces in terms of their storage characteristics. Traditional storage practices were developed in the context of location-specific environmental, climatological and social conditions and according to the specific characteristics of the varieties grown. When new varieties do not store well, the traditional storage techniques *per se* may not be at fault; rather, the increase in volume of production or the changes in the varieties' storage needs may have made existing facilities inadequate (Greeley, 1982).

As most developing countries have little new land to put under cultivation, and yield increases resulting from breeding and improved agronomy have levelled off for many crops, the post-harvest area has been seen by some scientists and policy makers as a promising means of increasing food availability. Booth and Burton (1982), for example, estimate that if post-harvest losses in potatoes can be cut by 50%, which is a reasonable possibility in many developing countries, this would be equivalent to a 12.5% increase in yields. In many cases, it costs less to obtain this yield equivalent through improved post-harvest technology than through the increased use of fertilizer, pesticides or other inputs. A major problem in attaining these post-harvest gains, however, is the development of novel approaches that are accessible to small farmers.

This chapter describes a case of on-farm research (OFR) in Peru that focused on potato storage. A major factor in this case was that the research team first had to understand the farmers' *perception* of the problem before arriving at a working solution.

SOCIOECONOMIC CONTEXT OF POST-HARVEST TECHNOLOGY

The storage of any crop for consumption, processing or planting must be understood as an integral part of the overall production process. Particular storage needs are, to a large extent, determined by specific consumer requirements and the magnitude, duration and frequency of harvests. For example, potato production and marketing systems, as well as the demand patterns, determine storage needs and must be fully understood before attempts are made to improve existing storage techniques or introduce new ones. Roots and tubers, which have a much higher moisture content than grains (for example, 80% for potatoes, 12% for rice), require special consideration in the development of post-harvest technologies.

In the development of appropriate technologies, careful consideration must also be given to the specific system for which the basic technology is being adapted (Shaw et al., 1982). Farmers, with their long-term understanding of local storage conditions, socioeconomic reality, crops and markets, assume the status of experts in this case (Rhoades et al., 1985). Building upon traditional practices, rather than replacing them, is the route to successful problem-solving in the post-harvest area. Many post-harvest projects, however, have operated under the assumption that farmers have little of value to say to technicians. Traditional post-harvest food handling systems in most developing countries have evolved over hundreds or thousands of years. Before introducing new technologies it is necessary to understand the total system and to appreciate how changing one part of the system will affect other parts of the system (Bourne, 1982).

In the Peruvian Andes, for instance, farmers regard the storage of food, seed and tools as a domestic activity. The security of the farmer's house and the flexible use of space within the house would not necessarily be compensated for by the possible technical advantages of having a specialized storage facility physically separate from the house (Rhoades, 1984). Within the house and the walled compound, potatoes and other assets can be kept from thieves and from the evil eye (*mal de ojo*) of envious neighbours, and at the same time be conveniently close at hand for cooking and processing (Werge, 1977).

Multipurpose storage involves considerable compromise in design and management to avoid excessive storage losses of individual commodities. Home storage facilities are generally less technically effective than buildings constructed specifically for potato storage. However, the traditional home storage system may be the most appropriate and efficient when the whole farming system and social context is considered. In addition, in many potato producing regions, capital is simply not available to construct and effectively operate separate storage facilities (Shaw and Booth, 1981).

DIAGNOSTIC STUDIES

Since the late 1960s, the Peruvian Government and various development agencies operating in the country have sought technical solutions to help control the flow of potatoes into the Lima market. The government built five large storage complexes, with a combined capacity of 2.0×10^4 t, in various parts of Peru. The rationale for building these stores was to hold potatoes during surplus periods and then release them into the Lima market to take advantage of rising prices. This would benefit not only the producers but also the consumers, who would pay lower prices during the 'critical months' of potato scarcity. Despite technical efficiency and good intentions, however, these stores and others built later have stood empty since they were constructed, largely because the designers did not fully understand the post-harvest system of potato agriculture in Peru. Such mistakes are not unique to Peru. Similar potato storage complexes, technically sound but empty, can be found throughout the developing world (Booth and Rhoades, 1985).

When post-harvest research at the International Potato Center (CIP) began in the early 1970s, the aim was to reduce storage losses of consumer potatoes by designing improved storage structures and systems. Initial interest centered on farmers in Peru's Mantaro Valley, the site of CIP's main highland research station and the main source of potatoes for the Lima market. Potato specialists first assumed that the major storage problems stemmed from inadequacies in farmers' traditional storage practices, leading to heavy losses as a result of rotting, insect attack, shrinkage and sprouting. All CIP's early storage experiments were conducted on the experiment station, with little contact with farmers; in 1975, however, an interdisciplinary team made up of an anthropologist and two post-harvest technologists was formed to work closely with national scientists. They began their research by looking at farmer storage practices and their rationale within the local context.

Although Andean farmers have been successfully storing potatoes for thousands of years, this was one of the first serious attempts by agricultural scientists to study these practices. Early Spanish references report the existence of a highly organized storage network in which the state concentrated supplies at strategic centres (Morris, 1981; D'Altroy and Hastorf, 1984). The quantities of goods stored amazed the Spanish, who drew extensively from state warehouses. The Incas were knowledgeable about insolation, insulation and ventilation, important principles in storing crops, especially maize and potatoes. They were particularly adept at evaluating and selecting areas where natural conditions favoured the storage of different commodities (Morris, 1981). These pre-Hispanic traditions continue today in Andean villages and farms (Rhoades et al., 1988).

By *beginning* with the traditional knowledge of farmers and heeding their advice, it became clear that CIP scientists had perceived the storage problem differently from the farmers. The team was guided by three basic questions: how were farmers presently storing potatoes; why did they store this way; and what did they perceive their storage problems to be. Initially, the team sought answers to these questions through open-ended dialogue with the farmers. No questionnaire was used and no prior assumptions were made. To the team's surprise, when farmers were asked about losses they responded that they had no 'losses'. They said that potatoes that shrunk or suffered insect and disease attack were simply removed and fed to pigs; these potatoes, already the less desirable part of the harvest, were considered necessary as feed for livestock. In addition, some women claimed that small, shrivelled potatoes tasted sweeter and sometimes had better cooking quality.

Further dialogue with the farmers, however, revealed that storage problems did exist but not in a way that the team members had originally perceived. The farmers said that nothing was wrong with their traditional stores; rather, that the problem lay with the improved, commercial potato varieties they had adopted over the past decade. They argued that stored seed potatoes of new varieties lost considerable weight under traditional storage management and produced extremely long sprouts (storage of these new varieties in darkness does result in excessive sprout development, sometimes reaching as long as 80 cm). This forced farmers to invest heavily in family labour to desprout the tubers before planting; they stated that an average of 5 man-days was needed to desprout the seed required for 1 ha of land. They expressed less concern about consumer potato storage problems, the focus of most of the earlier research efforts. Thus, the emphasis of CIP's on-station research shifted towards solving the farmers' perceived problems of storage of improved seed potatoes.

RESEARCH ACTIVITIES

On-station research

After the diagnosis studies, CIP scientists set out to design a low-cost technology that would solve the problems of excessive sprouting and moisture loss during the storage of seed tubers. The conventional approach of controlled low-temperature storage (refrigeration, forced-air ventilation) for preventing excessive sprout growth was clearly not appropriate for most of these small farmers.

Fortunately, some scientific knowledge already existed that indicated that storing seed potatoes in diffused light (not direct sunlight) would reduce sprout growth and improve seed quality. This suggested that scientists should develop a simple technology based on natural diffused light that would reduce sprouting and result in higher yields. The post-harvest research team began to experiment on-station with storing seed potatoes in thin layers in trays exposed to natural diffused light conditions. The results from these experiments indicated that diffused light storage did offer a potential solution. Table 11.1 shows that the average mean length of sprouts in dark storage was 18.8 cm compared to an average of 1.4 cm in diffused light, and that there was less weight loss in diffused light storage. Furthermore, when the seed stored under different conditions was planted in the field, the seed that had been stored under diffused light had higher rates of emergence and higher yields, as shown in Table 11.2.

The diffused light storage technique developed on the experiment station appeared to offer multiple benefits over traditional dark storage methods: the seeds lost less weight; the sprouts were shorter and stronger; field emergence was faster; there was greater resistance to pest attacks; seeds could be conserved for a longer period of time; and yields were higher. Despite these results, however, it was not known how widely applicable or acceptable the diffused light storage principle would be to farmers or how it could be adapted to specific local storage needs. Although modest by earlier potato project standards, the first diffused light stores on the experiment station were large, expensive structures (from the farmers' point of view).

Table 11.1   Effect of natural diffused light during storage[1] on storage behaviour of seed potato tubers, central Peru

Storage behaviour	Antarqui	Cuzco	Mi Perú	Mariva	Renaci- miento	Revolu- ción	Rosita	N-578-4	Average
Mean length of sprouts (cm):									
dark stored	22.5	20.6	12.1	33.7	10.3	19.8	19.3	11.8	18.8
light stored	2.1	1.5	1.1	1.2	1.2	1.6	1.8	1.1	1.4
Mean number of sprouts/tuber:									
dark stored	1.2	1.7	2.0	1.2	2.1	1.1	1.1	2.6	1.5
light stored	2.5	3.5	5.0	5.1	3.0	2.8	3.9	6.1	4.0
Mean storage loss[2] (% initial weight):									
dark stored	16.3	23.3	12.2	22.8	8.7	11.8	15.3	12.3	15.3
light stored	8.8	10.7	6.3	11.2	5.5	9.7	6.7	8.3	8.4

Notes: 1. 180 days storage with average monthly storage maximum temperature of 18.9°C and average monthly storage minimum temperature of 5.7°C.
          2. Includes tuber weight loss and weight of sprouts removed prior to planting.
Source: CIP, 1979

Table 11.2 Effect of natural diffused light during storage[1] on field emergence and yield, central Peru

Emergence and yield	Antarqui	Cuzco	Mi Perú	Mariva	Renaci-miento	Revolu-ción	Rosita	N-578-4	Average
% field emergence after 20 days:									
dark stored[2]	1.5	0.0	3.1	0.0	0.0	0.0	0.0	4.7	1.2
light stored	71.8	64.8	54.7	67.2	37.5	46.9	5.0	78.1	58.8
% field emergence after 30 days:									
dark stored	90.6	68.7	100.0	76.6	89.0	96.8	87.5	95.3	88.0
light stored	100.0	100.0	100.0	100.0	95.3	100.0	96.8	98.4	98.0
% field emergence after 40 days:									
dark stored	100.0	92.0	100.0	100.0	100.0	100.0	100.0	100.0	99.0
light stored	100.0	100.0	100.0	100.0	100.0	100.0	100.0	100.0	100.0
Yield (t/ha)[3]:									
dark stored	23.7	25.8	29.8	26.6	18.1	28.6	18.9	25.7	24.6
light stored	26.4	30.2	35.1	29.4	19.6	34.8	24.1	31.1	28.8

Notes: 1. 180 days storage with average monthly storage maximum temperature of 17.0°C and average monthly storage minimum of 5.7°C.
2. Desprouted prior to planting.
3. Average of 4 replicates each with 66 plants.

On-farm trials

In order to test the relevance of the diffused light technique, the CIP team initiated on-farm trials. Over time, the trials led to a simplification of the technology which made it more relevant to resource-poor farmers without altering the technical benefits which had been observed on the research station. Initially, this effort involved two main activities — on-farm testing and adaptations to local conditions. In practice, these two activities were highly interdependent.

Potatoes in Peru are produced in the highlands (the major zone for both commercial and subsistence production) and the coast (where production is geared towards the Lima market during winter, when potato prices are higher). The CIP team decided to tackle on-farm storage research in both areas (*see* Figure 11.1 *overleaf*).

Sierra region

In the Sierra there are two planting seasons: an early planting under irrigation, in June and July, and a main rainfed planting from September to December. This results in varied storage needs at the farm level. During 1979 and 1980, the CIP team built numerous small test stores with farmers in the central highlands in Potaca, Churcampo and Pampas (Department of Huancavelica) and Comas (Department of Junin). The stores were

Figure 11.1 Southern Peru, showing the coastal and highland areas involved in the study

built of readily available local material and placed in a covered patio or courtyard. Some were collapsible, free-standing structures that could be moved easily by two or three men.

Examples of research results obtained in these stores are given in Table 11.3. In general, storage losses were reduced by half, the number of sprouts per tuber increased and the sprouts remained short, avoiding the need for much desprouting. In the field, tubers stored under diffused light conditions gave earlier and more uniform emergence and yield increased by about 10%. On the basis of these results, the Peruvian national agricultural research system began to develop additional test stores throughout the highlands, and a programme of training for extension agents was launched.

Coastal region

The needs of Peru's coastal region differed from those of the highlands, given that little traditional storage technology had existed for potato production. Coastal potato producers rarely stored seed but instead acquired it from the Sierra. Seed costs accounted for 35-50% of total production costs. The availability of seed often determined whether farmers decided to plant potatoes or opt for other crops.

In the coastal region there is usually one production season, with early plantings in March and April and a main planting in May and June. The CIP team and the national researchers felt that the adoption of the diffused light technology on the coast could help farmers to keep their highland seed for one or more multiplications on

Table 11.3   On-farm storage trials, Sierra region, 1979-81

Year	No. of cultivars	Storage system	Length storage (months)	Weight losses (%)	No. of sprouts	Sprout length (cm)	Yield (t/ha)	Yield increase kg	%
1979-80	9	DLS[1]	3.3	9.76	3.32	1.08	13.944	1180	9.24
		dark	3.3	17.60	1.94	8.83	12.764		
1980-81	9	DLS	3.9	9.92	3.06	0.96	16.791	1567	10.29
		dark	3.9	18.40	1.58	11.31	15.224		

Note: 1. DLS = diffused light store.

the coast. This would make ealier planting, in March or April, more feasible, enabling farmers to obtain mean higher prices at harvest, and would reduce seed transportation costs from the Sierra to the coast.

The research was initially conducted at two coastal locations: Barranca (north of Lima) and Cañete (south of Lima). As in the Sierra, the CIP team travelled through the region, talking to farmers, and storage surveys were conducted in Barranca and Cañete. Between 1979 and 1982 three stores were built, two with large co-operatives and the third on a medium-sized private potato farm. Each store had a capacity of 15 t, was designed according to local needs and used readily available materials such as eucalyptus poles, cane, bamboo and straw mats. The stores were built in shady locations over irrigation ditches to keep them as cool and humid as possible.

Results over a 3-year period indicated that seed tubers could be successfully stored from one season to the next under coastal conditions with storage losses not rising above 2% per month. These losses were less than half of those experienced using the traditional methods practised on the coast and only slightly higher than those obtained in costly refrigerated storage. Yields were between 10% and 20% higher than those obtained from traditionally stored seed tubers and comparable with those obtained from either cold-stored tubers or seed tubers brought from the highlands. Three examples of the types of results obtained are given in Table 11.4.

Table 11.4   On-farm storage trials, coastal region, 1979-82

Year	Place	Variety	Length of storage	Weight loss (%)	% yield increase over traditionally stored tubers
1979-80	Cañete	Revolución	5 months	6.4	12.3
1980-81	Barranca	Revolución	5 months	8.3	18.0
1981-82	Barranca	Revolución	5 months	10.4	13.9

Demonstration stores

Following the successful OFR effort, Peru's national agricultural research institute, Instituto Nacional de Investigación Pecuaria e Agrícola (INIPA), in collaboration with CIP, other technical assistance agencies and a small FAO team working on storage in the highlands, concentrated on greater diffusion of the technology through the construction of more demonstration stores and increased training. By 1982 there were 28 demonstration stores in the Sierra region and 13 stores in the coastal region, representing a total capacity of 316 t.

Over the following 6 or 7 years these efforts were expanded, with testing and demonstrations being initiated in various parts of the country. In a project supported by the International Development Research Centre (IDRC), eight diffused light stores with a capacity of 1-4 t were constructed in Huancane, Ilave and Sandia (Department of Puno), and in the Department of Cuzco the new technology was incorporated into various development projects.

TECHNOLOGY ADOPTION AND ADAPTATION

The demonstration of a new technology is not a sufficient condition for immediate farmer adoption. Small-scale subsistence farmers invariably face a number of constraints that may not be apparent to the outsider. Adoption could be slowed or even stopped because of simple factors such as lack of space. Alternatively, in the case of the diffused light technology, the need for food during periods of scarcity could restrict adoption of the technology because potatoes being saved as seed might need to be consumed as table potatoes. Once potatoes have been 'greened' in diffused light stores they become bitter and are not suitable for human consumption.

Thus, continuous efforts are required to monitor diffusion and adoption in order to identify potential technical and socioeconomic problems at an early stage. Two major surveys on farmers' response to the diffused light technology were carried out in the 1980s, the first in 1981 and the second in 1985.

1981 adoption survey

The CIP survey was conducted in the Mantaro Valley. The team interviewed 31 farmers and managers of cooperatives (each cooperative had several hundred members) and found that the farmers' response to the new technology was positive and that no farmer interviewed had abandoned either the trial or demonstration store. Several farmers had 'copied' the original model proposed, while the rest had understood the diffused light principle and had modified their traditional storage systems accordingly, many of them making use of the corridors or verandas which are common in farm houses in the region; some small farmers had combined their new diffused light store with a cage for their guinea pigs and rabbits.

A survey of cooperatives in the coastal region was also undertaken. In Barranca, as a result of the OFR, the cooperative had built a diffused light store with a capacity of 100 t; in 1981 this had reduced cooperative's dependence on highland seed by 50%. Interviews with cooperative managers revealed the following benefits of using diffused light storage:

- earlier planting and earlier harvest to avoid *Mosca minadora*, the major pest in the coastal areas which causes severe damage later in the season;

- higher prices because the potatoes were marketed before those produced from the main harvest;

- lower cost of production as earlier planting resulted in earlier emergence, which minimized pest and disease damage and reduced expenditure on pesticides and fungicides;

- flexibility, enabling the cooperative to plant when it wished rather than be dependent on time of arrival of highland seed;

- *criolla* (local) seed stored in diffused light gave uniform, early emergence;

- tubers stored using the diffused light principle were 20% cheaper than highland seed which, with higher market prices following earlier harvest, made the whole operation more profitable;

- cooperative members did not have to pay interest on borrowed money, as was the case when they bought seed from the highlands.

1985 adoption survey

The aim of the second survey, conducted in the Mantaro Valley with support from the IDRC, was to evaluate the actual levels of technology adoption. The specific objectives were: to verify if a real diffusion was taking place of either the diffused light stores as demonstrated or the diffused light principle; to determine the individual process of adoption; to identify the advantages (or disadvantages) of the system, as perceived by the farmers; and to identify the reasons that might prevent farmers from adopting and/or adapting the diffused light principle.

The survey, based on a questionnaire, initially involved 45 farmers. These farmers were asked if they knew of anyone else who was using the technology, and this resulted in 106 more names being added as users of the diffused light principle. Of the 45 farmers surveyed, 43 were now storing their seed tubers in diffused light while only two farmers continued to store in the dark. Of the 43 adopters, 28 had constructed new diffused light stores and 15 had adapted an existing farm building (*see* Table 11.5).

The survey showed that, whereas the small farmers (those with less than 1 ha) had a surplus of seed in the installed diffused light stores, the larger farmers did not have enough diffused light storage capacity for their planting needs. This is partially explained by the fact that these farmers bought some of their seed at planting time from highland seed producers or from coastal potato producers with whom they maintained commercial relationships.

An interesting feature of the survey was the farmers' knowledge of the characteristics of the different varieties they cultivated and stored. Seven farmers (16%) reported not storing all the varieties they cultivated

Table 11.5   Surveyed farmers' farm size and storage facilities, Mantaro Valley, 1985

No. of farmers	Cultivated land (ha)	Separate diffused light storage facility	Diffused light principle[1]	Dark store
5	<1	2	2	1
23	1<5	15	7	1
7	5<10	6	1	—
10	>10	5	5	—

Note: 1. Farmers using corridors, second-floor rooms, terraces, sheds, etc.

in diffused light because of lack of space. Two of the farmers who planted both local and improved varieties continued to store three local varieties in the dark, while the other five stored the variety Yungay (a semi-early improved variety with a longer post-harvest dormancy period) in the dark, and used the diffused light store for early maturing improved varieties such as Revolución, Mariva and San Roque. Four farmers treated the seed intended for early planting differently from the other 41 farmers. They left this seed in sacks (although they had diffused light stores), which promoted the rapid sprouting of tuber seeds. These practices illustrate considerable farmer knowledge and show how farmers' own research and testing efforts were used to adapt storage technologies, including the new diffused light technology, to specific requirements.

Several farmers reported that they had used the practice of 'greening' their seed tubers before the period of formal diffusion by INIPA and before CIP started its research on the technology in 1977. Five farmers indicated that they had been aware of the effects of diffused light as early as 1973-75. In most of these cases, the discovery had been accidental. For example, farmers had sometimes left a few potatoes in the field at harvest; when they came back to prepare the field for planting, they found that these potatoes were green because of the sunlight they received. As the potatoes appeared to be in good condition (no pest attack was noticed), the farmers had decided to use them as seed, and had noticed differences in performance between 'green seed' and the seed stored traditionally in the dark. In other cases, in years of over-production potatoes had been left outside because of lack of space in the house. It is worth noting also that some farmers in the region use what seems to be an ancient practice — greening an Andean tuber known locally as 'olluco' (*Ulucus tuberosus*); this knowledge of the principle of greening, albeit with another crop, should help facilitate the adoption of greening for potatoes.

Follow-up studies on the diffused light storage technology were conducted in other countries, such as Sri Lanka and the Philippines, and showed that although farmers often claimed that they had used the technology prior to the CIP programme, they were in fact 'early adopters' who had learned about the principle during a field day or a visit by a technician. Thus, while it is difficult to verify exact dates of adoption by farmers, the pattern of accelerated adoption of diffused light storage is clear. Also, it is known that in a number of countries some farmers were indeed making use of natural diffused light before CIP started active research on the topic. CIP's research, however, has resulted in a fuller understanding of the technology and, with the collaboration of many national programmes, is responsible for accelerated adoption.

Farmer adaptation and experimentation

It has been suggested that if a new technology is to succeed it must be technically, economically and culturally 'competitive' and possess an evolutionary capacity for continued improvement (Moran, 1978). This view is supported by the CIP studies on farmer adoption of the diffused light storage technology.

Adoption in all countries began by either a few individuals or the community as a unit taking the initial risks of experimenting with the new technology. Farmers did not blindly accept technicians' judgements, and few of them attempted to make all improvements at one time, supporting the need for evolutionary capacity. They preferred to make alterations slowly, as they began to understand the principles involved and as they saw the success of their neighbours. The variation of these adaptations ranged from simply spreading potatoes in front of a window to constructing diffused light stores with a capacity of 100 t (Rhoades et al., 1983).

From the 1985 survey in Peru, it appeared that there were two different patterns of adoption. Farmers working directly with INIPA extension agents tended to progress directly from storage in darkness to specially built diffused light stores, whereas those who attended a course or field day, or saw the rustic seed stores at neighbours' houses, tended to show more interest in the principle of diffused light than in building copies of what they saw, often adapting the most suitable parts of their dwellings or other farm buildings. Some farmers were found using second floor rooms (such as attics), terraces and sheds for diffused light storage. Investigation has shown these adaptations to be quite adequate, with uniform greening of seed potatoes observed in most cases. The diversity of adaptation does not stem from a lack of building materials but, more importantly, from the farmers' understanding of the diffused light principle for which only indirect light and good ventilation is needed. Worldwide, this principle has been translated into an amazing array of farmers' versions of potato stores, each with a particular cultural flavour.

Follow-up by the CIP scientific team corresponds to the final evaluation stage by the farmer when the technology is totally under the farmer's management. This stage is necessary to understand farmer response so that the technology might be improved or reconsidered, or if rejection has occurred, to begin the research cycle once again with the farmer. The crucial point is that research must *begin* with the farmer and *end* with the farmer, and that it must be a continuous, interactive and cyclical process.

Benefits of adoption

In the 1985 follow-up study, farmers were very clear about the benefits of the new technology (*see* Table 11.6). None of those who had tested the technology had abandoned it. As a result of this survey and other experiences with the technology, it seems that the most important benefit of this technology is that it ensures that good quality seed is available at planting, protecting the farmer from the high seed prices prior to the planting season. In addition, seed tubers stored in diffused light remain in a condition suitable for planting for a longer period than those stored in most traditional systems, which provides extra flexibility and reduces the risk to the farmer.

Table 11.6   Perceived benefits of diffused light storage[1]

Main benefits in storage	% (N = 102)	Main benefits in the field	% (N = 77)
Less water loss	14	High yield	28
Less weight loss	24	High resistance	38
Less sprouting	16	Early emergence	34
Better management of seed	25		
Lower insect attack	21		

Note: 1. Based on field survey data, July 1985.

The present potato marketing system in Peru is complex and is the subject of a separate CIP research project (Scott, 1985). Some commercially produced seed tubers are sold at harvest time to coastal potato producers while others are either sold directly or stored for varying periods and sold to highland producers. Seed tuber prices generally increase with increased storage time and reach a peak shortly before major planting periods. From the Peruvian data it is not clear whether farmers would be willing to pay more for 'green' seed than for traditionally stored seed tubers. In the 1985 survey, 44% of farmers interviewed reported that although there was a saving, as no desprouting was required, the diffused light technique required more labour inputs during storage than traditional systems. In surveys in other countries, such as the Philippines, farmers were reported to be willing to pay more for diffused light stored seed tubers.

The concept of seed quality in Peru refers more to the size of the seed (1st, 2nd, 3rd quality) than to 'green' seed or seed of a specific phytosanitary or physiological quality. This situation may change with the continued efforts of INIPA to improve the real quality of seed tubers. It is worth mentioning here, however, the case of a large potato producer to whom many neighbouring small farmers go in search of seed. Several of these small farmers do not have the money to pay for the size graded 1st, 2nd, or even 3rd class seed tubers. In these cases the large producer 'lends' them very small seed (not even of 3rd class) he has left outside in sacks that in effect has been stored in diffused light. Now, more and more farmers pass by his farm looking for this class of seed which has to be 'paid back' with 1st class produce at harvest time.

Non-adoption

Farmers in the 1985 survey were asked about the non-adoption of the diffused light storage technology. The reasons they gave for non-adoption can be grouped into three categories:

- *Lack of knowledge:* This is the consequence of distance, inaccessibility and lack of human resources in the formal channels of diffusion. The lack of well-trained staff to handle the diffusion of new technologies is a serious problem. The INIPA extension agents, for example, are sometimes in charge of areas that are too large and have poor communications, which tempts them to concentrate their efforts on a few of the most interested and accessible farmers and to leave aside others who are isolated or who do not appear to be very motivated. Nevertheless, an extensive training and extension effort has been undertaken by INIPA.

- *Economics:* For some of the more marginal farmers, lack of materials (or funds to purchase materials) is an important limiting factor for the acceptance of the diffused light technology and any other technology. However, recommended materials can be easily replaced by other materials available in the region. Thus, to enable farmers to adapt the principles of the diffused light technology to their own needs, efforts must continue to focus on making farmers aware of these principles rather than concentrating on specific construction designs of the rustic stores.

- *Sociocultural factors:* The most common reason given by farmers for non-adoption was 'custom' or 'tradition'. However, this is not so much an indication of farmers' resistance to change as a caution against unproven and possibly maladapted ideas or technologies. It is clear from the Peru case and others that farmers are constantly searching for alternatives that could help them optimize production in the context of their total operations.

Lastly, while some small farmers were willing to use the diffused light storage technique in cooperative or community stores where potatoes were being produced commercially (four such examples were surveyed), the same small farmers were not adopting the technology for their own use where only very small amounts of potatoes were being produced for household use. It is possible that at this level of potato production the concepts of storage losses, quality seed tubers, security and even yields are viewed differently and need to be studied and understood better before attempts are made to introduce changes.

FOLLOW-UP RESEARCH

Several problems with the diffused light technology and its transfer to farmers have been identified. From the technical point of view, the two major problems are loss of tuber quality through the spread and build-up of aphid-transmitted viruses during storage and, in warmer climates, excessive storage losses associated with damage caused by the potato tuber moth. These topics are presently being researched at CIP and information is already being relayed to national scientists who are again assisting in testing these research findings.

From the socioeconomic point of view, the major bottleneck has been the lack of detailed knowledge about the prevailing seed distribution and marketing systems, which has led to difficulties in accurately diagnosing storage needs and problems. A few attempts have been made to introduce the diffused light technology into situations where it was not an appropriate solution. To remedy these problems additional technical and, in particular, socioeconomic training is required.

CONCLUSION

The generation and transfer of diffused light technology for storing seed potato tubers in Peru illustrates how the 'farmer-back-to-farmer' approach functions (Rhoades and Booth, 1982). The data clearly indicate that farmer adoption through their own research and adaptation is well under way. Of equal importance, perhaps, is the fact that the case provides an illustration of successful institutional linkages, greatly facilitated by training at several levels and by follow-up activities, which enabled an international centre, a national programme and supporting technical assistance projects to pull together in helping to solve a farmer-identified problem.

Probably the main conclusion to be drawn from this study is the vital role the farmers themselves played in the technology generation and evaluation process. They participated right from the first step of problem diagnosis and definition and, with the help of social scientists, continued to interact with international and national scientists in generating, transferring, evaluating and adapting the technology.

Although it is considered unrealistic at this time to attempt to measure the impact of this technology in terms of tons of potatoes saved or produced, the available data suggest that a significant impact is occurring as more and more farmers learn about and use the technology. More important, however, is the fact that the technology is easily adapted to specific farmer needs as it increases farmers' flexibility and reduces risk, two aspects which are of vital concern to many small potato farmers throughout the world.

By 1984, about 4000 cases of adoption had been reported. Most of this had occurred in a small number of countries — Peru, Sri Lanka, Philippines, Guatemala and Colombia. There were several factors underlying the successful generation and transfer of the technology in these countries (Rhoades and Booth, 1983):

- a need for improved seed storage practices was correctly diagnosed;

- there was a strong commitment on the part of national scientists to improve seed storage as an integral part of an overall seed potato programme;

- national scientists did not unquestionably accept the technology from CIP but tested it under their conditions together with local farmers;

- researchers, extension agents and farmers cooperated closely in the technology testing stage;

- local promoters were knowledgeable about village conditions, had an excellent rapport with farmers and used effective demonstrations which showed a wide range of possible ways of adapting the technology;

- scientists and extension agents conducted follow-up studies, especially in the early adoption stage, to refine and improve the technology.

The farmer-back-to-farmer model was derived from experiences with post-harvest technologies at CIP. The model offers suggestions on how the points of view of the biological scientist, the social scientist and the farmer can be combined to generate acceptable technology. It assumes that no single discipline has a corner on truth or that the farmer has all the answers. In the Peruvian potato storage case, when an attempt was made to adopt an interdisciplinary approach and to fully involve farmers, 25 years of failure in potato storage work came to an end. Although the technology has not been adopted by all farmers and, indeed, is not considered appropriate for all farmers, it is clear that many farmers have profited from this innovation.

References

Booth, R.H. and Burton, W.G. 1982. Future needs in potato post-harvest technology in developing countries. *Agricultural Ecosystems and Environment* 9(1983): 269-80.

Booth, R.H. and Rhoades, R.E. 1985. Requirements and recommendations for improving potato storage in Third World countries. Paper presented at International Symposium for Potato Storage, Michigan State University, USA, June 1985.

Bourne, M.C. 1982. Guidelines for post-harvest food loss reduction activities. Draft. New York, USA: Cornell University.

CIP. *Annual Report*. 1979. Lima, Peru: CIP.

D'Altroy, T.N. and Hastorf, C.A. 1984. The distribution and contents of Inca Stea Stonehouses in the Xauxa region of Peru. *American Antiquity* 49(2): 334-49.

Greeley, M. 1982. Pinpointing post-harvest food losses. *Ceres* January/February 1982.

Moran, M.J. 1978. Transfer of post-harvest technologies to small farmers. *Desarrollo Rural en las Américas* X(3): 143-51.

Morris, C. 1981. Tecnología y organización inca del almacenamiento de viveres en la Sierra. In Lechtman, H. and Soldi, A.M. *La Tecnología en el Mundo Andino*. El Batan, Mexico: Universidad Autonoma de Mexico.

Rhoades, R.E. 1984. Tecnicista vs. campesinista: Praxis and theory of farmer involvement in agricultural research. In Matlon, P., Cantrell, R., King, D. and Benoit-Cattin, M. (eds) *Coming Full Circle. Farmers' Participation in the Development of Technology*. Ontario, Canada: IDRC.

Rhoades, R.E. and Booth, R.H. 1982. Farmer-back-to-farmer: A model for generating acceptable agricultural technology. *Agricultural Administration* 11: 127-37.

Rhoades, R.E. and Booth, R.H. 1983. Acceptance of diffused light potato seed storage technology in developing countries. In Hooker, W.J. (ed) *Research for the Potato in the Year 2000*. Lima, Peru: CIP.

Rhoades, R.E., Booth, R.H. and Potts, M.J. 1983. Farmer acceptance of improved potato storages practices in developing countries. *Outlook in Agriculture* 12(1): 12-15.

Rhoades, R.E., Batugal, P. and Booth, R.H. 1985. Turning conventional agricultural research and development on its head: The farmer-back-to-farmer approach. Paper presented at the Seminar on Agricultural Research and Development for Small Farms, sponsored by IFDC, PCARRD and SEARCA, Los Baños, Philippines, 12-14 May 1985.

Rhoades, R.E., Benavides, M., Recharte, J., Schmidt, E. and Booth, R. 1988. *Traditional Potato Storage in Peru: Farmers' Knowledge and Practices*. Potatoes in Food Systems, Report No. 4. Lima, Peru: CIP.

Shaw, R.L. and Booth, R.H. 1981. *Introduction to Potato Storage*. Lima, Peru: CIP.

Shaw, R.I, Booth, R.H. and Rhoades, R. 1982. On the development of appropriate technology: A case of post-harvest potatoes. *Food Technology*. October 1982: 114-16.

Scott, G. 1985. *Mercados, Mitos e Intermediarios*. Lima, Peru: Universidad del Pacifico.

Werge, R.W. 1977. *Potato Storage Systems in the Mantaro Valley Region of Peru*. Lima, Peru: CIP.

Part 3

FUTURE DIRECTIONS FOR ON-FARM RESEARCH

Planned Change in Farming Systems
Edited by R. Tripp
© 1991 R. Tripp
A Wiley-Sayce Co-Publication

12

The Limitations of On-Farm Research

R. TRIPP

The cases presented in Part 2 demonstrate how on-farm research (OFR) can deliver improved technologies to farmers. They show how a detailed consideration of farmers' conditions and priorities, combined with careful experimentation on representative fields, can lead to the development of useful innovations for farmers. They provide well-documented evidence that OFR can work. In addition, many of these cases are examples of success after previous attempts at technological change with the same farmers had met with failure. Rather than simply being done in the right place at the right time, the research described in these cases strongly supports the hypothesis that a fresh approach to developing and transferring agricultural technology is needed.

Yet the cases described, while not unique, are certainly in the minority. A review of efforts in OFR, farming systems research (FSR) and similar approaches to technology development would show that in most instances little agricultural change can be credited to those initiatives.[1] The purpose of this concluding section is to explore why change has been so limited and to suggest ways of making OFR more effective. This chapter outlines some of the main problems, and the following chapters provide detailed suggestions for possible improvements.

Many critics have been too harsh in judging OFR, perhaps to take advantage of current shifts in fashion among funding agencies or to encourage jealousies deriving from the excessive attention lavished on OFR in the past. After all, if the standard for judging research is the successful delivery of useful technology to resource-poor farmers in developing countries, few research strategies win plaudits. Agricultural research of any type takes time and includes many twists, detours and dead ends. OFR is still a relatively new approach. Its already significant contributions to understanding farming practices and developing methods for diagnosis, experimentation and analysis are beyond dispute (Tripp et al., 1990). To say that a particular project has not brought new technology into the hands of farmers is not to say that the project is necessarily a failure, nor to disparage the quality of the work carried out.

But even when the relative youth and accomplishments of OFR are considered, its effectiveness and future contributions give cause for concern. Why, in spite of so much investment and interest, have the tangible results of OFR been so modest? The explanations are many and varied. Part of the blame can be placed on the institutional context within which OFR has operated in developing countries. National research and extension services are often poorly funded and badly organized, and frequently OFR has been imposed upon these institutions by outside agencies with competing interests and approaches. Another factor is the quality of OFR carried out by both national and international institutions. Too much research under the name of OFR or FSR has been poorly conceived and implemented. In addition, even where strong OFR programmes have operated, they have been noticeably isolated from the rest of the research organization. This lack of integration has denied adaptive research access to the technical support it needs and prevented it from contributing to the establishment of research priorities at the national level. Additional symptoms of this isolation are the fact that OFR has often been carried out with inadequate links to extension services and little consideration of the policy environment. Each of these factors is explored briefly here and in subsequent chapters.

THE INSTITUTIONAL CONTEXT

A variety of national agricultural research and extension institutions in developing countries have conducted OFR. Many of these institutions are poorly funded, inadequately managed and subject to frequent shifts in personnel and administration (SPAAR, 1987; Lele and Goldsmith 1989). These problems explain much of the low productivity for OFR and for most other research strategies and programmes that these institutions have taken up. In a review of OFR in Africa, Collinson (1988) cites researchers' low morale and weak motivation as one factor explaining the slow progress in implementing a farming systems perspective.

In addition, national programmes often lack experience in defining or identifying research targets. Such targeting is, of course, an essential requirement for successful OFR. Collinson (1988) points to several factors inhibiting the effectiveness of many national programmes. Sometimes political leaders are wary of encouraging popular participation in developing technology, maintain a 'top-down' orientation that gives only lip service to accepting grassroots views, and adopt a paternalistic approach to farming populations. Many national agricultural researchers and extension agents neither know, nor care to know, much about the problems and aspirations of most of their country's farmers. Their training has scarcely stimulated their interest in the farming population; even if they are interested, the institutions that employ them do nothing to reward their curiosity. When research and extension proceed with little or no accountability to the majority of the farming population, there is scant hope that an OFR project will be anything more than an exercise in spending donors' money.

However, to say that many national research and extension programmes are poorly organized to accommodate OFR is not to suggest that the institutions attempting to introduce OFR are themselves paragons of research management. Heinemann and Biggs (1985) comment on the paradox of introducing a research approach based on understanding client farmers' needs into national programmes through organizations that pay little attention to the characteristics of their own clients and make little effort to adapt OFR to the needs or capacities of national institutions. A recent review of the experience of the United States Agency for International Development (USAID) with FSR projects concludes that few people in the agency understood the approach they assiduously promoted (USAID, 1989). In these FSR projects there was often little consensus on what methodology to follow and no explicit policy to define the role that FSR would play at the institutional level.

Large amounts of funding encouraged efforts to develop, package and promote distinctive approaches to OFR. The resulting variety of research methods and vocabulary can be seen to a certain extent as a logical outcome of healthy competition, but it has added to the confusion within national and international institutions alike. Only recently, for instance, have the international agricultural research centres (IARCs) been able to sit down and talk about their differences with respect to OFR (ICRISAT, 1987). Attempts to identify common features in the competing research methods and philosophies are beginning to be made (Harrington et al., 1989).

Another symptom of this rush to define and defend an approach to OFR has been the tendency to develop overly elaborate research methods. It is a small step from saying that a broad, holistic approach to research is required to allowing a wide variety of scientists carte blanche to pursue their individual interests. An attractive and well-funded research movement such as FSR held out great incentives to scientists in international and academic institutions. Much of the criticism of the FSR movement is motivated by a perception that, rather than improving the efficiency of the research process, the process became an end in itself (Arnon 1989; Eicher 1990). Although some of the work related to OFR undoubtedly stemmed from efforts to establish careers and reputations, much of the work was certainly well motivated. But the end product was an outrageously varied body of research projects, methods and jargon that too often lost sight of the client. Fascinating descriptions of small-farm management fill the literature, but much less attention is paid to impact.

The growing emphasis on methods and information, rather than on developing technologies, meant that OFR practitioners began to forget that until data on farmers' conditions are translated into real improvements for those farmers, OFR cannot defend itself from its critics. One of the main problems in assembling cases for this book, for instance, was the limited number of adoption studies conducted after OFR recommendations had been made. This lack of attention to monitoring and follow-up calls into question OFR's supposed concern for reaching farmers.

THE CONDUCT OF ON-FARM RESEARCH

Despite these institutional problems, many national programmes have accepted and established OFR in a range of settings, deploying researchers and extension agents in projects ranging from initiatives managed and directed by international organizations to those undertaken by local institutions receiving little or no outside funding or advice. Yet in many of these cases progress has been limited. One explanation is simply the low quality of the research. It is not too much of an exaggeration to say that one of the main problems with OFR is that it has rarely been done properly. Researchers often initiate work with inappropriate client groups, develop an inadequate conception of research priorities and strategies, and conduct OFR poorly in the field.

Defining client groups has been a major problem. Targeting research and extension towards resource-poor farmers has always been difficult (Röling, 1988), and although OFR has been developed specifically to address that concern, much of it proceeds with little attention to client groups. Researchers may be able to describe the farming practices in their area, but even so they often locate trials on unrepresentative fields which may produce results of little relevance to the majority of farmers. In addition, for a research approach that claims to emphasize farmers' point of view, a surprisingly high proportion of OFR is carried out with only superficial interaction between researchers and farmers. This has provoked increasing criticism of OFR and generated numerous proposals for more substantial farmer participation (Chambers and Jiggins, 1986; Farrington and Martin, 1988). The scope for expanding the farmer's role in OFR is further explored in Chapter 14.

A second problem of OFR has been poor diagnosis. Although extensive (and often poorly focused) diagnostic surveys are a feature of many OFR projects, there may be little connection between the data generated and the research priorities that eventually emerge. The story is told of a project that devoted a year to intensive, multiple-visit diagnostic surveys covering everything from household labour supply to marketing practices, only to enter farmers' fields in the following year with two types of experiments: new maize varieties intercropped with local beans and new bean varieties intercropped with local maize!

Faulty trial design and implementation has been a third problem for OFR. On-farm experiments are often badly managed. They may be so few and planted under such variable conditions that little useful information can be extracted. Insufficient agronomic data may be collected to interpret the results of the experiments.

A fourth problem is lack of focus. Much OFR takes the overly enthusiastic approach of addressing a wide range of production problems simultaneously. A few experiments are devoted to each problem, denying the possibility of developing sufficient data for progress in any one of them. Discontinuity in personnel and research direction often compounds the problems of poorly focused research, so that research themes enter and exit the stage in rapid succession and no conclusions are ever reached. While the lack of recommendations emerging from OFR projects is sometimes the result of a hard-headed assessment of the performance of technologies under difficult conditions, more often it is simply a manifestation of poorly conceived research.

In essence, the management of OFR frequently fails to implement appropriate field methods and address the demands of the farming environment. Most farming systems are heterogeneous, complicated and constantly evolving. An OFR programme run by poorly trained and under-funded staff, featuring a partially analysed survey, a series of mechanically conceived experiments in farmers' fields and a set of unclear research goals, will not diagnose problems correctly, much less develop technological innovations appropriate to the system. The complicated interaction between research methods and research management in OFR has been addressed only recently. A summary of the most comprehensive study to date is presented in Chapter 15.

THE INTEGRATION OF ON-FARM RESEARCH

Although many OFR projects have been poorly implemented, there are still many examples of high quality work. But even where a particular OFR project has been managed well, OFR has rarely been successfully integrated as a strategy within a research and extension system. One of the main explanations for this failure is at the same time one of the very strengths of OFR: location specificity. Because OFR stresses the need to develop technologies for well-defined groups of farmers, it tends to be conducted in an atomistic, isolated fashion. Most commonly, small teams of researchers conduct OFR in particular locations. The frequently poor communication between these teams and with the larger institution has two serious implications. First, OFR has not mustered the technical support it needs to carry out its work. Second, it has cut itself off from being able to provide feedback to the larger research institution and from the possibility of helping to set research priorities.

OFR must look to other types of research for sources of technology. Few OFR programmes have produced technological innovations themselves; rather, OFR is a method for adapting existing technologies to the needs and conditions of particular groups of farmers. One problem with seeing OFR merely as a process in which researchers select 'on-the-shelf' technologies and then modify them is that the shelf is sometimes bare or at least inadequately stocked for local researchers. Many of the production problems faced by resource-poor farmers are exceptionally difficult and admit to no simple solutions (Carr, 1989). Research on weed control in Zambia (Vernon and Parker, 1983), for instance, demonstrated a significant gap between yields obtained under the farmers' practice and yields obtained with more timely weed control. Yet opportunities for improving practices were not immediately obvious, taking account of the labour requirements for extra weeding, the cost of available herbicides and the variability in field conditions.

Although OFR may identify production constraints at the local level, considerable support from experiment station trials is often required to develop technologies that make a significant contribution to farm productivity. OFR often experiments with such modest changes (for example, adjustments in fertilizer rates) that farmers are not motivated to take notice. The kinds of technology available must be analysed more carefully and applied research must be directed towards important constraints identified in the field (Chapter 13). Valuable resources can be wasted by establishing OFR where production problems offer little promise of short-term solutions (Eicher, 1986). The idea of a self-sufficient OFR team working independently in a location is as impractical as the notion of a completely self-sufficient family farm.

Another consequence of the isolation of OFR is that research may not take account of trends in the regional or national economy or policy environment that seriously affect the viability of any recommendation. FSR needs to balance its interest in the internal dynamics of local systems with more attention to the interplay between systems and their environment. Systems are changing constantly, and OFR is indeed similar to 'hitting a moving target' (Maxwell, 1986). This is especially true for the economic environment, and if local researchers lack access to good information about expected trends in labour costs, input prices or government policies, their research results may be well off the mark (Barlow et al., 1986). Pachico and Borbón (1987) describe how changes in land availability in Costa Rica, combined with the release of new bean varieties, led to a significant replacement of traditional bean production practices with modified practices. They show how research on the traditional planting system which did not take account of these local trends in resource availability would have been unproductive.

The relative isolation of OFR also affects its capacity to provide information relevant to setting research priorities at the national level. OFR has yet to fulfill its promise of bringing the farmer's voice to research decision making. Even when location-specific work yields useful technologies for farmers, few national programmes maintain a mechanism that allows the lessons learned from OFR to be transferred to similar research being done elsewhere.

More important, if OFR is to depend on technologies developed by other parts of the research organization there needs to be a way of channelling and synthesizing local-level data which ensures that relevant priorities are set. For instance, location-specific data on varietal requirements (maturity, grain type, agronomic characteristics) is developed by OFR at considerable cost. If a variety with appropriate characteristics is available 'on-the-shelf', it will enter on-farm testing. But if such information is presented to breeders in an uncoordinated fashion through individual reports, breeders are unlikely to be able to collate and compare it and decide which characteristics warrant an investment of research time and resources. The same is true for setting priorities for agronomic research. Not all agronomic research needs to be done in farmers' fields, but it does need to respond to an assessment of the most pressing production problems and constraints of significant numbers of farmers.

By establishing itself as a separate activity, or at times as a separate discipline (Jha, 1987), OFR has cut itself off from the mainstream of researchers that it hopes to influence and needs for support. If OFR is seen more as a perspective on research and a set of research methods (Byerlee and Tripp, 1988; Tripp and Woolley, 1990), it can be better integrated into the work of research and extension organizations. Such integration will not be easy to achieve, but it is necessary if OFR is to do anything more than nibble at the edges of agricultural development (Chapter 16).

LINKS WITH EXTENSION AND THE POLICY ENVIRONMENT

Two other factors are sometimes responsible for the lack of impact of OFR: weak links between OFR and extension, and inadequate attention to the policy environment. Extension and policy are admittedly easy scapegoats and at times are blamed for failures that are really the responsibility of research. But in a number of cases, more emphasis on extension or more awareness of the policy environment would have made a significant contribution to OFR.

OFR has often been presented as an ideal way to link extension to research, but this link has rarely been realized. With few exceptions, the attention given by donor and international organizations to strengthening national research capacity has rarely been complemented by an equivalent effort towards extension. Meanwhile, extension organizations continue to perform a wide variety of functions, usually under the auspices of inadequately funded ministries of agriculture subject to various political pressures. The most significant amount of attention paid to extension services has been through the World Bank-sponsored Training and Visit (T&V) programmes (Benor and Harrison, 1977). Although T&V can be seen as potentially complementary to OFR (Moris, 1989), that potential has not been realized. In addition, the information developed by OFR often leads to recommendations that represent 'best-bet' compromises rather than rigid packages, and traditional extension programmes may have trouble accepting such recommendations (Chapter 13).

It is clear that good technologies do not necessarily sell themselves and high-quality OFR has not always been matched by an adequate extension mechanism. Any large-scale restructuring of extension must be contemplated with caution, however. Recent studies indicate that the best way to improve the efficiency of OFR in relation to extension is to develop a series of links between research and extension institutions, such as the informal cooperation of staff at the field level, joint planning and analysis, formal committees and staff secondment schemes (Ewell, 1989). The possibility of forging links with other organizations active at the community level, particularly non-governmental organizations (NGOs), is also receiving increased attention (Thiele et al., 1988).

Although the FSR movement has occasionally seen itself as playing an important role in formulating agricultural policy (Davidson, 1987), its principal contribution would seem to be providing field-level data for conventional policy analysis. Examples are emerging in which OFR has developed information on input use and requirements that has been used to modify policy at the local level (Yates et al., 1988). The contribution of such information to policy formulation at the national level, however, suffers from the same problem of synthesizing data that affects research planning. Unless a critical mass of information from a significant cross-section of farming populations can be presented to policy makers, there is little chance that OFR will be able to contribute in any significant degree to policy formulation.

Before it can make an impact on policy makers, however, OFR needs to better understand the policy environment. Too often, OFR promotes varieties without sufficiently analysing seed production and distribution capacities, tests inputs anticipating that their prices will fall, or proceeds independently of local farm credit agencies. An analysis of the potential of adaptive research in Mali (de Frahan, 1990) showed that the returns would be low unless there were accompanying changes in fiscal policy, marketing systems and credit and input supply policies.

AN EXAMPLE: EARLY MATURING MAIZE IN ECUADOR

The multiple requirements for successful OFR can be illustrated by an example from Ecuador, where several years of OFR on an early maturing maize variety led to only modest adoption. The research is more fully described in Tripp (1985) and Moscardi et al. (1983).

The research was part of an OFR effort undertaken by the International Maize and Wheat Improvement Center (CIMMYT) and the Instituto Nacional de Investigaciones Agropecuarias (INIAP). One of the first locations where OFR was established in Ecuador was in a highland area, 2200-3200 m above sea level, in the province of Imbabura. The major crops were maize intercropped with climbing beans (*Phaseolus vulgaris*), but a wide variety of other crops was grown, including wheat, barley, potatoes and broad beans (*Vicia faba*). The main growing season began in September with the first rains. Local maize varieties took 9 months or more to mature and were not harvested until June or later. This long growing season appeared to be one of the principal research opportunities, and research focused on the possibility of introducing early maturing maize varieties that would allow a second crop (such as peas or potatoes) to be planted in rotation, thus giving two harvests a year.

Informal and formal surveys were carried out in 1976 and a programme of on-farm experimentation began in 1977. These experiments addressed several topics besides the varietal issue, including fertility requirements, insect control and weed control, but as the programme evolved, the main emphasis was on identifying early maturing maize varieties appropriate for the farming system. Experiments examined several possible varieties, investigated their compatibility with local climbing beans (and tested other bean varieties that might be planted with the new maize variety), and established reasonable levels of fertilization for the early maize. Once a variety had been identified, further experiments looked at possible rotations, particularly early maize followed by peas.

By 1981 seed of the new maize variety, INIAP-101, was being sold by the national seed company. The demonstrations organized by INIAP for farmers generated much enthusiasm and raised expectations that demand for INIAP-101 would increase rapidly. Initial follow-up studies indicated that a number of farmers had planted the new variety and had followed it with a second crop of peas, potatoes or beans (Villafuerte, 1983). But a more comprehensive adoption study in 1986 showed that adoption of INIAP-101 was quite low (Crowder, 1990). Farmers who did plant it tended to use it for green maize, sowing an average of only 12% of their maize area to INIAP-101. Most farmers followed the early maize with peas, potatoes or barley.

The adoption study uncovered several reasons for the low acceptance of INIAP-101 that provide lessons for the management of OFR and constitute an inventory of most of the requirements for successful OFR. The farmers' main complaint was the lack of seed. The national seed company could provide seed only in 100-lb bags, a size which far exceeded the average farmer's requirements. Initially, INIAP arranged for the seed to be sold in smaller quantities through extension offices, but this was soon discontinued. The extension service, already burdened with a range of administrative duties, had participated only minimally in the OFR that led to the release of INIAP-101, and showed little inclination to promote it or to consider alternative seed production or distribution strategies. The INIAP-101 seed that farmers continued to plant was maintained by farmers who had hosted experiments or demonstrations or was obtained through local seed exchange among farmers.

But even if seed of INIAP-101 had been easily available, farmers would have experienced serious problems in using it. The fact that farmers harvest most of the INIAP-101 green and express appreciation that the maize is available for consumption early in the season calls into question the validity of researchers' original goal of intensifying the cropping system. Although farmers follow early maize with another crop, they usually plant very small areas. For many of these farmers, subsistence food needs may take precedence over the possibility of growing more marketable crops after maize. Although farmers' opinions regarding the new maize were carefully sought during the course of the research (Tripp, 1985), no forum was established where farmers could contribute to decisions regarding research priorities. In addition, little research was done to explore the feasibility of marketing peas, potatoes or other crops that would be grown in rotation with INIAP-101.

Other concerns that farmers voiced about INIAP-101 related to its agronomic problems. Some of these problems became obvious in the early years of on-farm experimentation, but others were never addressed. As might be expected, a maize that matures 2-3 months earlier than the local variety will sacrifice some yield, and this trade-off was always made clear to farmers. But INIAP-101's fertility requirements are higher than those of local maize, and as soon as a farmer decides to plant INIAP-101 on more than a small portion of particularly fertile land near the house, chemical fertilizer may be necessary. Equally important, an analysis of historical rainfall data showed that in one area of the province where INIAP-101 was recommended its flowering period corresponded to the time when mid-season drought was most likely to occur. The local, late maturing maize varieties did not flower then and were better able to withstand dry years.

Further problems, which were uncovered relatively early in the research process but were not completely addressed in the INIAP-101 breeding programme, concerned the cooking qualities of the new variety and its acceptability in the most common maize preparations. The lack of a concerted strategy to address both the agronomic and consumer problems of INIAP-101 is indicative of perhaps the most important problem in the

OFR programme. Although there were informal relations between the OFR programme, the experiment station crop breeding programmes and the disciplinary departments in INIAP, no formal mechanism existed for presenting and discussing the results of OFR. Although INIAP-101 was tested in other OFR projects in the Ecuadorian highlands (e.g., Kirkby et al., 1981), little effort was made to compare results across areas. Maize breeders took an interest in the progress of the OFR, and the information generated certainly influenced some of their breeding objectives, but farmers' concerns with the new variety represented difficult breeding challenges regarding such characteristics as grain type and rooting depth. Breeders would reasonably demand adequate justification before committing resources to a reorientation of their programme. Unfortunately, information from several OFR projects looking at similar problems could not be channelled and used in setting priorities for breeding programmes or agronomic research.

The research on INIAP-101 was part of an attempt by INIAP and CIMMYT to develop a more effective capacity in adaptive research. The aim of reviewing that work here is to illustrate the extraordinary problems that OFR must face in a research and extension system. The research in Imbabura was well focused and competently managed. It identified an important research opportunity and developed a promising technology. Yet the expected impact never materialized. It is easy, with hindsight, to speculate about what should have been done, but clearly the research was deficient in several ways. More emphasis needed to be placed on input supply (in this case, on working within the limits of existing seed production capacity or encouraging alternative strategies). More commitment was required from the government extension service, and more support should have been given to the cooperatives and local NGOs that were important collaborators in OFR for Imbabura. More socioeconomic and agronomic research would have helped identify limitations and prospects for early maturing maize, and a stronger role for farmers in the research process would have helped clarify research priorities. Lastly, the OFR should have been integrated better with the rest of the research system, and this system itself needed to improve its internal communication links and to articulate a consistent strategy for setting research priorities.

CONCLUSION

The implications of the Ecuador example are sobering. For OFR to be effective, it needs to be well connected with extension; to operate where input availability is assured; to direct good-quality agronomic and socioeconomic research at a wide range of issues that may arise; to enlist farmers' participation in the research process; and to be well integrated with the rest of the research system. This list may be long, but it differs little from well-established thinking regarding the need for a broad approach to agricultural development (Mosher, 1966). For present purposes, a shorter corollary may be more useful: on its own, OFR will rarely make an impact.

As pointed out in Chapter 1, small-farm agriculture has changed, and will continue to change, with or without OFR. The promise of OFR for bringing about more rapid and effective technological development must thus be balanced against the prerequisites for success. In other words, the investment in establishing a field presence, developing information about farmers' conditions and carrying out a programme of experiments on farmers' fields must be matched by the formidable requirements listed above. To be effective, OFR must be carefully targeted. It must be directed towards problems that promise technically feasible solutions and for which a critical mass of research resources is available. And it must be aimed at groups of farmers who represent those sectors of the rural poor that can realistically benefit from technical change in agriculture (de Janvry and Sadoulet, 1989).

A strong capacity for conducting on-farm adaptive research that is carefully targeted should both complement and stimulate the development of an improved system for gathering information about farmers' conditions. In many instances, the information may be used to set priorities for research that is more applied or strategic than adaptive. The nature of production problems and the needs of farming populations in a given country will determine the balance between adaptive OFR and more long-term research efforts. But even if the locus of research lies far from farmers' fields, there must be constant feedback between researcher and farmer. It is here that OFR is indispensable. Experience shows that research that separates itself from farmer input, even if based on a careful analysis of farmers' conditions, easily goes astray (Starkey, 1988). Researchers need to constantly test their ideas against the reality of farmers' conditions, to be ready to make the compromises demanded by complex farming systems and to be challenged to direct high-quality research towards the development of 'suboptimal solutions' (Carr, 1989).

The demands for developing better information about farmers' conditions, and using that information to direct a research programme that includes a strong OFR capacity, are considerable. Meeting these demands is particularly daunting for national research systems in developing countries. But there is no alternative. As private sector agricultural research expands its role in developing countries, it will assume greater responsibility for providing technology to more affluent groups of farmers. The mass of resource-poor farmers in these countries demand and deserve attention, and public sector research must provide it. Stronger farmer organizations and the work of NGOs will help articulate the demand and develop farmers' own capacities to participate in the research process. The challenge and the responsibility, however, rest squarely on the shoulders of national agricultural research systems.

Notes

1. As noted in Chapter 1, the term 'on-farm research (OFR)' describes several related approaches to adaptive research, while 'farming systems research (FSR)' refers to the broader movement that engendered these research strategies.

References

Arnon, I. 1989. *Agricultural Research and Technology Transfer*. London, UK: Elsevier.

Barlow, C., Jayasuriya, S.K., Price, E., Maranam, C. and Roxas, N. 1986. Improving the economic impact of farming systems research. *Agricultural Systems* 22: 109-25.

Benor, D. and Harrison, J.Q. 1977. *Agricultural Extension: Training and Visit System*. Washington DC, USA: World Bank.

Byerlee, D. and Tripp, R. 1988. Strengthening linkages in agricultural research through a farming systems perspective: The role of social scientists. *Experimental Agriculture* 24: 137-51.

Carr, S. 1989. *Technology for Small-Scale Farmers in Sub-Saharan Africa*. World Bank Technical Paper No. 109. Washington DC, USA: World Bank.

Chambers, R. and Jiggins, J. 1986. *Agricultural Research for Resource-Poor Farmers: A Parsimonious Paradigm*. Discussion Paper No. 220. Sussex, UK: IDS.

Collinson, M. 1988. The development of African farming systems: Some personal views. *Agricultural Administration and Extension* 29: 7-22.

Crowder, L.V. 1990. Agents, vendors, and farmers: An assessment of agricultural knowledge transfer and utilization in Imbabura Province, Ecuador. PhD thesis. New York, USA: Cornell University.

Davidson, A.P. 1987. Does farming systems research have a future? *Agricultural Administration and Extension* 24: 69-77.

de Frahan, B.H. 1990. The effects of interactions between technology, institutions and policy on the potential returns to farming systems research in semi-arid north-eastern Mali. PhD thesis. East Lansing, Michigan, USA: Michigan State University.

de Janvry, A. and Sadoulet, E. 1989. Investment strategies to combat rural poverty: A proposal for Latin America. *World Development* 17: 1203-22.

Eicher, C. 1986. *Transforming African Agriculture*. Hunger Project Paper No. 4. New York, USA: The Hunger Project.

Eicher, C. 1990. Building African scientific capacity for agricultural development. *Agricultural Economics* 4: 117-43.

Ewell, P. 1989. *Linkages between On-Farm Research and Extension in Nine Countries*. OFCOR Comparative Study No. 4. The Hague, Netherlands: ISNAR.

Farrington, J. and Martin, A. 1988. *Farmer Participation in Agricultural Research: A Review of Concepts and Practices*. Agricultural Administration Unit Occasional Paper No. 9. London, UK: ODI.

Harrington, L.W., Read, M.D., Garrity, D., Woolley, J. and Tripp, R. 1989. Approaches to on-farm client-oriented research: Similarities, differences, and future directions. In Sukmana, S., Amir, P. and Mulyadi, D.M. (eds) *Developments in Procedures for Farming Systems Research*. Jakarta, Indonesia: AARD.

Heinemann, E. and Biggs, S.D. 1985. Farming systems research: An evolutionary approach to implementation. *Journal of Agricultural Economics* 36: 59-65.

ICRISAT. 1987. *Proc. of Workshop on Farming Systems Research*. Patancheru, India: ICRISAT.

Jha, D. 1987. Strengthening agricultural research in Africa: Some neglected issues. *Quarterly Journal of International Agriculture* 26: 265-75.

Kirkby, R.A., Gallegos, P. and Cornick, T. 1981. *On-Farm Research Methods: A Comparative Approach. Experiences of the Quimiag-Penipe Project, Ecuador*. Cornell International Agriculture Mimeograph No. 91. New York, USA: Cornell University.

Lele, U. and Goldsmith, A.A. 1989. The development of national agricultural research capacity: India's experience with the Rockefeller Foundation and its significance for Africa. *Economic Development and Cultural Change* 37: 305-43.

Maxwell, S. 1986. Farming systems research: Hitting a moving target. *World Development* 14(1): 65-77.

Moris, J. 1989. Extension under East African field conditions. In Roberts, N. (ed) *Agricultural Extension in Africa*. Washington DC, USA: World Bank.

Moscardi, E., Cardoso, V.H., Espinosa, P., Soliz, R. and Zambrano, E. 1983. *Creating an On-Farm Research Program in Ecuador*. CIMMYT Economics Program Working Paper 01/83. El Batan, Mexico: CIMMYT.

Mosher, A. 1966. *Getting Agriculture Moving*. New York, USA: Praeger.

Pachico, D. and Borbón, E. 1987. Technical change in traditional small farm agriculture: The case of beans in Costa Rica. *Agricultural Administration and Extension* 26: 65-74.

Röling, N. 1988. *Extension Science*. Cambridge, UK: Cambridge University Press.

SPAAR. 1987. *Guidelines for Strengthening National Agricultural Research Systems in Sub-Saharan Africa*. Washington DC, USA/The Hague, Netherlands: World Bank/ISNAR.

Starkey, P. 1988. *Practical Agricultural Research: Lessons from Thirty Years of Developing Wheeled Toolcarriers*. Agricultural Administration Research and Extension Network Discussion Paper No. 25. London, UK: ODI.

Thiele, G., Davies, P. and Farrington, J. 1988. *Strength in Diversity: Innovation in Agricultural Technology Development in Eastern Bolivia*. Agricultural Administration Research and Extension Network Paper No. 1. London, UK: ODI.

Tripp, R. 1985. Anthropology and on-farm research. *Human Organization* 44: 114-29.

Tripp, R. and Woolley, J. 1990. Training in farming systems research: Review and prospects. *Experimental Agriculture* 26: 247-62.

Tripp, R., Anandajayasekeram, P., Byerlee, D. and Harrington, L. 1990. Farming systems research revisited. In Eicher, C.K. and Staatz, J.M. (eds) *Agricultural Development in the Third World*. (2nd edn). Baltimore, USA: Johns Hopkins University Press.

USAID. 1989. A review of AID experience: Farming systems research and extension projects, 1975-1987. *AID Evaluation Highlights* 4. Washington DC, USA: USAID.

Vernon, R. and Parker, J.M.N. 1983. Maize/weed competition experiments: Implications for tropical small-farm weed control research. *Experimental Agriculture* 19: 341-47.

Villafuerte, J.A. 1983. Evaluación de la utilización del maíz INIAP-101 en la Provincia de Imbabura. Thesis, Ingeniero Agrónomo. Quito, Ecuador: Universidad Central del Ecuador.

Yates, M., Martínez, J.C. and Sain, G. 1988. *Fertilizer in Les Cayes, Haiti: Addressing Market Imperfections with Farm-Based Policy Analysis*. CIMMYT Economics Program Working Paper 88/01. El Batan, Mexico: CIMMYT.

Planned Change in Farming Systems
Edited by R. Tripp
© 1991 R. Tripp
A Wiley-Sayce Co-Publication

13

On-Farm Research in Southern Africa: The Prospects for Achieving Greater Impact

A.R.C. LOW, S.R. WADDINGTON and E.M. SHUMBA

In the early 1980s most countries in southern Africa introduced on-farm research (OFR) into their national agricultural research systems and over the following few years OFR activity in the region expanded dramatically, most of it initiated with donor funds. OFR was accommodated within the organizational structures of national systems in three different ways:

- addition of OFR mandates and scientists to commodity teams, institutes or research stations (Swaziland, Zimbabwe);

- a set of OFR teams operating with regional mandates to adapt technology from commodity research (Zambia, Malawi);

- ad hoc OFR projects or units added to the research system (Botswana, Zimbabwe).

This chapter reviews the performance of OFR within national institutions across the region. As the institutional settings in which OFR operates differ considerably from country to country, the emphasis here is on drawing lessons from common experiences with the implementation of OFR in southern Africa by focusing on maize, the major food crop in the region. The output from 6-8 years of OFR is discussed and an assessment made of its impact on maize productivity. From this analysis, some general constraints on the effectiveness of OFR are identified and ways in which these constraints might be overcome are suggested.

Despite considerable effort, the impact of OFR in the region has not been as great as expected. An assessment of the reasons for this indicates that improved methods of implementation, though necessary, are not enough to ensure success. Even well-conducted OFR will have limited impact unless it is well integrated with commodity and disciplinary research, links effectively with extension, and addresses serious deficiencies in the input service sectors.

ON-FARM RESEARCH OUTPUT

The discussion here on the improvements in maize production technology resulting from OFR work in southern Africa looks first at the research opportunities identified from diagnostic work and then outlines the outcomes of OFR programmes based on the identified opportunities. A more detailed analysis of opportunities and outputs from OFR with maize in southern Africa can be found in Low and Waddington (1990).

The diagnostic aspect of OFR has contributed to a better understanding of maize production systems in southern Africa and of the reasons why farmers use different production technologies. While this information, of itself, can be regarded as an output from OFR, its purpose was to identify potential opportunities for experimentation which would in turn lead to improved production technologies. In general, the diagnostic part of OFR has not been limiting. Most programmes in the region had no difficulty in identifying clear sets of research opportunities from initial diagnostic studies based on formal and informal surveys (e.g., CIMMYT, 1978; Shumba, 1983, 1988; Kawanga et al., 1984; Seubert, 1988).

Research opportunities

Improved germplasm

Long-season (140 days or more, at an altitude of 1200 m or more) hybrid maize with a high yield potential (and, to a lesser extent, open pollinated varieties) has been adopted by smallholder farmers only in certain areas of southern Africa, including Zimbabwe, Swaziland and parts of Zambia. On-farm trials and diagnostic studies have identified the main reasons why hybrids are not universally preferred over local unimproved varieties:

- because of their long cycle, most maize hybrids perform poorly in zones with a short erratic rainfall season (Whingwiri and Harahwa, 1985; Mataruka and Whingwiri, 1988) or when late planted in wetter subhumid areas (Shumba, 1988, 1989; Waddington and Kunjeku, 1989);

- the hybrids offer little yield advantage over local varieties when grown by smallholders who weed late and use little or no fertilizer (Kydd, 1989);

- most hybrids have soft endosperm (dent grain) and thus are not suitable for the local pounding methods used in Malawi and eastern Zambia to de-hull the grain for food preparation (Kydd, 1989);

- most hybrids have loose sheath cover and this, together with soft endosperm, means that they do not store well in traditional storage structures; they are not preferred for on-farm storage and consumption in Malawi (Kydd, 1989) and parts of Zambia (Waterworth and Muwamba, 1989).

These problems have led several OFR programmes in the region to examine the suitability of other maize varieties in good and marginal environments, using criteria important to farmers.

Late planting

Traditionally, maize production recommendations are based on the assumption that the crop will be planted early, but surveys conducted by OFR have indicated that many farmers are forced to plant late, which reduces grain yields (Shumba, 1984, 1988, 1989). The results from a survey and from trials conducted over 4 years (1982-83 and 1985-86) in Mangwende Communal Area, Zimbabwe, showed that the grain yield of SR 52 and R 215 hybrids was reduced by 1.3% per day for the period from the first effective planting rains until some 50 days later (Shumba, 1989; Waddington et al., 1991). About 33% (175 000 ha) of the subhumid maize area in Zimbabwe is planted 4 or more weeks after the first planting rains (Mudhara and Waddington, 1990).

Research is therefore needed to develop maize materials and practices suited to later plantings (Waddington and Kunjeku, 1989; Waddington et al., 1991). Suggested research includes breeding maize that has a shorter development cycle and an ability to tolerate low soil fertility, weedy seedbeds, the deep or shallow seeding associated with rushed planting, low radiation levels, and waterlogging early in crop development. It is also neccessary to assess the advantages of bringing forward the time of first weeding and fertilizer topdressing in order to accommodate the faster rate of crop development that is associated with late planting.

Draught power shortage

In ox-drawn mouldboard plough systems in southern Africa draught power shortages are common (Shumba, 1984; Baker, 1988; Masi et al., 1988; Seubert, 1988). Households without draught animals are forced to wait until draught owners have completed their own initial land preparation before being able to hire. This results in delayed and hurried land preparation after the onset of the rains, which gives low emergence percentages, increased early competition by weeds and lower maize yields. Reduced or zero tillage options have been examined to address these problems (Shumba, 1989; Waterworth and Muwamba, 1989).

Plant population density

In high-potential subhumid areas, optimum maize plant population densities are well known (about 40 000 plants/ha). However, for semi-arid areas or in late planting or low input situations in subhumid areas, optimal maize plant population densities (and their interactions with rainfall, variety and soil fertility) have not been well established (Waddington and Kunjeku, 1989). Surveys and observations on small farms indicate that densities of 15 000-20 000 plants/ha are common where farmers are aiming at the recommended 35 000-44 000 plants/ha. Poor-quality seedbeds, insect attack, dry spells, soil capping and inappropriate planting equipment and methods are among the reasons for low plant populations.

Fertilizer management

Current recommendations on fertilizer rates are based on responses from early planted, improved hybrids (or varieties). As a significant proportion of the smallholder maize area is planted to local and improved varieties 30-60 days after the recommended planting dates on sandy soils, OFR is needed to establish appropriate fertilizer inputs for these conditions.

Most smallholder farmers apply basal fertilizer 1-3 weeks after crop emergence, rather than at planting as currently recommended. Farmers delay applications of basal fertilizer because of the shortage of labour at planting, the risk of wasting fertilizer if the plant does not emerge, and fertilizer burn inhibiting germination (e.g., Shumba, 1988, 1989). Fertilizer topdressings are also often applied late, around tasselling, because of labour constraints (mainly weeding of other crops or other maize plantings). OFR has looked at appropriate rates and timing of fertilizer applications for non-hybrid maize grown in semi-arid areas or planted late.

Weed control

Late and inadequate weeding is widely recognized as a common feature of smallholder maize production in the region. Factors which contribute to this problem include the priority given to expanding the area planted, the time needed for hoe weeding and the shortage of draught oxen.

The importance of reducing early weed competition during the first 20 days after emergence has been established from experiment station work and generally holds in on-farm work. In Zambia, OFR has found this weed competition to be especially critical if it coincides with early dry periods and in the absence of inorganic fertilizer (Waterworth, 1989), and has looked at ways of combining fertilizer application with mechanical or chemical weed control to reduce labour and draught requirements in Central and Eastern Province (Waterworth, 1989; Waterworth and Muwamba, 1989).

Mixed cropping

In hoe-plant systems on ridges, such as those in Malawi, many farmers plant grain legumes between maize plants on the ridge. However, they often vary the placement and number of grain legume plants between maize plants. Generally, they aim to achieve some yield from the legume without sacrificing yield from maize. OFR recognized a need to assess the effect of intercropping with grain legumes on maize yield and establish the best legume arrangement within the maize crop on-farm, given farmers' resource constraints.

Research outcomes

Broadly, OFR in maize production in southern Africa has had one of four outcomes:

- *Led to adoption of new technology by farmers.* The major example of farmer acceptance and adoption has been of shorter-season maize germplasm with flint or semi-flint grain type in Zambia and short-season dent hybrids in Swaziland. In many cases, farmers preferred these materials for reasons other than yield: provision of food during a mid-rainy season hunger period; flint grain type suitable for pounding and on-farm storage; good performance with no fertilizer; and use for late planting (Seubert, 1989; Waterworth, 1989; Waterworth and Muwamba, 1989).

- *Demonstrated advantages but not led to general adoption.* Several technologies extensively tested on-farm over a number of years have been shown to have yield or cost advantages over current practices, but have not been widely adopted. These include: zero tillage in Zambia; reduced tillage (on-row tine tillage) in Zimbabwe; and herbicide weed control and combined weed and fertilizer management in Zambia and Zimbabwe (Shumba, 1989; Waterworth and Muwamba, 1989).

- *Confirmed the superiority of farmer practice over standard recommendations.* The technologies that fall into this category include management and application rate of inorganic fertilizer. On-farm trial results from many locations have shown no clear advantage of basal fertilizer application at planting over the farmers' practice of applying fertilizer 1-2 weeks after emergence. Similarly, trials showing the advantages of lower fertilizer rates for late planted maize compared with early planted maize are consistent with farmer practice.

- *Shown technologies to be ineffective on-farm or too demanding of resources.* An example of these technologies is stalk bending, which was tested as a way to avoid cobrot in Malawi. Inversion of the cob was expected to reduce moisture accumulation at its base. However, yield advantages in most years were minimal and the practice required considerable extra labour. Again, heavy initial liming of maize to correct pH gave little immediate yield benefit but demanded considerable labour and cash.

A general analysis of the outcome of on-farm work (on maize, as well as on sorghum, cotton, beans and sunflower) in Zambia (Luapula, Central and Eastern Provinces), Zimbabwe (Mangwende Communal Area) and Swaziland gave the results presented in Figure 13.1. The table shows that about one third of the original research themes formulated from identified opportunities resulted in technologies being adopted. Most of these

Figure 13.1 Analysis of the progression from on-farm research initiatives to farmer adoption, in Swaziland, Zambia and Zimbabwe

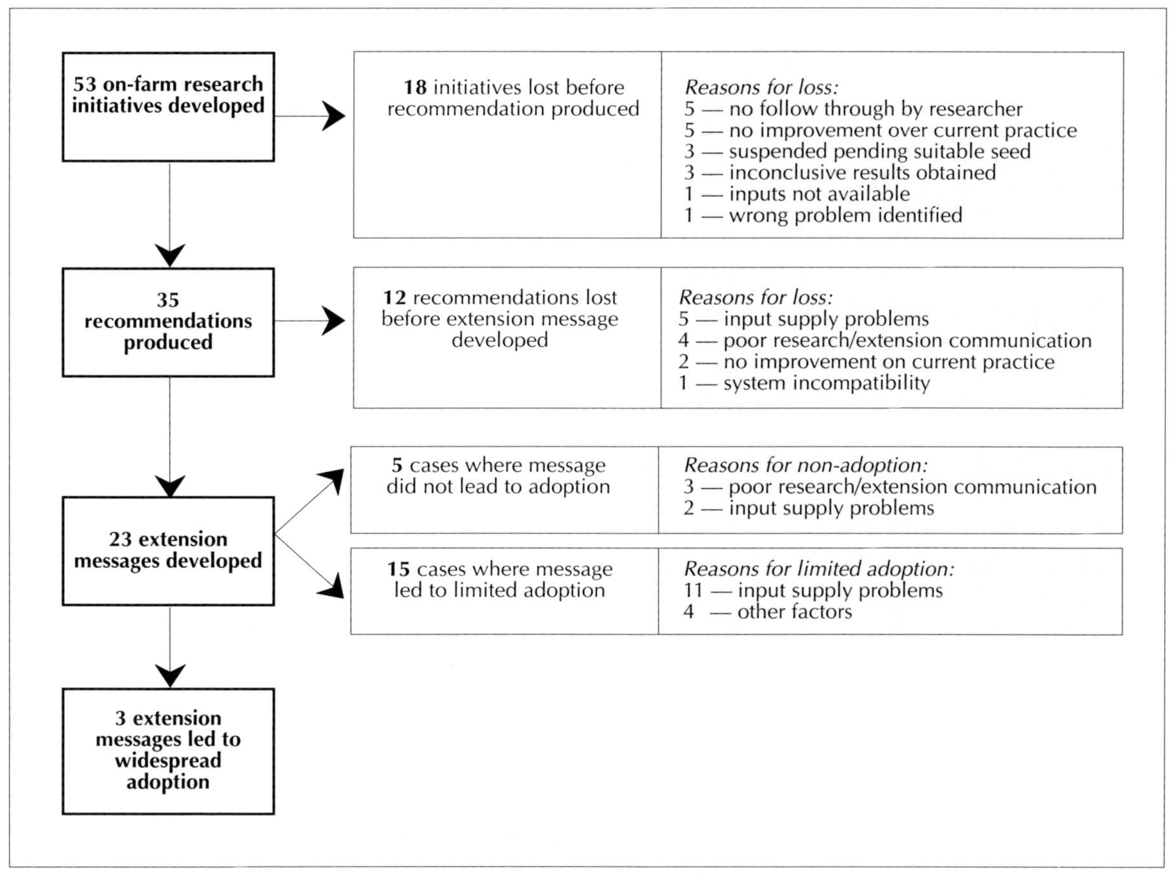

were adopted only partially or by limited numbers of farmers (less than 100, generally representing a proportion of the farmers cooperating in the research). Of the three technologies that were widely adopted, all of them in Zambia, two were crop varieties (a new bean variety and short-season maize MMV400) and the other a system of cultivation on receding flood waters. Figure 13.1 indicates the points at which there was a breakdown between the identification of research opportunities and adoption, and gives the reasons why the OFR effort was lost at these points.

ASSESSMENT OF FACTORS LIMITING ON-FARM RESEARCH IMPACT

The factors which have limited the impact of OFR in southern Africa can be grouped into four categories: implementation deficiencies; technology source; technology delivery; and links with input supply services. This categorization does not imply that these issues are unconnected (Low and Waddington, 1990). The discussion here looks at what needs to be done to improve performance; it is based on a review of many OFR teams and the recommendations are therefore generalized. The shortcomings highlighted in this discussion do not apply in all cases, but there is no programme in the region that does not suffer from at least some of these shortcomings.

Implementation deficiencies

In some cases the planning and implementation of OFR programmes has not been done thoroughly. Deficiencies have included the following:

* *Superficial diagnosis of problems and causes.* Often, the diagnosis has been based on a few days of informal surveying in the field because of a perceived urgency to get to the trial stage, resulting in some inappropriate trials. Inadequate attention has been paid to agronomic aspects of diagnosis, identifying which farmers face a given problem and targeting solutions.

* *Poor implementation of trials.* In many OFR programmes too little attention has been given to selecting farmers and field sites, training field assistants and supervising experiments. This has often been caused by a tendency to implement more trials than researchers are able to manage effectively. Poor implementation has led to doubts about the validity of the results of many on-farm experiments.

* *Inadequate analysis and interpretation of trial results in relation to the implications for farmers and the planning of the next cycle of trials.* In many cases this has occurred because the purpose of the trial or its place in the overall research programme has not been clearly defined. A common experience has been that the initial diagnosis fixes the content of trial programmes for many years.

* *High turnover of donor-funded expatriate staff as well as national staff.* This problem has exacerbated the problems outlined above. An important factor here is that there has also been little good documentation of research rationale and results to mitigate the high staff turnover.

About 50% of the losses shown in Figure 13.1 can be attributed to these various implementation deficiencies. One clear implication is that there is still a considerable need for further training and assistance in the region to improve on the quality of the OFR work being carried out. However, even where OFR has been well implemented, the three other factors discussed below have limited its impact.

Technology source

A major reason for limited adoption of maize technology by farmers and for several research opportunities remaining unaddressed by on-farm researchers is the lack of suitable technology available for field testing. It is clear that, historically, most maize commodity research in southern Africa has been geared towards larger-scale farmers with few resource constraints and has often neglected addressing smallholder needs directly. Thus OFR has had to work with a reduced portfolio of suitable technology. What is required is a more reliable way of providing OFR with technologies (particularly for maize) that are likely to offer clearer benefits to smallholder farmers.

Benefits from integrated research

If OFR had been in a better position to provide information to commodity researchers and if commodity researchers had been more receptive, gaps in appropriate maize technologies may have been filled more quickly. One way of helping to ensure that appropriate technologies are available for OFR is for on-farm researchers to participate in the development of those technologies, along with component researchers. The establishment of strong links between OFR and component research is essential for effective OFR, but in practice such cooperation has proved difficult to initiate and maintain. The introduction of OFR as a separate activity from the rest of already established research is a major reason why integration has not been achieved.

Cooperation is essential for two main reasons. First, OFR has to adapt technologies developed by component research to the needs of smallholder farmers. Second, through a problem focus and close contact with farmers and extension staff, OFR is in a unique position to help orientate component research agendas towards developing technologies that stand a good chance of being useful to farmers.

By providing appropriate information and guidance to component researchers on research needs, OFR and extension can facilitate a flow of appropriate new technologies for adoption and dissemination. Although this has long been recognized as one of the main roles for OFR (e.g. Collinson, 1986; Merrill-Sands and McAllister, 1988; Tripp and Woolley, 1989), it has been neglected. Indeed, of the five OFR-component research linkage functions identified by Merrill-Sands and McAllister (1988), the one involving feedback from OFR to component research was considered the least successfully implemented. The need for orientation is urgent, given the long period involved in researching most new technologies.

Degree of integration required

The information to be shared between commodity or disciplinary researchers and on-farm researchers needs to cover more than a mere identification of the problem(s) to be addressed. OFR should establish the reasons for the problem in the context of the farming system and suggest which types of technologies would be more appropriate, given farmer circumstances. Component research can help by demonstrating why a new technology is a valid candidate for further OFR, in terms of its potential benefits to particular groups of farmers.

Given the detail required, close involvement by commodity and disciplinary researchers and on-farm researchers in both technology development and adaptation is needed, as this allows joint development and appreciation of ideas and needs, avoids misunderstandings and obviates the requirement for OFR to spend a great deal of effort on developing and presenting a case to component researchers and *vice versa*. For an important crop such as maize, joint involvement should be encouraged to a level where the same researchers are responsible for both component research related to technology development and for OFR.

Examples of integration

There are some examples in southern Africa where OFR has provided useful information and guidance for component research and where structures (formal and informal) have been developed to facilitate interaction between OFR and component research. Since 1986 there has been good interaction between Adaptive Research Planning Teams (ARPT) and Commodity Research Teams in Zambia (Kean and Singogo, 1988). ARPTs have contributed to reorienting the research priorities of the maize team towards flinty, early maturing varieties. Based on a better understanding of farmer conditions, the maize team provides ARPTs with maize materials for on-farm testing. Joint planning and implementation of on-farm and on-station trials, joint visits to trials and joint presentation and review of results are undertaken. In Malawi, after initial difficulties, links have been developed between the adaptive research teams and the maize commodity team (e.g., Tinsley, 1988) involving joint participation in diagnostic surveys and joint planning, implementation and assessment of trials focusing on such issues as the yield response of local varieties to low levels of fertilizer inputs under farmer management.

Integration has also resulted from a broadening of production research activities done on-farm towards basic research issues related to small farm productivity over the longer term (6-25 years). Concerns for 'sustainable agricultural systems' have support across the spectrum of agricultural research and are leading to the development of integrated agronomic research in southern Africa that includes a key OFR component. For example, work on the agronomy and productivity of continuous maize production systems in Malawi is in progress and involves researchers from the maize commodity team, adaptive research, grain legume commodity team and the soils team in problem-driven, on-station intercropping and rotation trials, with promising outputs to be examined on-farm later.

Technology delivery

The standard research/extension model in southern Africa is of the type described as 'transfer of technology' (TOT) (Chambers and Ghildyal, 1985). Inherited from colonial times, this model has the following features:

- researchers develop production recommendations for new crops and varieties for specific agroecological regions; they conduct this work mainly on research stations with the aim of developing an 'ideal' set of production practices that maximizes the yield of the crop in question (in some cases economic optimization is the criterion);

- extension agents are trained to know these technical production requirements and how to communicate them to farmers; the style and content of extension messages are influenced by the administrative and regulatory roles of parent institutions and emphasize the promotion of correct husbandry methods (Blanckenburg, 1988);

- success is judged on the basis of how many farmers are contacted or demonstrations held (Kennan, 1980), and on the development of 'elite', non-representative farmer groups (Drinkwater, 1987; Moris, 1989); these farmers are supposed to be the innovators, whose example will eventually stimulate others to do things correctly.

Adherence to this model means that the content of extension messages is quite uniform, regardless of different ways of organizing message delivery. A review of five different extension approaches in Zimbabwe concluded that, in all cases, standardized messages were delivered repetitively by rote and had little relevance for the majority of poorer and less progressive farmers (Chipika, 1987).

Some of the outputs generated by OFR, such as better adapted varieties or recommendations adjusted for specific agroecological regions, pose no real conflict with this traditional extension model. But many OFR-generated outputs imply recommendations that are suboptimal, with input levels or management below those shown to give good returns at high yield levels, or conditional on natural, economic or seasonal circumstances.

Suboptimal recommendations take account of farmers being unable to manage all factors of production at optimum levels, because of resource constraints or conflicts. In these cases, recommendations are not concerned with 'the best way' to grow a crop, but with reducing management conflicts or improving management practices or resource use within the given constraints. For example, a suboptimal recommendation involving delayed application of basal fertilizer to maize was developed in Central Province, Zambia. Researchers found that labour shortages were forcing farmers to delay weeding until the crop was about 70 cm tall, at which time the farmers combined a single handweeding with topdress fertilizer. OFR results showed that earlier weeding, at 20 cm, increased yields by 17% and that a further yield increase was obtained by also bringing forward the timing of topdressing. As the timing of basal fertilizer had no significant effect on yield, it was concluded that delaying basal fertilizer application and making a combined (basal and topdress) fertilizer application in conjunction with weeding at 20 cm would not only increase yields but also reduce labour requirements by 6 man-days/ha during peak periods for labour (Waterworth and Muwamba, 1989). Three seasons of trials confirmed the economic benefit of the combined fertilizer and early weeding practice.

However, extension staff misinterpreted the results of these trials. In the extension demonstrations set up to verify the trial results and elicit farmer response over the following 2 years, basal fertilizer was applied at planting (as currently recommended), thus missing the additional labour saving of mixing basal and topdressing applications. Farmers were not enthusiastic about the demonstrations and recommendations were never issued on the OFR trial findings.

Some argue that conditional recommendations will become more common as we move away from the Green Revolution era, based on new varieties (Byerlee, 1987; Lele, 1989), and as research becomes more farmer- and problem-oriented (Baker and Norman, 1987; Chambers, 1988). These types of research output seem to generate conflicts with traditional TOT-based extension approaches in southern Africa. In addition, the conditional recommendations seem to be equally difficult for extension staff to digest. OFR results in Luapula Province, Zambia, led to the conclusion that maize variety recommendations should be conditional on whether fertilizer is to be applied or not. Consistent OFR results indicated the superiority of an open-pollinated improved variety over hybrids when no fertilizer was applied (Waterworth and Muwamba, 1989). However, the extension messages concerning maize varieties and fertilizer rates related only to hybrids and recommended 60 kg/N/ha; the option of using no fertilizer and non-hybrids was not included.

These examples illustrate the problem that extension staff have with handling management-based OFR findings that do not conform to accepted 'technical' ideals. Acceptance of the utility of suboptimal and conditional recommendations tends to conflict with the technical training and in-service experience of most extension staff and requires the development of new skills in making conditional judgements about what input levels or management practices are appropriate for which farmers. About 40% of the losses between research recommendations and adoption shown in Figure 13.1 were attributed to communication problems between research and extension. In order to build stronger and more effective links between OFR and extension, adjustments are required on both sides.

Extension adjustments

The adjustments to be made in extension fall into two main categories. The first category concerns relating extension methods to extension content, while the second focuses on training in the use of a problem-oriented approach to extension.

Under existing extension systems, particularly the Training and Visit (T&V) method, demonstrations and crop production packages — quarter hectare ('Lima') Crop Recommendations in Zambia, Crop Packages in Zimbabwe, Crop Production Guidelines in Swaziland — are the standard extension tools. Although demonstrations are an effective tool for extending new varieties and the inputs to go with them (Borlaug, 1976; Russell, 1981), they do not allow for the flexibility required to transfer information about suboptimal or conditional techniques (Sutherland, 1988). There is a need to develop a wider range of extension methods that are geared to the types of information being extended and a diversity of clients (Sagar and Farrington, 1988; Gentil, 1989).

Extension of suboptimal and conditional recommendations requires dialogue with farmers, in groups or individually, about the choice of options or the reasoning behind the management compromises. It suggests a collaborative rather than a teacher-pupil relationship between extension agent and farmer. Much is said about developing a two-way flow of information between the two groups but, in reality, extension training and perceived roles in southern Africa ensure that there is a one-way flow from those who know (research and extension) to those who supposedly do not (smallholder farmers). For example, the current Zimbabwean training course on extension methods teaches participants how to educate farmer leaders; in its list of available extension methods, the course manual does not mention consultation and dialogue, and individual discussion about production problems is discouraged except with elected leaders (AGRITEX, 1986).

Alternative extension methods suggested by Sutherland (1988) in the Zambian context include: field meetings at strategically selected local farms; group discussions on common problems; individual informal visits to innovative farmers; and more focused T&V messages on priority crops and problems.

With regard to adjustments in extension training, in southern Africa standard technical crop production packages still form the basis of most extension advice and extension training at all levels emphasizes 'correct crop husbandry methods'. In-service training courses focus on upgrading and updating technical knowledge, which makes it difficult to accept deviations from the accepted norms.

In Zambia (Sutherland, 1988) and Botswana (Baker, 1988) it was observed that extension agents could not easily see how resource and socioeconomic factors related to technical aspects of production. Training in a management perspective was advocated to enable them to vary their advice, taking account of particular sets of circumstances, such as competing demands for cash and labour among different crops. In Swaziland researchers also recognized that extension agents needed to have more experience in taking the farmers' views into consideration. In 1988, farmer participation was incorporated into the extension training sessions conducted by researchers. The extension agents, guided by farming systems researchers, visited farmers and performed their own resource and constraints analyses based on discussions with the farmers. In Zimbabwe, Agricultural Extension (AGRITEX) staff have been very receptive to training in field diagnosis techniques which emphasize a problem-oriented approach. The idea of starting with an understanding of what farmers are doing, and why, and developing messages around identified problems and opportunities, was well received, and informal field diagnosis will become the standard technique used to develop the content of extension messages (Hakutangwi, pers. comm.).

Through OFR, research scientists have embraced a problem-oriented approach to the development of their research agendas. Substantial resources have been spent on demonstrating the benefits of this approach and training researchers on how to use it. Some of this has spilled over into extension connected with OFR projects (Sagar and Farrington, 1988). In southern Africa, however, OFR has remained primarily a research activity and extension agents continue to look to research to provide the recipes they should be teaching farmers. Logic, as well as experience from Asia (Pickering, 1985), suggest that OFR should be as much an extension as a research activity. Adoption of an OFR approach in extension implies moving away from packaged recipes towards providing farmers with options and advice on how to improve production. This change in the role of extension not only requires new skills on the part of extension staff at all levels (Denning, 1985) but also needs sanction and support from directors of research and extension.

Adjustments in OFR

It has been observed that OFR methodology is weak on incorporating its findings into the extension system (Moris, 1989). Ewell (1989) suggests that lack of good information about research findings has particularly limited the transfer of technologies other than improved varieties. Attempts in southern Africa to overcome this shortcoming in OFR have taken two broad forms. First, formal structural or organizational changes designed to facilitate information flows from OFR to extension have been introduced. Second, attempts have been made to develop information formats designed to make research findings more relevant and usable by extension.

Three very different structures to facilitate information flow from OFR to extension have been set up in Malawi, Zimbabwe and Zambia. In Malawi, adaptive research teams have been located within, and operate under the control of, area-based extension programmes (Agricultural Development Divisions). The teams are coordinated and administered by a section of the research department, located at the central research station, Chitedze. General professional support and training is provided from research, but the teams' work programme and research content is agreed jointly with ADD management and the research service. This arrangement has led to conflicts about the content and coverage of the adaptive research programmes. ADD management sees the teams as an additional resource for implementing extension and development programmes, whereas the adaptive research coordination unit sees their role as identifying research opportunities in specific locations and feeding this information back to commodity teams or following up through on-farm testing and evaluation. This has overstretched the adaptive research team capacities, with the result that neither role has been effectively performed.

In Zimbabwe, the extension branch (AGRITEX) and the research branch (Department of Research and Specialist Services) of the Ministry of Agriculture and Lands both conduct on-farm trials and/or demonstrations. To improve coordination of this on-farm work, the Committee for On-Farm Research and Extension (COFRE) was established in 1986 (Shumba et al., 1991). This committee appoints sub-committees (by commodity grouping), who are responsible for vetting all OFR or demonstration proposals related to their commodity grouping. COFRE keeps research and extension informed of each other's on-farm activities and has had some impact on the quality of planning and field implementation of on-farm trials and demonstrations.

In Zambia, the ARPT has appointed research/extension liaison officers (RELOs) to improve information flows between the adaptive research teams and extension. RELOs are expected to monitor technology adoption, train extension agents on new recommendations, organize interactive meetings between adaptive researchers and extension, and ensure that extension literature on new recommendations is available. Although they are administratively responsible to the extension branch, they are technically responsible to the provincial ARPT coordinator. Over the past 6 years the actual roles of the RELO have changed and have varied between provinces. Extra activities have included the production of regular (monthly) bulletins, the updating of all recommendations for the province and the organization of extension demonstrations. On the other hand, some of the intended activities for RELOs, such as producing extension training materials and monitoring adoption and feedback, have hardly been touched.

The experiences of the formal structures suggest that their success depends on a mutual appreciation of the need for collaboration between research and extension. Where this is lacking, it needs to be developed through the formal structures themselves, through training and through providing appropriate incentives for research and extension to work together.

With regard to developing information formats, efforts have been made in Zambia and Swaziland to devise formats suited to the types of output being produced by the OFR teams. The monthly T&V bulletin put out by Eastern Province, Zambia, started by being technically oriented and crop specific. Recently, more economic information with a management orientation has been included. This is an attempt to give some background and rationale for the conditional recommendations to help extension staff apply them in different circumstances.

In Swaziland a similar need was recognized, and researchers developed Field Support Guides in which the extension messages differ from previous extension support materials in several ways: they focus on a particular topic (often a constraint, or a specific part of the production system) rather than on all the recommended practices for a crop; they provide basic background information on the topic (including social, economic and biological factors) to help extension agents understand the rationale for the recommendations; and they provide guidance on when to apply the recommendation or how to modify it to suit particular farmer conditions. The format of the Field Support Guides also differs significantly from previous materials in that they use simple language, explain new technical terms, are short and easy to read, use illustrations to improve understanding and are easy to carry in the field because of their small size (Seubert, 1989).

Links with input supply services

It is clear from Figure 13.1 that input supply problems are often to blame for the lack of impact from research recommendations. Although the quality of input supply, infrastructure and organization varies from country to country, timely access by small farmers to appropriately packaged technology has restricted the adoption of some recommendations throughout the region. Input supply problems have included the following:

- in a number of countries the wrong type of input has been supplied (for example, a compound fertilizer with low nitrogen and high phosphate content was supplied to Eastern Province, Zambia, despite repeated requests for a compound with higher nitrogen and lower phosphorus which was preferred by farmers and shown by research to be more effective and economic);

- the late availability of topdress fertilizer restricted the adoption of earlier application recommendations in Zambia and Swaziland, and the late supply of short-season maize varieties inhibited uptake of new varieties in Zambia;

- the inability to supply herbicides in appropriate pack sizes slowed uptake of herbicides in Swaziland and Zimbabwe;

- insufficient supplies of minimum tillage equipment in Zimbabwe reduced uptake rates;

- credit for fertilizer tied to hybrid maize seed inhibited fertilizer use in Malawi;

- short-season varieties of maize, which farmers preferred over hybrids for late planting or for use as an early 'hunger gap' crop, were not supplied at all to some provinces in Zambia.

Although researchers are not in a position to rectify these problems directly, they are well placed to highlight them and to point out their significance to policy makers at the district, provincial or national level. In an attempt to do just this, Zambia's Eastern Province OFR team developed what they called 'Management Briefs'. These are addressed to managers of input supply or marketing organizations responsible for the relevant service. They attempt to make use of the research data collected to provide an estimate of the extent of the problem and an idea of the likely benefits for the individual farmer as well as for the province as a whole (Waterworth, 1989).

To date, the Management Briefs do not seem to have been very effective in overcoming supply problems in Eastern Province, but they do provide an example of the potential role that field level data can play in informing policy makers about such problems. Economists on OFR teams are in a good position to make use of these data and to communicate them to other economists in planning ministries or supply and marketing organizations. This is a role that OFR economists are now beginning to develop. They need guidance, encouragement and perhaps some training to help them identify opportunities and conduct appropriate analyses.

CONCLUSION

In southern Africa, OFR started with strong donor support. Indeed, much of the early activity was initiated by donor organizations and met with some resistance from established research systems. However, OFR has now become institutionalized and forms an integral part of the research system of most countries.

It is clear that OFR has contributed significantly to the identification of relevant research opportunities. There is increasing evidence that it has also generated an awareness among commodity/disciplinary research units and extension services of the criteria influencing farmers' decisions and the importance of non-technical factors in the adoption of technology. Both research and extension staff now have a better appreciation of the variation in circumstances between groups of farmers. The notion that it can be useful to develop a range of suboptimal technical options to suit different circumstances, priorities or capacities of farmers is beginning to gain ground. We need to build on this by paying more attention to the development of effective links between OFR on the one hand and commodity research and extension on the other. Some progress has been made in both these directions in southern Africa. The lessons from these experiences should provide a basis for further progress. Increasingly, OFR programmes in the region are having to manage with lower budgets than they enjoyed previously, as donor interest and enthusiasm shift in other directions. One adjustment that can be made to accommodate these lower budgets is to better integrate the work of OFR, commodity research and extension, which would provide the rare prospect of simultaneously reducing costs and improving effectiveness.

The success and effectiveness of OFR as assessed in the mid-1990s will depend as much on how it links with and influences other actors in the technology development and transfer process, as on how well its procedures and methodology are implemented in practice. It should be remembered that OFR was initially seen as a linking mechanism (e.g., Andrew and McDermott, 1985), not as a branch of research or extension in its own right.

References

AGRITEX. 1986. *Extension Methods Resource Manual.* Harare, Zimbabwe: Training Branch, AGRITEX

Andrew, C.O. and McDermott, J.K. 1985. *Farmer Linkages to Agriculture and Research: A Question of Cost Effectiveness.* Mimeograph. Gainesville, Florida, USA: Farming Systems Support Project, University of Florida.

Baker, D.C. 1988. *Draught Power Arrangement in Shoshong and Makate.* ATIP Working Paper No. 16. Gaborone, Botswana: Department of Agricultural Research, Ministry of Agriculture.

Baker, D.C. and Norman, D.W. 1987. *The Farming Systems Research and Extension Approach to Small Farmer Development.* ATIP Miscellaneous Paper No. 87-11. Gaborone, Botswana: Department of Agricultural Research, Ministry of Agriculture.

Blanckenburg, P. 1988. *Agricultural Extension Systems in Some African and Asian Countries.* FAO Economic and Social Development Paper No. 46. Rome, Italy: FAO.

Borlaug, N.E. 1976. The Green Revolution: Can we make it meet expectations? *Proc. of American Phytopathological Society*, 3: 6-31.

Byerlee, D. 1987. *Maintaining the Momentum in Post-Green Revolution Agriculture.* International Development Paper No. 10. East Lansing, USA: Michigan State University.

Chambers, R. 1988. Farmer first. *International Agricultural Development* November/December 1988: 10-12.

Chambers, R. and Ghildyal, B.P. 1985. Agricultural research for resource-poor farmers: The farmer-first-and-last model. *Agricultural Administration* 20: 1-30.

Chipika, S. 1987. A brief analysis of extension approaches within the National Extension Framework Study. Paper presented at 5th Annual Agricultural Extension and Research Project Review Conference, June 1987, Zimbabwe.

CIMMYT. 1978. *Demonstration of an Interdisciplinary Approach to Planning Adaptive Agricultural Research Programmes.* Miscellaneous Report No. 3. Nairobi, Kenya: CIMMYT.

Collinson, M.P. 1986. On-farm research with a systems perspective: Its role in servicing technical component research in maize, with examples from eastern and southern Africa. In Gelaw, B. (ed) *To Feed Ourselves: Proc. of First Eastern, Central and Southern Africa Regional Maize Workshop.* El Batan, Mexico: CIMMYT.

Denning, G.L. 1985. Integrating agricultural extension programs with farming systems research. In Cernea, M.M, Coulter, J.K. and Russell, J.F.A. (eds) *Research-Extension-Farmer.* Washington DC, USA: IBRD/UNDP.

Drinkwater, M.J. 1987. *Exhausted Messages: Training and Groups, A Comparative Evaluation of Zimbabwe's Training and Visit System.* Working Paper AEE 2/87. Harare, Zimbabwe: Department of Agricultural Economics and Extension, University of Zimbabwe.

Ewell, P. 1989. *Linkages between On-Farm Research and Extension in Nine Countries.* OFCOR Comparative Study No. 4. The Hague, Netherlands: ISNAR.

Gentil, D. 1989. A few questions on the training and visit method. In Roberts, N. (ed) *Agricultural Extension in Africa. A World Bank Symposium.* Washington DC, USA: World Bank.

Kawanga W.T, Bulla, G.M. and Chisambiro L.J. 1984. *Exploratory Survey, Background Information and Experimental Results: Ntcheu Rural Development Project.* Chitedze, Malawi: Adaptive Research Section, Ministry of Agriculture.

Kean, S.A. and Singogo, L.P. 1988. *Zambia: Organization and Management of the Adaptive Research Planning Team (ARPT), Research Branch, Ministry of Agriculture and Water Development.* Special Series on the Organization and Management of On-Farm Client-Oriented Research (OFCOR), Case Study No.1. The Hague, Netherlands: ISNAR.

Kennan, P.B. 1980. Agricultural extension in Zimbabwe, 1950-1980. *Zimbabwe Science News* 14: 183-86.

Kydd, J. 1989. Maize research in Malawi: Lessons from failure. *Journal of International Development* 1: 112-144.

Lele, U. 1989. *Managing Rural Development in Africa: Lessons of Experience for Governments and Donors.* World Bank MADIA Study Paper. Washington DC, USA: World Bank.

Low, A.R.C. and Waddington, S.W. 1990. On-farm research on maize production technologies for smallholder farmers in southern Africa: Current achievements and future prospects. In *Proc. of Third Eastern and Southern Africa Regional Maize Workshop.* Nairobi, Kenya: CIMMYT.

Masi, C., Mumba, A., Sutherland, A. and Valades, N. 1988. *On-Farm Diagnostic Survey of Small Scale Farmers in Choma and Monze Districts, Southern Province, Zambia.* Mimeograph. Choma, Zambia: Mochipapa Regional Research Station.

Mataruka, D.F. and Whingwiri, E.E. 1988. Maize production in communal areas and the research options for improving productivity. In Gelaw, B. (ed) *Proc. of Second Eastern, Central and Southern Africa Regional Maize Workshop.* El Batan, Mexico: CIMMYT.

Merrill-Sands, D. and McAllister, J. 1988. *Strengthening the Integration of On-Farm Client-Oriented Research and Experiment Station Research in National Agricultural Research Systems (NARS): Management Lessons from Nine Country Case Studies.* OFCOR Comparative Study No. 1. The Hague, Netherlands: ISNAR.

Moris, J. 1989. Extension under East African field conditions. In Roberts, N. (ed) *Agricultural Extension in Africa. A World Bank Symposium.* Washington DC, USA: World Bank.

Mudhara, M. and Waddington, S. 1990. *Study of Maize Planting Dates in Natural Region 2 and 3 Communal Areas of Zimbabwe, 1989/90 Season.* Harare, Zimbabwe: CIMMYT.

Pickering, D.C. 1985. *Sustaining the Continuum.* In Cernea, M.M., Coulter, J.K. and Russell, J.F. (eds) *Research-Extension-Farmer.* IBRD/UNDP Symposium. Washington DC, USA: IBRD/UNDP.

Russell, J. 1981. Adapting extension work to poor agricultural areas. *Finance and Development* 18(2): 30-33

Sagar, D. and Farrington, J. 1988. *Participatory Approaches to Technology Generation: From the Development of Methodology to Wider-Scale Implementation.* Agricultural Administration Unit Network Paper No. 2. London, UK: ODI.

Seubert, C. 1988. Background and overview of agronomy farming systems research in Swaziland. In Curry, J.J. and Seubert, C. (eds) *Report of a Review of On-Farm Research in Swaziland: 1982-1988.* CIMMYT Network Report No. 12. Harare, Zimbabwe: CIMMYT.

Seubert, C. 1989. *On-Farm Research and Extension Linkages: Experience from Swaziland.* CIMMYT Network Report No. 19. Harare, Zimbabwe: CIMMYT.

Shumba, E.M. 1983. *A Diagnostic Survey of Mangwende Communal Area, Zimbabwe for On-Farm Research Planning.* Research Report No. 6. Harare, Zimbabwe: DR&SS, Ministry of Agriculture and Lands.

Shumba, E.M. 1984. Reduced tillage in the communal areas. *Zimbabwe Agricultural Journal* 81: 235-39.

Shumba, E.M. 1988. Maize technology research in Mangwende, a high potential communal area environment in Zimbabwe, Part 1: Developing a research agenda. *Farming Systems Newsletter* 34: 12-34.

Shumba, E.M. 1989. An agronomic study of appropriate maize (*Zea mays*) tillage and weed control technologies in a Communal Area of Zimbabwe. PhD draft thesis. Harare, Zimbabwe: Faculty of Agriculture, University of Zimbabwe.

Shumba, E.M., Waddington, S.R. and Navarro, L.A. (eds). 1991. *Research and Extension Linkages for Small Holder Agriculture in Zimbabwe.* Proc. of Workshop on Assessing the Performance of the Committee for On-Farm Research and Extension (COFRE). Harare, Zimbabwe: DR&SS/AGRITEX/IDRC.

Sutherland, A. 1988. Extension and farming systems research in Zambia. In Howell, J. (ed) *Training and Visit Extension in Practice.* Agricultural Administration Unit Occasional Paper No. 8. London, UK: ODI.

Tinsley, R.L. 1988.The adaptive research programme in Malawi. Paper presented at the annual LADD Senior Staff Seminar Ntcheu RDP, July 1988, Malawi.

Tripp, R. and Woolley, J. 1989. *The Planning Stage of On-Farm Research: Identifying Factors for Experimentation.* El Batan, Mexico/Cali, Colombia: CIMMYT/CIAT.

Waddington, S. and Kunjeku, P. 1989. Potential technology and research needs for rainfed maize production in drought-prone environments of southern Africa. *Farming Systems Bulletin, Eastern and Southern Africa* 3: 28-41.

Waddington, S.R., Mudhara, M., Hlatshwayo, M. and Kunjeku, P. 1991. Extent and causes of low yield in maize planted late by smallholder farmers in subhumid areas of Zimbabwe. *Farming Systems Bulletin (Eastern and Southern Africa)* 9.

Waterworth, J.V. 1989. *On-Farm Research and Extension Linkages: Experiences from Eastern Province, Zambia.* CIMMYT Network Report No. 20. Harare, Zimbabwe: CIMMYT.

Waterworth, J.V. and Muwamba J. 1989. *Report of a Study on the Use of On-Farm Research Results by Extension in Zambia and Swaziland.* Mimeograph. Harare, Zimbabwe: CIMMYT.

Whingwiri, E. and Harahwa, G. 1985. Maize varieties for the low-yielding environments in Zimbabwe. *Zimbabwe Agricultural Journal* 82: 29-30.

14

Adopters and Adapters: The Participation of Farmers in On-Farm Research

J.A. ASHBY

An international audience of experienced on-farm researchers, extension agents and policy makers watches a sequence of slides showing farmers in their fields interacting with officials, the groups easily distinguished by their dress. When asked to interpret what is happening in each slide, members of the audience agree that 'the agronomist is teaching the farmer...', 'the extension agent is giving a recommendation...', 'the visitor is showing the farmers how to overcome a problem in their crop...', and so on in a similar vein. These interpretations reflect an assumption common to on-farm researchers: that farmers are recipients of technologies and of recommendations taught to them by researchers and extension agents. In other words, farmers are primarily *adopters* of ideas and practices generated by outside experts.

This chapter compares the role of farmers as adopters with another function they perform — that of experimenting with and adapting recommendations to suit their own resources and purposes. Adoption is understood to mean the application of a technology as it is recommended. Adaptation occurs when farmers apply the principles or inputs embodied in a new practice, but in a novel way. A good example is provided by the diffused light storage case discussed in Chapter 11. While some farmers constructed potato storage facilities according to researcher recommendations, others used the principle of diffused light storage on which the recommendations were based to modify existing facilities or build innovative structures (Rhoades and Booth, 1982).

Technology adaptation by farmers can be merely observed and tolerated as incidental to the mainstream on-farm research (OFR) effort, or it can be stimulated and integrated into this research process. In this chapter, the advantages and disadvantages for OFR of placing more emphasis on facilitating experimentation and adaptation by farmers are analysed. It will be suggested that the greater degree to which an OFR effort emphasizes the role of farmers as *adapters* of technology, the more participatory is the research style and often — though not always — the more immediate is the pay-off for farmers from the research.

TECHNOLOGY ADOPTION AND ADAPTATION

Notional and developed technologies

An important element in the distinction between adoption and adaptation is the idea of 'notional' vs 'developed' technology. Notional technology is the concept or principle embodied in an innovation at an early stage of testing and development. Developed technology has evolved from notional technology, and has been tested to provide formal recommendations about its use (Menz and Knipscheer, 1981).

The relationship between notional/developed technology and adoption/adaptation is illustrated in Figure 14.1. Farmers are usually expected to adopt developed technology. However, they often adapt developed technology, altering its components and application so as to achieve better results (from the farmers' point of view) than those obtained with the researchers' recommendations (Biggs and Clay, 1981).

An example of farmers' adaptation of developed technology is provided by Doorman's (1983) study of rice technology adoption in the Dominican Republic. The technology package consisted of 10 components (double cropping, new planting dates, improved varieties, seedbed preparation, plant density, herbicides, fungicides, mechanized land preparation, fertilizer applications and insecticides). Doorman found that none of the farmers in the study had adopted the complete package. Some had experimented on their own initiative on a small area, applying different amounts and types of fertilizer over two or more seasons, and so had determined the correct amount and type to suit their conditions. Although the recommendation had been to apply fertilizer evenly over the plots, farmers were applying it only on those parts of the plots where plants were poorly developed because of inadequate drainage and levelling. Another adaptation was to apply higher levels of fertilizer than recommended when transplanting rice late in the growing season. This practice of transplanting, as opposed to the recommended practice of direct seeding, was in itself an adaptation aimed at addressing a number of

Figure 14.1   Adaptation or adoption of notional or developed technology

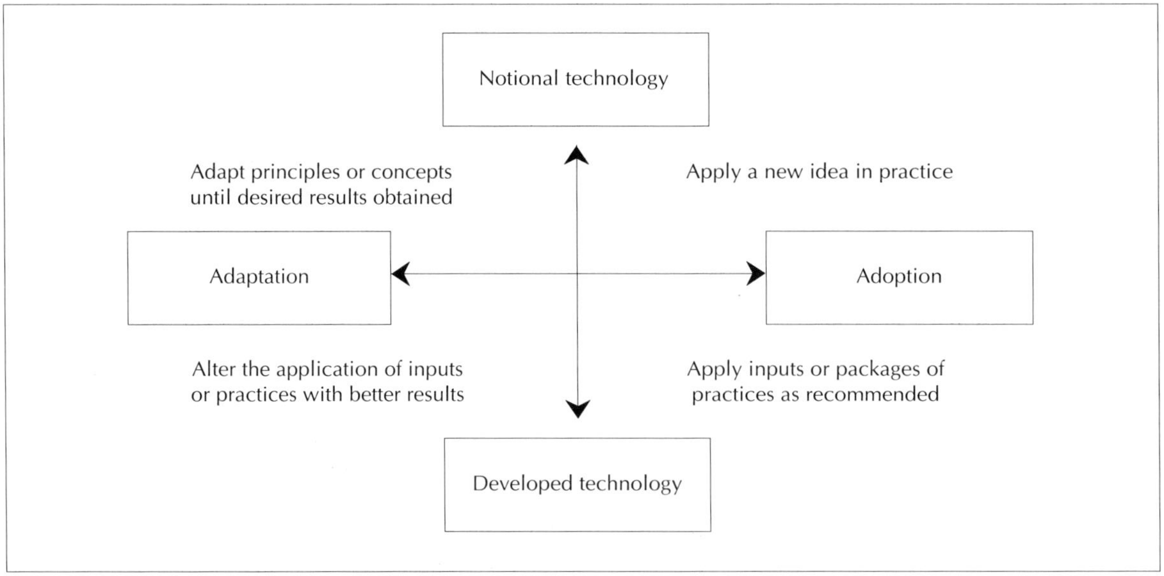

problems, such as pest damage, excess water and poor germination, which had not been anticipated in the design of the technology package. Overall, Doorman identified seven components of the package which had not been adopted; five different adaptations; and only three components which had been adopted in less favourable environments characterized by poor drainage and levelling (representing the majority of farms).

As indicated in Figure 14.1, farmers may adopt notional technology, taking a new idea not yet developed into a well-defined technology and finding a way to make it work in practice. For example, contrary to researchers' expectations, from an exploratory on-farm trial conducted in Sumapaz Province in Colombia, local farmers adopted the idea of estimating the level of insect infestation in snap bean plots before choosing the appropriate control measure for white-fly. However, they developed their own classification of the white fly life cycle and a scale for measuring infestation levels that was easier to use, and to communicate to other farmers, than the one researchers had used in the trial (Cardona et al., 1991).

Farmers may also adapt some of the principles embodied in a notional technology to suit their own purposes. In an experimental artesanal seed production enterprise in Colombia, farmers took part in planning seed production trials. They determined that the recommended specialized seedlots were not as advantageous as the multipurpose production plots where farmers grew grain which was selected for seed, domestic consumption or animal feed, depending on its quality. The multipurpose plots enabled farmers to 'play' the markets for different products, financing the extra cost of selecting and storing seed from the sales of grain for consumption. In this instance, the cultural practices recommended for specialized seedlots were adopted, but the basic principle of specialization was adjusted to achieve the goals of multipurpose production (Roa et al., 1991).

Technology diffusion

In any farming community there will be some members who experiment with new technologies, whether notional or developed, and others who wait and see what the experimenters' conclusions are before they take up an innovation. Small farmers in developing countries seldom adopt a new practice or input without first testing it on a small scale. Although this trial period slows down the rate of adoption, it is very necessary because of the risk involved in adopting an inappropriate recommendation (Biggs, 1980).

The classic S-shaped curve of most adoption studies shows that a few farmers in the community first test an innovation to see if it suits their conditions. If they are satisfied that this is so, only then does the diffusion process 'take off' as the rate of adoption increases rapidly, levelling off when most farmers have adopted (Rogers, 1962). When full diffusion does not take off, it shows that in the initial testing period farmers found the innovation to be unsuitable (Ashby, 1982). But what is overlooked is the fact that in many instances the initial testing period enables farmers to *adapt* the original recommendation to their own purposes. The adapted recommendation may well diffuse widely even though the original recommendation is abandoned by farmers.

When farmers participate in research as adapters of technology, the farmers' trial period (an integral feature of the diffusion of any innovation) will merge with the period of formal technology testing by researchers. When farmers are involved principally as adopters of developed technology, then the farmers' trial period will occur after the formal on-farm trials. Farmers will want to tinker with the technology to see what adaptations are desirable because their participation has been late in the game and of a limited type and duration. OFR which perceives farmers only as adopters often faces situations like the one described here:

> At the end of the on-farm trials, farmers were asked for their opinion of the improved maize variety and
> its effect on their total system. Only about 5% of them indicated that they would grow the early maize

variety ... because of the increased income. Most farmers were not interested in the early maize harvest because at that time they would be harvesting and processing palm kernels. Also, they preferred to use any extra time available to weed older cassava plots than to harvest early maize.

The trials indicated very clearly the need to consider farmers' requirements and their assessment of benefits in designing improved technology. (Ezumah, 1989).

Farmer participation which ensures that adjustments desired by farmers are integrated into the formal research process will improve research efficiency and increase the returns to public sector investments in agricultural technology development in several ways: by speeding up the rate of innovation; by reducing or eliminating wasteful research and extension on technologies eventually rejected by farmers; or by reducing the cost of 'fine tuning' technologies, putting this responsibility in the hands of producers.

TYPES OF FARMER PARTICIPATION IN ON-FARM RESEARCH

Farmer participation has always been an important concept in OFR. In early accounts of OFR, such as those by Harwood (1979), farmers were portrayed as colleagues in the research process. They were shown as active participants throughout the process, not just in testing developed technology. They helped to define objectives and the criteria for success and contributed to decisions concerning the inputs to experimentation and the management of trials. This type of researcher-farmer relationship tended to depend on researchers' personal style, rather than any explicit methodology, and was viewed as 'natural' and obvious.

However, as on-farm researchers began to develop formal methodology in order to teach and disseminate the approach, they lost sight of the importance of constructing a working relationship with farmers that would permit, and indeed encourage, farmers to be experimenters and adapters during the research. Instead, emphasis was placed on methods, data generation and experimental design, and thus on a researcher-managed process. Studying farmers became more important than interacting with them. As a result, adoption became the overriding goal, while adaptation was seen as something farmers did late in the technology generation process, only after a recommendation had been provided to them.

In addition, OFR became oriented towards tailoring recommendations for the use of developed technology in specific farming conditions. The mounting concern over the years to incorporate extension into OFR (hence 'farming systems research and extension' or 'FSR/E') reflected the definition of 'farmer participation' as 'farmer *adoption*' of developed technology. Research institutions that incorporated FSR/E typically defined its role as a refinement of agricultural extension. On-farm researchers were under pressure from production scientists who expected that FSR would increase the adoption of already developed technology. Thus the earlier approach, based on a respect for farmers' knowledge and experience and a recognition of their role as 'folk' researchers and inventors, was replaced by a transfer-of-technology model, developed in Europe and North America, in which the central players were professors, researchers and technocrats (Farrington and Martin, 1986; Chambers et al., 1989).

OFR directed at increasing the adoption of developed technology (that is, the transfer-of-technology approach) has developed a style of farmer participation which has been described as 'consultative' (Ashby, 1986; Biggs, 1989). A different style of farmer participation is required when OFR emphasises the hands-on involvement of farmers in technology design and development. This participation involves researchers and farmers in collaborative decision making, with shared control and responsibility (Ashby, 1986; Biggs, 1989). Consultative participation is by far the most predominant mode of involving farmers in OFR today (Biggs,

1989). There are many reasons for this, not least that decision-making participation is still in its infancy in terms of tried-and-tested methodological tools.

Consultative farmer participation usually involves on-farm researchers in direct interaction with farmers, which generates important information to orient the objectives and priorities of an OFR effort. For example, farmers are consulted as key informants in formal and informal surveys, their opinions of trials may be sought, and they are actively involved in trial management. However, the *researchers* set the priorities, design the trials and take most of the decisions. A good example of consultative participation is provided by Kean (1988), who describes a two-stage approach implemented by the Adaptive Research Planning Team (ARPT) in Luapula Province, Zambia. The approach involves reporting the results of on-farm trials to farmers, for them to discuss as a group, and then obtaining farmers' suggestions on how the trials can be improved for the following season's programme. First, small-scale field days are held at each cluster of on-farm trial sites, for farmers who are managing trials and for their neighbours. The trial farmers explain the trials and lead the discussions. At a later stage, a more comprehensive group of farmers meets in the local primary school. Different 'clusters' voice their opinions of the trials and make suggestions. The local extension agent makes a record of the farmers' ideas, and this record is consulted when researchers discuss trial proposals for the following season.

In contrast, decision-making participation is based on the perception of farmers as decision makers as well as informants. Farmers take part in the formulation of research objectives and priorities and in assessing the desired features of technology design. In collaboration with researchers, they decide what to test, and when and how to test it. The three main characteristics which distinguish this type of participation from consultative participation are: it enables farmers to take part in decisions about the content of the research programme throughout the research process; it promotes the early involvement of farmers, at the stage when a technology is still 'notional'; and it encourages farmers to take more responsibility for site-specific technology testing and adaptation.

Decisions about research content

In order to involve farmers in decision-making, participatory OFR typically requires farmers to take part in the generation and analysis of some of the information needed for the research. For example, various methods of group discussion have been used to enable farmers themselves to analyse problems in their farming situation, to identify research opportunities, and so to prioritise possible interventions (Lightfoot et al., 1988; Quirós and Ashby, 1988). This participation is qualitatively different from the type of group session in which farmers are simply informed about an OFR programme, and their support is sought. Group analysis also differs from the group interview, in which the group generates information but does not participate in synthesising or drawing conclusions from this information. Participatory group analysis advances the participants' own understanding as well as that of the researchers. A feature of decision-making farmer participation is that it involves farmers in making adjustments or even fundamental changes in the research agenda, in the design of testing procedures and in treatments or non-experimental variables in trials.

Various experiences show that it is feasible to take farmer participation beyond discussion, towards the involvement of farmers in decisions about the design of trials. In Peru, for example, proposed experiments on parasite control in animals in the highlands had to be modified when farmers were unwilling to treat only selected members of the herd or to sell slaughtered animals so that laboratory work could be carried out (Fernández and Salvatierra, 1989). By mutual agreement between farmers and on-farm researchers, entire family herds were used for each treatment, and animals destined for slaughter were replaced. Thus, the whole experimental procedure had to be redesigned to enable farmers to make their own assessment of the effectiveness of the different treatments.

In another example, Fujisaka (1989) reports that several adaptations of soil conservation techniques by farmers were subsequently incorporated into researchers' experimental work at Claveria in the Philippines. From a list of potentially useful technologies, the farmers decided to experiment with terracing using grasses, which they had observed established terraces faster than using trees. Researchers then added to their agenda a study of root-feeding insects which might be exacerbated by grass terraces. When the farmers explained that faster establishment of terraces to control erosion was important to them, researchers responded by shifting from an almost exclusive emphasis on improved nutrient cycling to more research on erosion control. One of the farmers improved upon the arrangement of napier grass and trees in a terracing experiment, leading to a change in the way these experiments were conducted. Farmer participation also led to changes in nursery establishment techniques, after researchers had observed that whereas there was poor seed germination in their tree nursery, farmers had experienced no such problems. Investigations on the timing of seed gathering and sowing were initiated to explain the farmers' success and to improve nursery establishment techniques. Overall, the farmers' adaptations of the trial technologies in this research effort stimulated researchers to reformulate their agenda.

Decision-making farmer participation in OFR occurs only when researchers are willing to cede some of their control over the conceptualization of an experimental programme (that is, its objectives, hypotheses and treatments of interest). A few cases have been reported in which farmers were given the opportunity to design formal trials in response to a problem they perceived as important. These trials built on the farmers' ongoing informal experiments aimed at solving the problem in question. For example, Lightfoot et al. (1988) report that farmers in the Philippines took researchers to sites where they could show that shading inhibited the growth of cogon grass, an invasive weed that farmers considered to be a major problem. Bibliographic research supported the validity of the principle of shading to control this weed, and formal trials were designed based on this finding. In Colombia, farmers who designed fertilizer trials based their design on the principle of mixing organic and chemical fertilizers in order to obtain greater efficiency in terms of costs (Ashby, 1986). Their informal experimentation involved testing the same principle of mixtures of different sources. The fertilizer mixtures used by the farmers proved to be economically competitive with the treatments used in the researchers' fertilizer trials.

It is seldom recognized that decision-making farmer participation requires involving farmers formally in the planning of a trial programme from the start. Instead, where decision-making participation does become a feature of the research effort, it is usually because researchers establish a good rapport along the way with farmers who manage on-farm trials. As a result of good communication, often facilitated by a social scientist, researchers become informed about farmers' preferences as trials are implemented. Colfer et al. (1989) describe such an experience in rice-based cropping pattern trials in West Sumatra, Indonesia. The trials examined the effects of four soil fertility treatments on crop yield, and in those treatments that included rock phosphate or lime, farmers were asked to hoe their fields in order to incorporate these inputs into the soil. Some farmers managing the trials viewed hoeing as essential, while others were reluctant to use it. Eventually, it became clear that hoeing was a more significant variable in terms of crop yield than any of the soil fertility treatments, and the trial design was modified accordingly.

The examples outlined above illustrate that farmer participation in designing tests of proposed technological innovations not only stimulates adjustments in technology design and the formal research programme, but also, typically, incorporates elements of informal farmer experimentation into formal trials.

Early involvement of farmers in developing technology

Early decision-making participation by farmers can make significant contributions to the direction of technology development. For plant breeding programmes this may involve doing more research at the early stages on-farm, or allowing farmers more access to the experiment station.

One method for moving more plant breeding research on-farm is a 'menu' approach, where farmers are given the opportunity to choose which innovations to test. This approach was used to screen improved rainfed rice varieties in Uttar Pradesh, India (Maurya et al., 1988). Up to seven improved lines were selected by the breeding programme because of their similarities to local varieties in a given village. Numerous farmers in the village were given samples of the improved breeding lines; each farmer managed up to two of these samples in split-plot trials alongside his/her own rice crop. When the plants reached maturity, the farmers assembled and visited all the plots to evaluate them. The breeding lines selected in this way outyielded the local varieties on 56 out of the 59 farms where this approach was used. In the national bean breeding programme in Colombia, larger numbers of advanced breeding lines have been offered in nurseries of up to 100 materials planted on-farm for evaluation by groups of farmers participating in the programme (CIAT, 1990). Materials selected by the farmers have been used as parents, as well as for breeding to introduce traits desired by farmers into the breeders' nurseries and into the lines for eventual release. In on-farm trials conducted over three seasons, varieties selected jointly by farmers and breeders outyielded varieties selected only by breeders or by on-farm researchers, or by farmers experimenting on their own.

This menu approach implies much earlier and more significant farmer involvement in the research process than the consultative participation characteristic of much OFR, where farmers test and evaluate only a few alternatives that have already been assessed in researcher-managed on-farm trials. This requires methods to encourage farmer evaluation. Methods such as ranking and group evaluation techniques help farmers and researchers arrive at an effective summary of farmer reaction to experimental treatments (Ashby, 1990).

Efforts to involve farmers in evaluating technology may also be done while the technology is still evolving on the experiment station. When small farmers in Rwanda were invited to participate in the experiment station's seasonal selection of advanced potato clones, researchers found that they had ignored several features of tuber colour, size and shape which were of interest to farmers (Haugerud and Collinson, 1990). Farmer participation in the station screening of varieties showed that researchers were more conservative than farmers and had been unnecessarily rejecting some acceptable materials. Women, who normally play the major role in bean seed selection in Rwandan farm households, regularly take part in evaluating hundreds of advanced breeding lines in the national bean research programme's on-station trials. The varietal lines selected by these women have consistently outperformed the varieties released so far by the breeding programme (Sperling, 1988; Pachico et al., 1989).

Farmers' experimentation

An important objective of decision-making farmer participation is to improve farmers' own capacity to carry out testing and experimentation. This objective is a central feature of the Farmer Innovation and Technology Testing (FITT) programme in The Gambia, in which farmers are given control over on-farm testing and take the initiative in requesting trials from research programmes. Farmer groups and development agencies — often non-governmental organizations (NGOs) — are provided with a set of innovations proposed by the research programmes; the farmers then decide which technologies to test and are encouraged to adapt technologies where they consider this is necessary. The menu of technologies offered by development agencies to farmers include, for example, alternative intercropping patterns, hand-drawn weeders, storage technologies, live fencing and seed treatment, as well as improved varieties of various cereals, root crops and fruit trees. As FITT evolves, it is expected that farmer groups will take more initiative in determining the menu of technologies proposed for testing (Mills and Gilbert, 1989).

Lightfoot (1987) has suggested a method to take advantage of farmer experimentation. The first step is to identify research topics that farmers are interested in and on which they are currently experimenting. Then participants are chosen, either from the farmers already engaged in experimentation or by selecting a sample group of farmers. The researcher distributes varieties or other inputs to be included in the experiment, but allows the farmers to manage them as they see fit. The farmers' research is then carefully monitored. This requires several visits to the field, informal conversations with the farmers during visits and the recording of key information on experimental management and results. When applied in the Philippines, this method yielded valuable information about varietal characteristics. In one case, farmer experimentation with sweet potato varieties showed the importance of criteria other than yield, such as rapid vining to prevent erosion and the capacity for prolonged sequential harvesting. In another case, farmers were able to adapt an upland rice variety to lowland fields.

Once farmers understand a problem, they may be the most appropriate source of solutions. In a project in Honduras (Bunch, 1990), trials were planted to investigate the use of mucuna (velvet bean) as a cover crop. The dilemma was that late planted mucuna did not produce enough biomass, but early planted mucuna required considerable management. Once farmers had seen the trials and were provided with seed, they developed the most appropriate way of managing the crop. The same farmers who carried out the experimentation were then used as extension 'agents' to communicate with other farmers in the project.

There are certain advantages and disadvantages to placing more responsibility for experimentation in the hands of farmers. A commonly voiced concern is the possible sacrifice in the quality of data, reducing researchers' ability to draw conclusions based on statistical analysis. This is an important concern, but it has not been proved that farmers necessarily compromise statistical validity when they set up and run farm trials. It must be balanced against the fact that a technology will ultimately be accepted or rejected by farmers not on the basis of statistical analysis but rather on farmers' own criteria and evaluation methods. Baker (1991) provides examples from Botswana and Cameroon that show that giving farmers flexibility in experimental design and management can provide additional information to researchers and in many cases still allow for rigorous analysis of experimental results.

IMPLEMENTING DECISION-MAKING FARMER PARTICIPATION

Philosophically and practically, decision-making participation in OFR is difficult to implement. It implies certain risks and costs for formal research. Farmer experimentation and adaptation of technology may appear intuitive and haphazard. Although many studies have shown that indigenous knowledge systems are based on careful observation, taxonomy and hypothesis testing (Brokensha et al, 1980), there are still serious gaps in farmers' information and their abilities to interpret it (Tripp, 1988; Bentley, 1989).

If farmers, working with incomplete information, are introduced to new ideas or notional technology of which they have no concrete experience, how can they participate effectively in designing, adapting and evaluating these ideas? At best, won't the farmer-participant bring ignorance to bear on the research? At worst, might farmer participation foster suspicions and prejudice that will jeopardize future acceptance and adoption? Researchers feel concerned that involving farmers in technology development at a stage when many applications are not well defined, when many options still need to be screened and when recommendations cannot be given will simply result in farmers losing confidence in the research systems.

There are also worries about the additional cost of managing a decision-making type of farmer participation if this involves researchers in more intensive interaction with farmers. How to scale-up farmer participation to achieve broad coverage of a large number of farmers without incurring excessive expenses and compromising the quality of participation is a key issue which has to be resolved.

Although at present there exists no empirical assessment of the cost-effectiveness of farmer participation in OFR, experience shows that when farmers participate in design and adaptation early on in the development of technology, the end product is more readily accepted by other farmers. In addition, there are several elemental components of a strategy for implementing farmer participation which contribute to efficient use of resources. These include: a reorientation of extension; a diagnosis of the cultural interpretation of participation at the research site; building the capacity for OFR in farmer organizations; and decentralising existing research systems to include cost-sharing with the private sector in particular in the local farm community.

Extension and education

One remedy for farmers' incomplete knowledge is to provide them with the information and skills they need to assess a new idea, principle or application. Effective farmer participation in OFR requires non-formal education and training of farmers. This training differs from conventional extension in that its objective is not to transmit information to farmers about a recommendation or developed technology but to provide them with insight into ideas, principles and methods which will help them test and evaluate potential innovations more effectively.

In his discussion on implementing this strategy, Bunch (1982, 1989) argues that teaching farmers how to experiment, to teach innovations to each other and to improve on those innovations by themselves is essential to sustainable agricultural innovation. He cites examples of farmer experimentation with contour bunds to control erosion, with planting distances, with intercropping, with pest control and with tools and machinery in programmes which have stimulated farmer-initiated research in this way.

Bunch's approach is based on the idea that researchers narrow down three to six innovations which are potential solutions to farmers' problems and have been identified by participatory group analysis. These alternative innovations are tested by farmers who are taught how to conduct experiments and keep farm records. Once these farmers gain experience, sophistication and confidence they are equipped to test another stage of technology introduction. Gradually, 'an inverted pyramid of technology' is built as further innovations are added for testing over time. Once an innovation is shown to be successful using this approach, Bunch recommends teaching small-scale experimentation as an extension technique because, in this way, more farmers learn an attitude of experimentation and a method of scientific enquiry, a cornerstone of self-sustaining farmer participation in agricultural development (Bunch, 1982). The Participatory Research in Agriculture (IPRA) project at the International Centre for Tropical Agriculture (CIAT) in Colombia is implementing a similar approach, involving committees of experimenting farmers formed by producers' organizations. Preliminary results from this project show that farmers with primary schooling can master the principles of controlled comparison, replication and random assignment and apply them in setting up and running their own trials (CIAT, 1991).

Another project which illustrates this approach is the Integrated Pest Control Program for Tropical Asian Rice. Important objectives of this project are to provide training for farmers in the principles of how and when to use pesticides, and to build knowledge and skills in the farming community for experimenting with various pest control methods (Biggs, pers. comm.). In the PROGETAPPS project in Guatemala, extension agents have made the shift away from teaching research recommendations to farmers towards stimulating farmer participation in decision making. Farmers still participate in field trials and field days where they acquire information about proposed innovations but, in addition to this, consultative groups (known locally as *grupos de consulta*) involve farmers in making decisions about the planning of these activities. These groups usually consist of about 30 people, all of whom are members of local farmer associations. The research-extension team which is responsible for a given region or area presents its proposed work-plan for the forthcoming season and the farmers' consultative group reviews and revises this plan in order to ensure its relevance to local problems (Ortíz et al., 1989).

Thus, decision-making participation involves a reorientation away from conventional concepts of agricultural extension. It requires farmers to be informed about possibilities and options, not only certainties, and to master principles of experimentation (such as controlled comparison) and elementary data collection and analysis skills. This is not as formidable as it sounds because numeracy is often widespread among even illiterate rural peoples; data management skills can be exercised by farmers using symbols and representational methods, to compare treatments in a trial, for example. Nonetheless, this approach does require a different style of extension. Extension agents must cease to be technology salespersons and become genuine educators if farmers are to take part in decision-making about research and the adaptation of technology, not just its adoption.

Culture and social protocol

As illustrated at the beginning of this chapter, the professional culture of on-farm researchers and extension agents defines farmers as adopters (Chambers and Jiggins, 1986). Indeed, in most of the cultures in which these professionals and their clients, low-income farmers, live and work there are long-standing traditions of autocratic and hierarchical social relationships, not necessarily compatible with participatory relationships. Most researchers, extension agents and farmers are accustomed to interact with each other as supervisors or subordinates, not as colleagues. Thus, decision-making participation can be seen very much as a western, democratic concept, alien to other cultures.

However, many rural societies have traditions of mutual self-help, co-operation and innovation in farming (Brokensha et al., 1980). Participatory research needs to be embedded in appropriate indigenous cultural and social institutions. In order to achieve this, a diagnosis of the cultural context, social protocol and local institutions needs to be carried out to determine how best to implement farmer participation. Stone (1989) reports such a study carried out in Nepal to assess the potential for community participation in resource conservation. The study found that there was a wide gap in the perception of community participation between villagers and development project staff. In particular, it identified the importance to the villagers of personal, hierarchical relationships based on mutual dependencies and exchanges of goods and services, which contrasted with the self-help, participatory development philosophy. Workshops for villagers and project staff were conducted to disseminate information, to initiate group planning and to make participants aware of the differences in their values and attitudes. As a result, villagers began to change their expectations and the types of demands made on the project.

Farmer organizations and costs of increased farmer participation

The need to provide non-formal education for farmers through reoriented extension approaches and to carry out a careful diagnosis of the sociocultural and institutional context imposes new burdens and additional costs on already overstretched public sector agricultural research and extension institutions. In spite of these formidable requirements, however, increasing farmer participation in OFR is advocated, for two reasons. First, the anticipated savings to public sector OFR and to farmers from more efficient technology development should compensate for these additional costs. The second, and generally more powerful, reason for promoting farmer participation is that it creates the possibility for public sector research and extension to externalise some of their costs to farmers and private sector institutions.

Decision-making farmer participation aims to mobilise farmer adaptation, and the resources of time and other inputs it represents, as early as possible in the development of a technology. Therefore, farmers are expected to continue to incur the costs of experimentation and adaptation, but as an integral part of the OFR process, instead of separate from it. This raises the question of who can then afford to participate in OFR. Won't this kind of participation become the property of the well-endowed farmer who can afford to experiment? Might this re-introduce a bias in favour of the rich, a bias that OFR was promoted to ameliorate?

Group participation is being used increasingly to overcome possible bias in favour of the better-off farmer. There are several examples of both formal groups (such as co-operatives) and informal groups (such as family networks, credit societies or neighbourhood associations) mobilising resources for group activity and spreading these resources among the poor as well as the better-off farmers. Farmer organizations in the highlands of Ecuador, for instance, have initiated their own research and extension programmes (Bebbington, 1991). Farm trials, field days, seed multiplication and input distribution is controlled and implemented by farmers. Several farmer organizations use their own resources to hire a formally trained agronomist, and conduct their own courses to train indigenous (farmer) extension agents. These activities are financed by commissions on inputs supplied through the farmer organization. Fujisaka (1989) describes how informal groups of farmers in one community in the Philippines host farmers from another community to teach them their innovations in soil conservation. Farmer-to-farmer training is seen as one means of achieving effective communication about new ideas and practices to stimulate adaptation of technology among farmers, at relatively low cost.

In addition to farmer groups, NGOs are becoming increasingly active in raising funds for farmer participation in all areas of rural development, including agriculture (Farrington and Biggs, 1990). Some non-formal education functions, and the socio-cultural diagnosis and sensitization cited earlier as requirements of decision-making participation, are being carried out by NGOs with farmers. An example of this is provided by the intervention of NGOs in agroforestry research in East Africa. Until recently, farmer involvement in formal agroforestry research programmes in this region took place only after researcher-managed on-farm testing had advanced the design of agroforestry practices to a virtually finished technology. In 1988-89, NGOs became extensively involved in monitoring farmers' own experiments and indigenous agroforestry practices, incorporating this information into NGO-sponsored trials and demonstration plots (Farrington, 1989). Whereas farmer experimentation and indigenous knowledge was isolated from formal agroforestry research, NGOs were quick to build these elements into their programmes. With alley farming, NGO projects have found that although farmers adopt the basic idea, they use species other than those which have been recommended and, instead of planting closely spaced rows of trees, have developed a practice of dispersed intercropping of trees which has proved to be much more widely adopted than the recommended practices. NGOs have enlisted farmers as 'local researchers' to help screen and adapt a menu of agroforestry species and practices (Kerkhof, 1990). This experience demonstrates how formal research systems could use their resources more efficiently by providing a menu of unfinished technologies to farmers via grassroot organizations which share with farmers the cost of the adaptive research to produce finished technology.

However, this approach still requires prior screening of technology components in order to develop a menu of manageable size. Here, one way of using farmer participation to help improve the efficiency of research is to provide farmers with the opportunity to contribute to the design of technology prototypes. Hands-on contact with prototypes by users is a strategy used to develop successful innovations in the private sector and can be applied in agricultural research (Nickel, 1988). Farmers could be invited to visit the experiment station to see plant breeders' ideotypes, for example, and be informed about the objectives of plant character improvement, with concrete illustrations given in terms they can understand (Ashby et al., 1987; Sperling, 1988). Giving farmers access to prototypes can be expensive, however, and prototype testing by users will require farmers and other interested parties to bear some of the costs of prototyope development.

An example of the benefits of involving farmers in developing prototypes is the small mechanical bean thresher developed by the IPRA project and the Seed Unit at CIAT in Colombia with farmer participation. The researchers' prototype was faster and more efficient than the one the farmers developed together with the engineers and it threshed the grain more cleanly. However, the farmers' prototype had a more maintenance-free structure, was easier to transport in hilly terrain and was cheaper (US\$ 420, as opposed to US\$ 1200 for the researchers' prototype). Most important, the farmers' prototype could cope with greener, moister bean pods for threshing (crucial to a rapid harvest in an area where drying is difficult) whereas the researchers' prototype rapidly came to a stop unless the bean pods were thoroughly dry. The farmers' prototype is diffusing rapidly to several countries (Roa et al., 1991).

Implementing increased farmer participation and responsibility in adaptive OFR is a means of devolving a significant proportion of this activity to institutions other than public sector organizations. To effectively integrate farmers into the process of technology adaptation, a devolutionary institutional model should facilitate interaction at many points between farmers, intermediate organizations, private industry, researchers and extension-educators (Merrill-Sands et al., 1989; Merrill-Sands and Kaimowitz, 1990). This is a challenging requirement, and evidence has yet to be obtained on how appropriate, sustainable and cost-effective such an institutional model is likely to be for research management.

CONCLUSION

Farmer participation aims to integrate farmers' ideas, opinions and adaptations into research. Participation in OFR has traditionally been limited to the consultative type, with farmers acting as informants and evaluators of relatively developed technology. But there are good reasons for involving farmers in making decisions together with researchers about what to test and how to test it. This does not imply turning all agricultural research over to farmers, whose objectives and short-term horizon for planning may be quite antithetical to the long-term needs of basic and applied research. Decision-making participation does, however, require the involvement of farmers in planning adaptive OFR and in pre-screening technology on experiment stations, an approach which has received relatively little attention. It also means that farmers must have access to information, skills and technology prototypes which are conventionally considered too sophisticated or risky for farmers to manage. The devolution of control of OFR implied by decision-making participation transfers some of the costs of research to farmers, but it offers the prospect of higher returns through more rapid development of relevant technologies.

References

Ashby, J.A. 1982. Technology and ecology: Implications for innovation research in peasant agriculture. *Rural Sociology* 47: 234-50.

Ashby, J.A. 1986. Methodology for the participation of small farmers in the design of on-farm trials. *Agricultural Administration* 22: 1-19.

Ashby, J.A. 1990. *Evaluating Technology with Farmers: A Handbook.* Cali, Colombia: CIAT.

Ashby, J.A., Quirós, C.A. and Rivera, V. 1987. *Farmer Participation in On-Farm Varietal Trials.* Agricultural Administration Network Discussion Paper No. 22. London, UK: ODI.

Baker, D. 1991. Reorientation, not reversal: African farmer-based experimentation. *Journal for Farming Systems Research-Extension* 2 (1): 125-47.

Bebbington, A. 1991. *Farmer Organizations in Ecuador: Contributions to Farmer First Research and Development.* Gatekeeper Series No. 26. London, UK: International Institute for Environment and Development.

Bentley, J.W. 1989. What farmers don't know can't help them: The strengths and weaknesses of indigenous technical knowledge in Honduras. *Agriculture and Human Values* 1989 (Summer): 25-31.

Biggs, S.D. 1980. Informal research and development. *Ceres* 13: 23-26.

Biggs, S.D. 1989. *Resource-Poor Farmer Participation in Research: A Synthesis of Experiences from Nine National Agricultural Research Systems.* OFCOR Comparative Study Paper No. 3. The Hague, Netherlands: ISNAR.

Biggs, S. and Clay, E.J. 1981. Sources of innovations in agricultural technology. *World Development* 9: 321-36.

Brokensha, D., Warren, D.M. and Werner, O. (eds). 1980. *Indigenous Knowledge Systems and Development.* New York, USA: University of America Press.

Bunch, R. 1982. *Two Ears of Corn. A Guide to People-Centered Agricultural Improvement.* Oklahoma, USA: World Neighbors.

Bunch, R. 1989. Encouraging farmers' experiments. In Chambers, R., Pacey, A. and Thrupp, L.A. (eds) *Farmer First: Farmer Innovation and Agricultural Research.* London, UK: Intermediate Technology Publications.

Bunch, R. 1990. *Low Input Soil Restoration in Honduras: The Cantarranas Farmer-to-Farmer Extension Programme.* Gatekeeper Series No. 23. London, UK: International Institute for Environment and Development.

Cardona, C., Prada, P., Rodriguez, A., Ashby, J. and Quirós, C.A. 1991. *Bases para Establecer un Programa de Manejo Integrado de Plagas de Habichuela en la Provincia de Sumapaz, Colombia.* Working Document No. 86. Cali, Colombia: CIAT.

Chambers, R. and Jiggins, J. 1986. *Agricultural Research for Resource-Poor Farmers: A Parsimonious Paradigm.* IDS Discussion Paper 220. Sussex, UK: IDS.

Chambers, R., Pacey, A. and Thrupp, L.A. (eds). 1989. *Farmer First: Farmer Innovation and Agricultural Research.* London, UK: Intermediate Technology Publications.

CIAT. 1990. *Farmer Participation in Technology Design and Transfer.* Report of a special project, April 1990. Cali, Colombia: CIAT.

CIAT. 1991. *Institutionalizing Local Leadership for Farmer Participation in Agricultural Technology Generation and Transfer in Rural Communities.* Report of a special project, July 1991. Cali, Colombia: CIAT.

Colfer, C.J.P., Agus, F., Gill, D., Sudjadi, M., Vehara, G. and Wade, M.K. 1989. Two complementary approaches to farmer involvement: An experience from Indonesia. In Chambers, R., Pacey, A. and Thrupp, L.A. (eds) *Farmer First: Farmer Innovation and Agricultural Research.* London, UK: Intermediate Technology Publications.

Doorman, F. 1983. *Adopción y Adaptación en el Uso de Tecnología en la Producción de Arroz. Resultados del Estudio de Casos en el Cultivo de Arroz entre Pequeños Productores en la Región de Nagua.* Leiden, Netherlands: IDC Microform Publishers, Wageningen Agricultural University.

Ezumah, H.C. 1989. Marketing improved varieties attractive to farmers. *IITA Research Briefs* 9(4): 6.

Farrington, J. 1989. Farmers participatory research: Institutions, approaches and methods. Paper presented at the CIMMYT East Africa Technical Network Workshop on On-Farm Research, Arusha, December 1989.

Farrington , J. and Martin, A. 1986. *Farmer Participation in Agricultural Research: A Review of Concepts and Practices.* Agricultural Administration Unit Occasional Paper No. 9. London, UK: ODI.

Farrington, J. and Biggs, S.D. 1990. NGOs, agricultural technology and the rural poor. *Food Policy* 15: 479-91.

Fernández, M. and Salvatierra, H. 1989. Participatory technology validation in highland communities of Peru. In Chambers, R., Pacey, A. and Thrupp, L.A. (eds) *Farmer First: Farmer Innovation and Agricultural Research.* London, UK: Intermediate Technology Publications.

Fujisaka, S. 1989. A method for farmer-participatory research and technology transfer: Upland soil conservation in the Philippines. *Experimental Agriculture* 25: 423-33.

Harwood, R. 1979. *Small Farm Development. Understanding and Improving Farming Systems in the Humid Tropics.* Boulder, USA: Westview Press.

Haugerud, A. and Collinson, M.P. 1990. Plants, genes and people: Improving the relevance of plant breeding in Africa. *Experimental Agriculture* 26: 341-62.

Kean, S.A. 1988. Developing a partnership between farmers and scientists: The example of Zambia's Adaptive Research Planning Team. *Experimental Agriculture* 24: 289-99.

Kerkhof, P. 1990. *Agroforestry in Africa. A Survey of Project Experience.* London, UK: Panos.

Lightfoot, C. 1987. Indigenous research and on-farm trials. *Agricultural Administration and Extension* 24: 79-89.

Lightfoot, C., de Guia Jr., O., Aliman, A. and Ocado, F. 1988. Participatory method for systems-problem research. Rehabilitating marginal uplands in the Philippines. *Experimental Agriculture* 24: 301-09.

Maurya, D.H., Bottrall, A. and Farrington, J. 1988. Improved livelihoods, genetic diversity and farmer participation: A strategy for rice breeding in rainfed areas of India. *Experimental Agriculture* 24(3): 311-20.

Menz, K. and Knipscheer, H.C. 1981. The location specificity problem in farming systems research. *Agricultural Systems* 7: 95-103.

Merrill-Sands, D. and Kaimowitz, D. 1990. *The Technology Triangle: Linking Farmers, Technology Transfer Agents and Agricultural Researchers.* The Hague, Netherlands: ISNAR.

Merrill-Sands, D., Ewell, P., Biggs, S. and McAllister, J. 1989. Issues in institutionalizing on-farm client-oriented research: A review of experiences from nine national agricultural research systems. *Quarterly Journal of International Agriculture* 28(3/4): 279-300.

Mills, B. and Gilbert, E. 1989. Agricultural innovation and technology testing by Gambian farmers: Hope for institutionalizing OFR in small country research systems. Paper presented at the FSR/E Symposium, Fayetteville, Arkansas, USA, October 1989.

Nickel, J.L. 1988. Excellence in agricultural research. *Agricultural Administration and Extension* 28: 43-58.

Ortíz, R., Ruano, S. and Juarez, H. 1989. Closing the gap between research and limited-resource farmers: A new model for technology transfer in Guatemala. Paper presented at the ISNAR Workshop on Making the Link between Agricultural Research and Technology Users, The Hague, Netherlands, November 1989.

Pachico, D., Voss, J. and Woolley, J. 1989. On-farm research. In Schoonhoven, A.V. and Voysest, O. (eds) *Bean* (Phaseolus vulgaris *L.*) *Production in the Tropics.* Palmira, Colombia: CIAT.

Quirós, C.A. and Ashby, J.A. 1988. *Pasos en una Metodología para Investigación Participativa en Agricultura.* IPRA Projects Working Paper No. 6. Cali, Colombia: CIAT.

Rhoades, R.E. and Booth, R.H. 1982. *Farmer-Back-to-Farmer: A Model for Generating Acceptable Agricultural Technology.* CIP Social Science Department Working Paper No. 1/1982. Lima, Peru: CIP.

Roa, J.I., Ashby, J.A., Gracia, T., Guerrero, M. and Quirós, C.A. 1991. *Investigación Participativa en la Producción de Semilla Mejorada por Pequeños Agricultores.* IPRA Projects Working Paper No. 12. Cali, Colombia: CIAT.

Rogers, E.M. 1962. *Diffusion of Innovations.* New York, USA: Free Press of Glencoe.

Sperling, L. 1988. Farmer participation and the development of bean varieties in Rwanda. Paper prepared for joint Rockefeller Foundation/CIP Workshop on Farms and Food Systems, Lima, Peru, September 1988.

Stone, L. 1989. Cultural crossroads of community participation in development: A case from Nepal. *Human Organization* 48(3): 206-12.

Tripp, R. 1988. *Farmer Participation in Agricultural Research: New Directions or Old Problems?* Discussion Paper 256. Sussex, UK: IDS.

15

Integrating On-Farm Research into National Agricultural Research Systems: Lessons for Research Policy, Organization and Management

D. MERRILL-SANDS, S.D. BIGGS, R.J. BINGEN, P.T. EWELL, J.L. McALLISTER and S.V. POATS

About a quarter of the world's population produces food under difficult agroecological conditions, using few improved technologies or inputs (Wolf, 1986). Production is low and unstable, relying on uncertain rainfall and fragile soils which are vulnerable to degradation. Farm resources for agricultural production are often severely constrained and infrastructures for input supply and marketing are inadequate. Yet, many analysts argue that it is these resource-poor farm families, raising a wide range of crops and animals across highly diverse ecological conditions, who will constitute the key to meeting food demands and reducing poverty in the future.

The population of the developing world will swell by almost 1 billion in the next 10 years and is expected to double in 35 years. During the past 25 years, escalating demands for food and agricultural products were met through the expansion of cultivated lands and the intensification of production through improved technologies. Dramatic increases in productivity were achieved for a few key food crops — primarily wheat and rice — in the more uniform, better-endowed and, in many cases, irrigated areas of the developing world. This was the basis of the Green Revolution which brought food self-sufficiency and improved welfare to many developing countries, particularly in Asia.

During the past decade it has become increasingly clear that the task of keeping pace with population growth now and in the future will be even more daunting. Possibilities for expanding cultivated land are limited and in many of the favourable environments grain yields appear to have reached their biological ceilings (Wolf, 1986; Byerlee, 1988; Pingali and Moya, 1989; TAC, 1990). Yield stability and efficiency of input use can still be improved in the more favourable environments, but much of the increased production needed to meet rising food demand and generate income for the rural poor has to come from improved productivity and stability of the complex farming systems characteristic of 'resource-poor agriculture'.[1]

DEVELOPING TECHNOLOGIES FOR RESOURCE-POOR FARMERS

Developing technologies appropriate for resource-poor agriculture is perhaps the greatest challenge facing national agricultural research systems in developing countries. The challenge is twofold: technological and institutional. The technological side is complicated by three factors: the agroecological environments in which resource-poor farmers strive to produce are diverse; researchers often have an inadequate understanding of these environments; and the associated production systems are more complex, unstable and varied than those typical of commercial agriculture in the more favourable regions. Moreover, these production systems are driven by multiple goals — both commercial and subsistence — which means that farmers use a range of different criteria for evaluating and adopting technologies (Merrill-Sands et al., 1986).

To serve this client group, research has to develop a different approach from that used for the more stable and homogeneous conditions associated with 'Green Revolution agriculture' (WCED, 1987). In the past, research has focused on producing more standardized technologies of wide applicability across a range of broadly uniform conditions. With resource-poor agriculture, research is expected to produce multiple products tailored to the identified needs of diverse client groups and production systems. Thus, when a research system makes a commitment to addressing the needs of resource-poor farmers, it must be understood that it will have to generate a greater number of different technologies, adapt these to a wide range of conditions and evaluate them according to the broader range of criteria that these farmers use.

Institutional factors also complicate the research process when addressing resource-poor agriculture. The issue is not just one of producing more and different types of technology; it also involves producing the *right type of technology*. This means that researchers must understand their production systems and be responsive to the goals and priorities of the farmers they are serving. For commercial or Green Revolution agriculture, this is not so difficult. The influence of the market makes farmers' decision-making more uniform (Piñeiro, 1989). Researchers, who often come from such farm backgrounds, understand the goal of production in these systems and criteria for evaluating technologies. Information flows relatively unencumbered through the technology system. Farmers are well enough organized to express their needs and demand specific products, services and information from research. Extension services and private sector input suppliers, as well as farmers who have ready access to information, credit and inputs, take on much of the responsibility for adaptive research — fine-tuning research recommendations to suit their specific conditions.

Such conditions do not obtain for resource-poor agriculture. These farmers have limited access, either directly or through extension services, to information generated by research. Moreover, their limited capacity to tolerate risk reduces their capacity to experiment with new technologies. Perhaps most important, they are rarely well organized or powerful enough to put pressure on research systems to meet their demands. If relevant technology is to be developed, these farmers have to be given a 'voice' in the research process (Norman, 1980). Strong institutional links between research and resource-poor farmers have to be developed.

On-farm research: Linking research and its clients

On-farm research (OFR) has been developed as a means for linking research more closely with its clients, particularly resource-poor farmers. It strengthens links in three basic ways. First, it makes research driven more by demand (responding to clients' needs) than by supply (reflecting scientists' interests). It feeds information on farmers' production systems and their primary concerns back into research priority-setting, planning and programming so that research focuses on developing solutions to farmers' most pressing problems. Second, it performs several tasks that are critical to successful technology generation and transfer: problem diagnosis, design of solutions, adaptive research and testing, technology evaluation and fine-tuning of recommendations. For commercial agriculture, research often does not need to perform these tasks because other institutions or farmers themselves carry them out effectively. Third, it monitors changing problems in the field or in farmers' situations, allowing research to take corrective action or respond more quickly to farmers' evolving needs.

To better meet the needs of resource-poor farmers, many national agricultural research systems in developing countries have undertaken major OFR efforts, of varying scope and intensity. Over the past 15 years at least two-thirds of the systems in Asia and sub-Saharan Africa, and about half the systems in Latin America, have made substantial institutional investments in OFR. Donors have also supported these developments actively.

Through the concerted efforts of scientists in national and international research institutions, significant progress has been made in developing and refining OFR methods, and innovative work continues (Byerlee et al., 1980; Zandstra et al., 1981; Rhoades and Booth, 1982; Shaner et al., 1982; Hildebrand and Poey, 1985; Collinson, 1987; Chambers et al., 1989; Tripp and Woolley, 1989). Experience has also shown, however, that while sound methods are a *necessary* condition for effective OFR, they are not a *sufficient* condition. Institutional and policy factors have had a major impact on the quality of the research conducted and on the development and sustainability of OFR within national research systems. The development of OFR within a research system must be accompanied by appropriate adjustments in the policy, organization and management of that system.

Institutional implications of on-farm research

By making research more demand-driven, OFR brings about basic changes in the research process and the organization of work. It expands the number and types of tasks a research organization performs and the number of products it develops. It brings in new types of information and knowledge. It alters the flow of information within the system and requires the development of new links between diverse actors within the research-technology transfer system. Horizontal communication across disciplinary and commodity programmes has to be strengthened, and the vertical flow of information has to be broadened. Feedback from the field becomes as important as the flow of information out from research to technology users. It requires more decentralization and flexibility in decision-making and wider dispersion of researchers and research activities. It creates increased demands for transport — scarce in many national programmes — and for logistic support for researchers in the field, often far from experiment stations. And, very importantly, it requires the development of human resources — training staff in new skills and research methods or hiring new staff with these skills.

Organizational theory tells us that such basic changes in the work process of any institution, whether a research institution or manufacturing firm, will have major implications for its organization, management and institutional culture (Woodward, 1965; Hage and Aiken, 1969; Van de Ven and Delbecq, 1974; Aldrich, 1979; Zey-Ferrell, 1979). For example, when private sector technology research and development companies, such as Hewlett Packard or Millikin, revamped their strategies to give top priority to client responsiveness, radical organizational and managerial changes were instituted (Peters and Waterman, 1984; Peters, 1987). In agricultural research, however, the institutional challenge of introducing research aimed at clients, particularly poor clients, has generally been underestimated. It is only in recent years that our attention has focused on the difficulties in the implementation and institutionalization of OFR in national agricultural research systems (Collinson, 1982, 1988; Norman, 1983; Chang, 1984; Biggs, 1985; Heinemann and Biggs, 1985; Norman and

Collinson, 1985; Abalu et al., 1988; Baker and Norman, 1988; Merrill-Sands, 1988; Chambers et al., 1989; Tripp et al., 1990).

National research managers have faced significant problems in organizing and managing this type of research. In some cases, institutional problems have resulted in the marginalization or stagnation of OFR efforts. In others, they have encouraged national scientists or donor representatives to keep OFR under the protection of externally funded donor projects, isolated from the mainstream of national agricultural research. In either case, the impact of OFR becomes limited and transitory.

In response to research managers' and donors' growing concern that institutional problems were undermining the effectiveness of OFR, the International Service for National Agricultural Research (ISNAR) has conducted a major study to analyse the institutional factors affecting the integration of OFR within national agricultural research systems.[2] The objective has been to identify common problems and develop guidelines for research managers on how to address the problems through improved research policies, organization and management. The study, funded by the Government of Italy and the Rockefeller Foundation, was carried out in collaboration with national agricultural research systems in nine countries (*see* Table 15.1).[3] In all these cases, OFR had been integrated into the national research institutions and had been carried out for at least 5 years. In addition to these criteria, the cases were selected to include a variety of organizational models and a balanced distribution across regions.

The study has generated several policy and institutional lessons about organizing and managing OFR so that it is effective and efficient in developing relevant technologies for resource-poor farmers and is incorporated as a stable and integral part of the research process. This chapter highlights the most important of these lessons. The discussion is not intended to be comprehensive. Each issue addressed is analysed in depth in publications from the study (Ewell, 1988, 1989; Merrill-Sands and McAllister, 1988; Biggs, 1989; Bingen and Poats, 1990). These papers, as well as the country case studies, also extract innovative solutions developed by managers and scientists in national agricultural research systems. Only some of these can be summarized here. The important point is that the lessons presented are based on a systematic and detailed analysis of over 25 specific experiences of conducting OFR in nine countries; they are not the product of armchair analysis or conventional wisdom.

The summary of lessons for integrating OFR successfully into national agricultural research systems is organized around four main themes:

- policy factors affecting OFR;

- organizational considerations;

- management of institutional linkages;

- management of research and resources.

POLICY FACTORS

Development policy commitment to resource-poor farmers

The ISNAR study confirmed that a strong national development policy commitment to resource-poor farmers is a key factor determining the successful institutionalization of OFR and its sustainability within national agricultural research systems (Merrill-Sands and McAllister, 1988; Biggs, 1989). One can argue that this is a

Table 15.1   Descriptive indicators of the nine OFR cases in the ISNAR study

Case studies	National agricultural research systems (NARS)		Organization of OFR	Years in operation[1]	Scale of OFR (scientist years)		
	Institutional type	Organization of research programme			OFR as % NARS human resources	Size of OFR effort
Ecuador	Semi-autonomous institute (INIAP)	Regional research stations/commodity programmes	Production Research Program (PIP)[2]: National programme with 2 coordinators and 10 teams based at regional research stations	9	6	14
Guatemala	Semi-autonomous institute (ICTA)	Regional research and commodity programmes	Technology Testing Department with 14 field teams in 6 regions and national socioeconomics department with limited regional representation[3]	14	34	65
Panama	Semi-autonomous institute (IDIAP)	Commodity programmes/ regional offices	National OFR plan identified target regions where OFR is implemented through special FSR projects or part-time OFR	7	16	24
Senegal	Semi-autonomous institute (ISRA)	Multi-commodity departments/ regional stations	OFR, located within Department of Production Systems Research and Technology Transfer (DRSP)[4], consists of 3 regional teams and a Central Systems Analysis Group	4	13	22
Zambia	Ministry (MAWD)	Commodity and factor programmes	OFR programme with national coordinator and 7 provincial teams at regional stations	6	20	38[5]
Zimbabwe	Ministry (MLARR)	Commodity- and disciplinary-based institutes and stations	OFR implemented by: 8 research institutes/stations with combined on-station/OFR programmes; and Farming Systems Research Unit (FSRU) based at central station with 2 regional teams	6	18	26
Bangladesh[6]	Semi-autonomous institute (BARI) of larger NARS with council	Disciplinary depart-ments/commodity programmes	On-Farm Research Division (OFRD), with Central Management Unit at headquarters and 24 teams deployed through BARI's network of regional stations, has official mandate for OFR; consolidation of previous OFR efforts	9[7]	12	104

Table 15.1   (continued)

| Case studies | National agricultural research systems (NARS) | | Organization of OFR | Years in operation | Scale of OFR (scientist years) | |
	Institutional type	Organization of research programme			OFR as % NARS human resources	Size of OFR effort
Indonesia[8]	Ministry, Department of Research (AARD) with many institutes and co-ordinating bodies	Commodity-based regional institutes	Two principal modes of implementation: research institutes conduct OFR as part of regular programmes; and OFR projects organized at AARD level with staff seconded from multiple institutes	11[9]	n.a	57[10]
Nepal[8]	1. Ministry	1. Commodity programmes/ disciplinary departments	1. Farming Systems Research and Development Division (FSR&DD) with 6 FSR sites, supported by Socio-Economics Research and Extension Division (SERED); and commodity programmes with multi-locational and outreach programmes	14[11]	n.a.	35[10]
	2. LAC and PAC, externally funded autonomous institutes	2. LAC: Multi-disciplinary research efforts PAC: Disciplinary departments	2. LAC and PAC, regional institutes with OFR as a generalized research strategy			

Notes:
1. Base year for all statistical data is 1986.
2. Programa de Investigación en Producción.
3. The Spanish names for these departments are Prueba de Tecnología and Socioeconomía.
4. Départment de Recherche de Systèmes de Productions et Transfer de Technologies en Milieu Rural.
5. Includes six research-extension liaison officers seconded from extension.
6. The case study is limited to the Bangladesh Agricultural Research Institute (BARI), the largest of the five institutes coordinated by the Bangladesh Agricultural Research Council (BARC).
7. Refers to national agricultural research systems (NARS). Several OFR programmes with complex histories operate within BARI. The oldest, the On-Farm Fertilizer Programme, dates back to 1957. This programme was reorganized in the late 1970s, about the same time that cropping systems research was established at BARI. The OFRD was not formally consolidated until 1984.
8. The data refer only to the subcase studies unless otherwise indicated; NARS-wide data are not available.
9. Refers to NARS. In 1973, multiple cropping research in the Central Research Institute for Food Crops (CRIFC) took on a systems orientation and was renamed cropping systems research; cropping systems research moved into farmers' fields in 1975.
10. Represents totals for subcase studies only. Not directly comparable to other NARS-wide data.
11. Refers to NARS. Cropping/farming systems research was initiated 9 years ago. On-farm research is 14 years old.

precondition for success in OFR. Development policy oriented towards resource-poor farmers exerts pressure on government organizations to redirect their research to meet this client group's needs and priorities. Indeed, in all the case study countries, OFR was either launched or strengthened in response to national policy pressures to target resource-poor agriculture in development.

Such pressures were seen to propel important changes in research policies favouring the integration of OFR within the case study institutions. In many cases, it stimulated a critical review of research priorities and recommendations. It provided incentives for researchers to understand resource-poor agriculture better and the means for them to move off the experiment stations to conduct on-farm work. And, with the emphasis on impact, it opened the door to the inclusion of social scientists in agricultural research and encouraged research organizations to develop stronger links with input supply and technology transfer agencies.

The evolution of OFR in Zimbabwe provides a good example of how development policy can foster OFR (Shumba and Fenner, 1989). Prior to independence in 1980, agricultural policy focused almost exclusively on the 4000 farmers in the large-scale commercial farming sector. With independence, the new government instituted radical changes and gave high priority to assisting the 850 000 farm families in the peasant farming sector. The Department of Research and Specialist Services (DR&SS) launched its OFR effort in direct response to this initiative. Within 6 years, the DR&SS had experienced a dramatic reorientation and important institutional changes (Avila et al., 1989; Shumba and Fenner, 1989). Nine of its 17 semi-autonomous research institutes or stations had developed major OFR efforts and about one-fifth of research staff time was now allocated to OFR. A small multidisciplinary Farming Systems Research Unit was established, social scientists were introduced into the DR&SS for the first time, and a high-level committee was created to coordinate research and extension activities conducted on-farm. Most importantly, a change in institutional culture was emerging. Researchers came to recognize resource-poor farmers as an important, and different, client group requiring different technologies.

Development policy changes can also have a negative impact on OFR. Vagaries in the policy environment and a government's attention to the resource-poor farm sector can destabilize OFR efforts and impede their effective integration in the research system. Dynamic OFR programmes in Ecuador and Panama floundered and eventually stagnated when governments shifted their development priorities away from this sector.

What can research managers do to sustain OFR efforts when policies shift away from resource-poor farmers? Although they may have limited influence on such policies, they do need to be ready to respond by: consolidating their programmes and network of field sites; using lower cost approaches; linking with other partners, such as non-governmental organizations (NGOs); working to maintain support to continue key research activities; and keeping a core staff of experienced researchers in place for when the policy climate becomes more favourable (Sompo-Cessay and Gilbert, 1989; Farrington and Biggs, 1990; Gilbert, 1990; Merrill-Sands and Kaimowitz, 1990). Perhaps most importantly, they can continue to lobby, using field data, for the importance of sustained attention to resource-poor agriculture and its potential contribution to development.

Cultivating support for on-farm research

Development policy can create a favourable environment for OFR, but it is senior research managers who have to translate such policies into research priorities and goals and promote these actively within their institutions. The case experiences show that without senior managers' commitment, OFR, still regarded as a new approach in many institutions, can easily become marginalized and reduced to little more than traditional technology testing. When research directors lose interest, senior researchers stop going to the field, vehicles are grounded, operating funds dwindle, on-farm researchers lose promotions and opportunities for training decrease.

The key lesson is that OFR leaders and scientists interested in sustaining vigorous programmes must give priority attention to cultivating and maintaining the support of senior managers in research and agricultural development institutions. They need to ensure that senior managers understand the role of OFR in the research process and its potential contribution. The OFR programme in Senegal did this by organizing institute-wide seminars and workshops (Faye and Bingen, 1989). In Ecuador, cultivating the support of senior managers, who changed frequently, was seen as a key responsibility of the national coordinator for OFR (Soliz et al., 1989)

In the longer term, however, it is *outputs and results* that convince policy makers, senior research managers and donors to continue supporting OFR. On-farm researchers have usually not paid enough attention to reporting their accomplishments. This has had high costs in loss of programme support. Some donors, operating under the erroneous assumption that OFR can deliver relevant technologies in 2-3 years, are now questioning its utility. More attention has to be given to demonstrating and reporting impact, whether intermediate impact in terms of reorienting the priorities of a commodity programme, or final impact in terms of farmers adopting new technologies. Gaining senior managers' support is especially important when OFR has initially been introduced through donor projects. Experience shows that, if internal support has not been cultivated, on-farm activities succumb to the 'special project syndrome' and are difficult to sustain when donor monies stop.

Several useful lessons for cultivating support can be drawn from the case studies:

- OFR efforts should focus on a limited and well-defined set of research problems to ensure high-quality research output;

- To ensure relevance, broad impact and efficiency in resource use, OFR should concentrate on farming systems or geographic target areas from which results can be extrapolated to a broad target group;

- In the early stages, it can be very useful to direct part of the OFR effort to solving easier research problems in higher-potential areas. Experiences from Guatemala, Panama, Nepal and Zimbabwe show that concrete results from adaptive research — revised recommendations or adopted technologies — produced in the short term can be crucial to developing broader support for the longer-term research efforts in the more complex environments (Martínez and Arauz, 1984; Ruano and Fumagalli, 1988; Avila et al., 1989; Kayastha et al., 1989; Cuellar, 1990). Great care must be taken, however, not to raise unrealistic expectations for adaptive research achieving short-term impact in the more marginal zones as well;

- OFR should generate specific products tailored to the needs of its various clients — policy makers, commodity or disciplinary researchers working on experiment stations, technology transfer agents and farmers. In Senegal, for example, the OFR staff cultivated policy support by distributing short, informal 'information notes' highlighting the policy implications of their research results (Faye and Bingen, 1989). In Zambia, annual research reports from on-farm teams included a specific section on the implications of experimental results and field observations for commodity research programmes;

- From the start, and throughout the research process, OFR needs to give higher priority to systematically monitoring farmers' reactions to new technologies and to conducting adoption studies.[4] The results should be communicated to senior research and development managers, national policy makers and donors. Such records of impact are especially important in institutions where senior managers change frequently.

Reliance on donor funding

Successful institutionalization of OFR is often equated with fully nationalized funding and staffing, but this is an elusive and, in many cases, unrealistic goal. Many research efforts, whether on the experiment station, in the laboratory or in farmers' fields, rely on external funding and research grants. Moreover, the financial situation, in terms of expenditure per researcher, is deteriorating in many national systems (Pardy and Roseboom, 1988). Donor support will continue to be important for the medium term; it is naive to assume that productive OFR can be carried out with local funds alone.

The goal should not be to free OFR from dependence on donor funds but to ensure there are sufficient funds and staff, whether national or external, to support a stable and productive programme which responds to national priorities and clients' needs. The key objective in institutionalization is that national scientists, not donors, control the research agenda. Obviously, this is not easy; donors have funds, power and clearly articulated interests. For OFR leaders to negotiate as partners, some basic conditions must pertain. They need to have well-defined research priorities based on a defensible assessment of clients' needs, coherent research strategies and methods, credible research results, a stable cadre of competent researchers and basic core funding from the national system.

The cases show that OFR managers have not given enough attention to shaping a research programme to attract donor funding and technical support; in too many cases they are running behind, moulding their research effort to meet donor interests. This is not to detract from the important role that some donor agencies have played in stimulating greater attention to resource-poor agriculture and supporting fledgling OFR efforts. Rather, it is to argue that national research managers must use donor projects not simply to implement research, but also to build the foundation for a strong and stable national OFR capacity. At the same time, donor projects should also have institution-building at the centre of their project objectives (Baker and Norman, 1988).

ORGANIZATIONAL CONSIDERATIONS

Because the organization of OFR is now recognized as central to its effective institutionalization in national research systems, the issue generates much debate (Collinson, 1982, 1986, 1988; Chang, 1984; Biggs and Gibbon, 1985; Norman and Collinson, 1985; Stoop, 1987; Abalu et al., 1988; Baker and Norman, 1988; Byerlee and Tripp, 1988; Gilbert et al., 1988; Tripp et al., 1990). The discussion centres around three key questions:

- Should OFR be organized as a separate programme or department, or incorporated as a research perspective which permeates the entire research system?

- Should researchers specialize in OFR or combine it with research conducted on experiment stations?

- Should researchers working on-farm reside at central research stations or in the target areas, closer to farmers?

No ideal model for organizing on-farm research

The answer to these questions is, unfortunately, 'it depends'. As unsatisfying as this may be, it is an important point and a key conclusion of the study. It means that there is no ideal model for organizing OFR (Heinemann and Biggs, 1985). As illustrated by the cases, various approaches have been used: establishing separate national departments or programmes, creating multidisciplinary regional teams, folding OFR into commodity programmes, or simply assigning OFR activities to individual scientists. Each model has distinct strengths and weaknesses, both in and of itself and within the institutional context. Organizational theory would predict this (Lawrence and Lorsch, 1967; Galbraith, 1977; Zey-Ferrell, 1979; Mintzberg, 1983); evidence from the cases and other experiences confirms it.[5]

When organizing OFR efforts, managers need to take into consideration:

- the *specific objectives and tasks* assigned to OFR and their relative importance;

- the *institutional context* of the research system in its policy, organizational and resource dimensions.

The objectives and tasks for OFR will determine which institutional links — those between livestock and crop scientists, those between on-farm and experiment station research, or field-level links between research and extension staff — should be given emphasis in organizational design. Grouping staff together within a unit facilitates communication and coordination and can be an important mechanism for linking tasks which are highly interdependent (Lawrence and Lorsch, 1967; Thompson, 1967; Mintzberg 1983; Bourgeois, 1990). It should be emphasized, however, that while different organizational models facilitate specific institutional links, managers can also use coordination mechanisms, such as joint planning meetings or collaborative research activities, to reinforce other important links (Merrill-Sands and Kaimowitz, 1990)

The institutional context is important because OFR, as a relatively new research approach, is generally being introduced into or strengthened within an existing research organization. Only in exceptional cases, such as Guatemala's Instituto de Ciencia y Tecnología Agrícolas (ICTA), has a research organization been designed around the concept of OFR. The institutional context affects the scale of activities and resources available, as well as the feasibility of integrating OFR. For example, experience shows that it is much more difficult to integrate OFR, which emphasizes the flow of information from the bottom up, into a highly centralized, hierarchical research institution which is organized precisely to facilitate the top-down flow of specialized information out to technology transfer agencies and farmers. Similarly, the way research programmes are organized, whether by commodities, disciplines or regions, also has an important bearing. The study showed, for example, that it is much easier to integrate OFR into regionalized research systems than into those organized along disciplinary lines. Although OFR, with its emphasis on multiple links and client contacts, never fits easily within a research organization, managers should try to make it as compatible as possible with the overall organization of research.

These general points can be illustrated by looking briefly at the comparative strengths and weaknesses of two common, but very different, models of organizing OFR: specialized regional on-farm teams, and individual scientists combining both experiment station and OFR activities in their work programmes.

Regional teams

The model of regional OFR teams, of varying size and complexity, has been widely used in national agricultural research systems. It has a number of key advantages: it promotes links at the field level with farmers and technology transfer agents; it facilitates geographical coverage; it encourages interdisciplinary collaboration; and it promotes the development of expertise and specialized OFR skills. Because regional teams can work with multiple commodities and across diverse farming systems, this model facilitates adaptive research in which component technologies are 'pulled down' to fit into specific farming systems, rather than 'pushed out' by commodity or disciplinary-based programmes (Norman and Collinson, 1985; Collinson, 1986).

The two most obvious disadvantages are that it sets up significant organizational barriers to integrating on-farm and experiment station research, and it can raise staffing problems unless clear incentives have been put in place. Senior researchers are often unwilling to live in remote areas and junior scientists may fear that such postings sidetrack them from normal career development paths within the institution.

The regional team model fits best in research organizations with broad mandates which have organized research by commodity programmes, regions or a combination of both. Also, research institutions should be large enough to have researchers specialize in OFR. It is often most appropriate for institutions which have technologies 'on-the-shelf' and need to define the demand for these technologies, adapt them to local conditions and promote transfer. In this context, the barriers to links with experiment station research are not as severe because although interaction with commodity or disciplinary specialists is necessary, it can be intermittent rather than on-going and intensive. Moreover, linkage problems with experiment station research can be minimized when on-farm teams are based at regional experiment stations with specialized commodity and disciplinary scientists.

Individual scientists

The second model, whereby scientists incorporate OFR activities into their on-going research programmes, has advantages and disadvantages which are almost directly opposite to those of the specialized regional team model. The principal advantage is that it promotes strong integration of on-farm and experiment station research. The scientist passes information efficiently back and forth between the field and the station. In addition, it does not require major institutional change and it can cultivate a stronger client-orientation throughout the entire research system (Biggs, 1983; Stoop, 1987; Byerlee and Tripp, 1988).

There are four main disadvantages. Unless strong coordination mechanisms are put in place, communication and collaboration across disciplines and commodities is inhibited. This undermines the use of a systems perspective in problem definition, priority setting and technology evaluation. Even more important, links with farmers and technology transfer personnel in the field tend to get slighted because of constraints on researchers' time and their lack of specialized skills, particularly in social science methods. Researchers' understanding of farmers' needs and priorities is often superficial or, at best, restricted to their particular discipline or commodity interest (Gibbon et al., 1989). Lastly, it can lead to duplication of effort and inefficiencies in the use of vehicles and operating funds because researchers, from numerous programmes or departments, each pursue their individual field activities. This is a cost few national programmes can afford.

The ISNAR studies indicate that this model is often most appropriate for research institutes with a narrow and clearly focused mandate, such as a single-commodity institute or regional research station with a well-defined regional mandate. The focused mandate facilitates coordination and reduces the volume of information and discrete OFR activities that have to be handled. The model can also be useful for institutions tackling difficult environments (Maurya et al., 1988), or addressing new commodities or client groups, for which few technologies are available. In this case, it is important to have senior researchers out in the field in regular contact with farmers early in the technology generation process. For small research systems, where staff resources are limited and specialization is a luxury they cannot afford, the individual scientists model is usually the only choice (Sompo-Cessay and Gilbert, 1989; Merrill-Sands and Kaimowitz, 1990).

For this model to work effectively, mechanisms have to be put in place to coordinate the diverse OFR efforts. Lumle Agricultural Centre, a small semi-autonomous regional research station in the highlands of Nepal, provides a good illustration. Of the approximately 16 researchers, 10 are involved in both on-farm and experiment station research (Kayastha et al., 1989). Their work is coordinated through several mechanisms: joint diagnosis, priority setting and planning in the field to design an integrated programme, implementation of trials at common field sites, monthly technical meetings for reviewing research, and periodic visits to the field (hiking together for 10-14 days!) to monitor on-going research activities. Informal communication among the scientists who reside together at the research station also greatly facilitates coordination.

Mixed strategy

For many medium-sized national agricultural research systems (150-300 researchers) with mandates covering a wide range of crops and agroecological zones, a model which mixes those described above shows considerable promise. OFR specialists, generally working in regional teams, are responsible for a broad range of farm-level activities with wide geographical coverage. At the same time, disciplinary and commodity scientists carry out a narrow range of on-farm trials near experiment stations and backstop the broader OFR efforts.

The key advantage of the mixed strategy is that it involves a large number of scientists, with varying levels of intensity, in OFR. This helps to foster a strong client-orientation in the research system as a whole. It can also minimize factions and reduce tensions which often arise between researchers working on farm and those working on station. Researchers working mainly on station become more aware of the complexities of conducting research on farm and more tolerant of the research methods, modes of analysis and evaluation criteria required. In turn, by promoting some specialization, the mixed strategy encourages researchers working mainly on farm to innovate, develop expertise and consolidate experience. While this model reinforces the strengths of both the regional team and individual scientist models, managers must bear in mind that it can be quite resource intensive and requires close management and strong coordination if it is to be effective.

Incremental build-up of on-farm research

Development policy, especially when reinforced by parallel donor interests, often exerts powerful pressure to expand OFR efforts rapidly in order to maximize geographical coverage. This problem emerged, sometimes acutely, in almost all the cases reviewed. Nothing can undermine an OFR effort more quickly. Research quality almost inevitably deteriorates as efforts become spread too thinly. Young researchers are sent to the field without supervision. OFR leaders spend too much time in administration and coordination to the detriment of scientific leadership. Moreover, expectations are raised which cannot be met. Policy makers, senior managers and donors can become impatient with the lack of progress and abandon OFR to seek new development approaches. Rapid expansion can also aggravate institutional jealousies as scientists in conventional research programmes perceive OFR to be siphoning away resources from their own efforts.

A key lesson is that research managers have to counteract pressures to produce 'too much too soon'. Research managers in Senegal and Zimbabwe, for example, were successful in taming the World Bank's overly ambitious plans for expansion (Avila et al., 1989; Faye and Bingen, 1989). They recognized the importance for the long-term sustainability of OFR of building up capacity gradually, consolidating skills and expertise, and giving priority to quality of research over quantity.

MANAGING INSTITUTIONAL LINKAGES

Successful OFR depends heavily on the effective management of four sets of institutional links: links with experiment station research, links with farmers, links across disciplines and commodities, and links with technology transfer. It is precisely this dependency which makes OFR a challenge to organize and manage.[6] This section highlights some of the key lessons to be drawn from the cases on managing these links. The discussion of links across disciplines focuses on integrating social science research, as this has been one of the most difficult aspects of OFR to institutionalize. A broader analysis of interdisciplinary collaboration is provided in Ewell (1988) and Bingen and Poats (1990).

Integrating on-farm and experiment station research

Interdependent research roles

A key lesson from the ISNAR study is that highest priority should be given to ensuring strong links between applied experiment station research and on-farm adaptive research. They have distinct roles in the research process but also depend on one another to be effective (Byerlee et al., 1980; Biggs, 1983; Fresco, 1984; Norman and Collinson, 1985; Collinson, 1987; Baker and Norman, 1988; Byerlee and Tripp, 1988; Merrill-Sands and McAllister, 1988). OFR should never be developed at the expense of applied research; a balanced build-up is essential.

OFR identifies niches for component technologies and adapts and tests them under the specific agroecological, socioeconomic and farm management conditions of particular client groups. It counts on applied research to provide a broad range of component technologies relevant for solving problems or exploiting opportunities identified in the field. In turn, applied research depends on OFR for evaluating technology performance under realistic farming conditions. On-farm and experiment station research also depend on each other for the exchange of specialized knowledge. OFR helps to ensure research relevance by feeding back information from its work with farmers in the field. Commodity and disciplinary specialists provide expertise for diagnosing farm-level constraints, designing solutions and analysing and interpreting research results (Merrill-Sands and McAllister, 1988).

Problems in managing links

That on-farm and experiment station research are interdependent is common sense. Nevertheless, the case experiences reveal that managers in national research systems have faced major problems in creating a strong partnership. Links have been easiest to forge with respect to adaptive research, the pulling down of technologies from the station and adapting them in the field. Some of the cases reported important successes in increasing the productivity of farming systems with adapted technologies (Ruano and Fumagalli, 1988; Kayastha et al., 1989). Nevertheless, these successes have tended to be confined to the more favourable environments for which technologies were available. Lack of appropriate technologies for marginal and more complex environments or the non-crop components of farming systems was cited as a major constraint in many cases.

OFR's vital feedback role in bringing clients' needs to bear on research priorities has been hardest to institutionalize. Feedback was assessed as weak in half the cases and serious conflicts had erupted during the development of OFR efforts in most of the cases reviewed (Merrill-Sands and McAllister, 1988). This finding is particularly disturbing as the cases represent relatively mature OFR efforts.

What are the sources of the tension that jeopardize this critical linkage? Conflicts erupt because of:

- differences in the research objectives and approaches;

- desire to control research agendas;

- poor understanding of the respective roles of on-farm and experiment station research (managers should anticipate these problems and develop strategies to address them).

Research approaches. The very factors which make on-farm and experiment station research complementary and interdependent create the potential for conflict (Arnold and Feldman, 1986; Merrill-Sands and McAllister, 1988). Differences in research objectives — for example, producing component technologies with wide adaptability across a range of environments versus adapting technologies to improve the productivity of a whole farm system in a specific environment — can lead to very different research agendas and priorities. Similarly, differences in methods, experimental designs, types of data collected and criteria for evaluating results can provoke fundamental disagreements about research priorities and what constitutes good science. Two common areas of controversy are the higher coefficients of variation and higher rates of trials loss typical of OFR. Disputes also often arise over the use in OFR of additional socioeconomic criteria to evaluate technological alternatives, such as returns to scarce labour, shortening of the hunger period, or simply farmers' assessments.

If managed effectively, such differences can lead to constructive debate which improves the quality and relevance of both on-farm and experiment station research (Rhoades and Booth, 1982; Arnold and Feldman, 1986; Kean and Singogo, 1988). Too often, however, such professional debate simply degenerates into unproductive infighting or studied avoidance.

Control over research agendas. Collaboration and exchange of advice, while leading to better quality research, also reduces researchers' scientific independence. Power struggles easily erupt when either feedback from OFR or technical support from experiment station research is allowed to become supervisory, rather than advisory. It is not uncommon, for example, for commodity researchers to fear that OFR, armed with clients' demands, will become the 'new boss' with the power to dictate research priorities. Similarly, if power shifts too much towards applied research programs, researchers working on-farm fear that they will be cast in a service role, responding to scientists' priority interests rather than to those of the farmers.

Defining responsibilities. Ambiguity fosters conflict. In over half the cases studied, tensions arose from applied researchers' false expectations about the role of OFR. The most common misconception was that OFR is simply a testing and demonstration activity, not a legitimate research activity. Such misunderstandings reflect management problems. In most cases, senior research managers had simply neglected to work with staff to define the roles, responsibilities and outputs of OFR, or to develop guidelines on how the two types of research should fit together.

Lessons for building stronger links

Several important lessons can be derived from the cases for minimizing conflict and strengthening links (Merrill-Sands and McAllister, 1988). First, managers have to ensure that OFR is not perceived as a corrective measure for past failures to generate relevant technologies. Too often, leaders of OFR efforts try to gain political support by highlighting the deficiencies of conventional research. Nothing undermines collaboration more quickly. Managers have to emphasize that OFR complements experiment station research and is critical to improving the capacity of research to respond to the diverse and challenging demands of resource-poor farmers.

Managers also have to manage collaboration actively. Unfortunately, this has been more the exception than the rule. Managers need to create incentives for scientists to work together. They should encourage informal consultations as an inexpensive linkage mechanism but, at the same time, they have to provide a regular forum for debate and sharing of information and set up mechanisms for joint planning and review of research. Regular joint visits to the field have proved very useful in stimulating stronger integration, as have small planning and review meetings based on single commodities or regional research programmes. In Zambia, conflicts abated and collaboration improved markedly once regional on-farm researchers and commodity scientists began going to the field together and holding joint programming and review meetings (Kean and Singogo, 1988).

Managers also need to recognize that links cost time and money. Collaborative activities have to be carefully selected and resources, both funds and staff time, have to be allocated explicitly to them. In addition, managers have to provide leadership and demonstrate their commitment by participating actively in linkage activities themselves. Lastly, and perhaps most importantly, they have to ensure the scientific credibility of OFR. Problems of scientific credibility have plagued fledgling OFR efforts in national research systems, undermining their effective integration with applied research efforts. Managers have two basic responsibilities here: to ensure the quality of OFR by assigning capable researchers and providing strong scientific leadership; and to encourage experiment station researchers to accept OFR as a viable research activity and clients' demands as a legitimate input into research planning and priority-setting. Fulfilling these responsibilities is a pre-requisite for the effective institutionalization of OFR's crucial feedback role.

Strengthening links with farmers

User participation is essential for relevant, innovative and efficient technology development, whether in agriculture or industry (von Hippel, 1978; Souder, 1980; Peters and Waterman, 1984; Gamser, 1988; Merrill-Sands and Kaimowitz, 1990; Röling, 1990). OFR has been used as a key mechanism for strengthening resource-poor farmers' participation in research.

Farmer participation in research

To what extent has this key objective of increasing farmer participation in research been achieved? The ISNAR study and others (Biggs, 1989; Byrne, 1989; Tripp et al., 1990) show that OFR has succeeded in increasing scientists' appreciation of farmers' priority needs and problems, particularly those of resource-poor farmers. Specific client groups have been identified, their farming systems described, and key constraints and research opportunities identified. In many cases, existing technologies have been adapted to specific farming systems and adopted by farmers, especially in the more favourable agroecological environments (Ruano and Fumagalli, 1988; Kayastha et al., 1989). In other cases, research priorities and agendas have also been changed to respond more effectively to problems identified at field level (Merrill-Sands and McAllister, 1988). In all cases, however, research managers and scientists have found that organizing and sustaining farmer participation beyond the initial diagnosis stage has been more demanding than they had anticipated.

Biggs (1989) identified four basic types of farmer participation in the ISNAR case studies: *contractual*, where scientists contract with farmers to provide land and services; *consultative*, where scientists consult farmers about their problems and then develop solutions; *collaborative*, where scientists and farmers collaborate as partners in the research process; and *collegial*, where scientists work to strengthen farmers' informal research and development systems. In over half of the 25 cases analysed, participation was consultative, with farmers playing a relatively passive role. In only a third of the cases had researchers set up mechanisms for more direct, intensive and continuous farmer participation. There were few documented examples of collegial participation.

The lack of active farmer participation partly reflects a lack of methods and training in skills required for a productive interaction with clients (Chambers and Jiggins, 1987; Norman, 1989).[7] It also reflects logistical constraints and in some cases managers' impressions that intensive farmer participation may be too costly.[8] However, at the root of much of the problem is, first, researchers' hesitance to reveal to farmers that they do not understand some problems nor have available solutions and, second, to expose their work to clients for fear of losing control over the research agenda. This problem is aggravated with respect to resource-poor farmers where differences of status, education and language often impose powerful barriers to effective partnership.

Lessons for strengthening participation

Our analysis of the experiences of national research systems reveals three important lessons for realizing the full benefits from strengthened farmer participation:

- researchers and managers need to be flexible in using different modes of participation;

- researchers need to use more rigorous procedures for selecting farmer collaborators;

- managers need to develop specific mechanisms for incorporating information from farmers into research planning.

Flexibility in using differing types of participation. Different types of participation are appropriate for different research objectives, institutional settings and the resources (both staff and funds) available. Each type also has distinct organizational and managerial implications.

A key lesson from the ISNAR study is that researchers should not pursue one type of farmer participation in a rigid or routine fashion to the exclusion of other approaches (Biggs, 1989). Different types can be used together or sequentially. Researchers may, for example, need to work closely with a group of farmers when a technical problem is poorly understood or involves complex interactions within the farming system. Here it would be important to involve farmers early in the technology generation process (Ashby, 1990). On the other hand, a less intensive survey mode may be most appropriate in areas where agroecological conditions are well understood and potential technological alternatives are available for adaptation and testing.

Farmer selection. Few of the OFR efforts reviewed used systematic and defensible methods for selecting farmers to collaborate in OFR (Ewell, 1988; Biggs, 1989). Although researchers had developed formal criteria for selecting farmers, they often did not apply them in the field. When faced with difficult logistical problems of contacting farmers, getting trials planted with the rains, keeping vehicles running or reporting results on schedule, researchers have often reverted to selecting farmer collaborators on an *ad hoc* basis. They have simply accepted suggestions of extension agents, taken volunteers at meetings or delegated selection to junior field staff with minimal guidance.

This is a persistent problem which requires research managers' close attention. Such *ad hoc* selection has been shown to bias samples considerably in favour of male, wealthy and politically active farmers (Sutherland, 1986; Kean and Singogo, 1988; Soliz et al., 1989). Moreover, lack of explicit methods for selecting farmers calls into question the quality and utility of the farm-level information being fed back into the system. Time and resource constraints often make it impossible to use rigorous sampling procedures. Nevertheless, whether working with farmer groups or individuals, a careful selection of farmers using explicit and defensible criteria is a minimal requirement. Such criteria should reflect the research objective and the type of participation sought (Biggs, 1989; Merrill-Sands and Kaimowitz, 1990; Ashby, 1990). If researchers are trying to mobilize farmers' knowledge, they should try to work with innovative and research-minded farmers (Abedin and Chowdry, 1989). If they are trying to obtain farmers' feedback on various technological alternatives in order to find out which should be formally recommended, they need to involve farmers who are representative of the targeted client group. On the other hand, if researchers are working with farmers to demonstrate and disseminate verified technologies, then involving respected community leaders may be the best approach (Ortíz et al., 1991).

Mechanisms for incorporating farmer feedback into research planning. The case experiences show that it is often difficult to institute mechanisms to ensure that information from farmers and field-level research is systematically fed into research priority-setting, planning and review processes (Ewell, 1988; Merrill-Sands and McAllister, 1988). Information from the initial diagnosis is generally analysed and used effectively. The wealth of knowledge gained through researchers' and technicians' on-going contact with farmers, however, has proved more elusive. Rarely is this information recorded systematically or written up in a form that can be communicated easily to other researchers, retained in institutional memory or fed systematically into planning processes. This problem affects both the efficiency and effectiveness of OFR. Considerable resources are being spent to put research staff in regular contact with farmers. If the knowledge they gain in the field cannot be mobilized fully to improve research relevance and impact, then these resources are not being well spent.

Some research managers in the cases studied had begun to experiment with various mechanisms to acquire this information. In addition to having technicians use field logs systematically, some managers had instituted regular research meetings with farmers to review on-going research and the plans for future work (Kean and Singogo, 1988; Ruano and Fumagalli, 1988; Abedin and Chowdry, 1989; Faye and Bingen, 1989; Kayastha et al., 1989; Cuellar, 1990; Budianto et al., forthcoming). If the conclusions are well documented, such meetings can provide a formal means for feeding information from farmers into the research planning and review processes (Norman et al., 1988; Ashby, 1990). A second solution, used in Nepal, Zambia, Zimbabwe and Bangladesh, is to involve field staff (junior scientists, technicians and field assistants) in annual planning and review meetings (Kean and Singogo, 1988; Avila et al., 1989; Jabber and Abedin, 1989; Kayastha et al., 1989). This is a low-cost option which helps to integrate knowledge gained through informal discussions with farmers into the research process. A third solution is to give more emphasis to formally monitoring farmers' reactions to and use of technologies being tested on-farm. Although there are some examples of on-going monitoring and feedback documented (Biggs, 1982), the general lack of such follow-up is a disturbing weakness in most national OFR efforts (Tripp et al., 1990).

Ensuring social science input

Sustaining the social science input in OFR activities has proved difficult for many national research systems. Although social scientists, primarily agricultural economists, had been instrumental in initiating OFR in all the cases reviewed, they had continued to play an important role in less than half of them (Ewell, 1988). Evidence from the cases challenges the widely held assumption that OFR is dominated by social scientists as, on average, they comprise only 20% of the on-farm researchers across the cases reviewed (Frankenberger et al., 1989).

Constraints to sustaining social science input

Several constraints to sustaining social scientific analysis in OFR were identified in the cases (Ewell, 1988; Bingen and Poats, 1990). The supply of appropriately trained national scientists is often limited, and thus on-farm programmes have often had to rely, at least initially, on foreign scientists with short-term contracts. When competent national social scientists have been brought into research institutions, they are often drawn away from field research into institutional planning and priority-setting functions. A related problem is the lack of established posts for social scientists. In Zambia, for example, 10 years after the initiation of a major OFR programme, the Research Department is still struggling to get posts officially established for its social scientists (Kean and Singogo, 1988).

A second constraint relates to the role of social scientists in agricultural research. While many research managers accept that social scientists can make a valuable input in the early stages of OFR — in defining client groups, characterizing farming systems and diagnosing key constraints — they have had a much harder time envisioning their potential contribution at later stages of the research process. In some cases, social scientists have seeded their own demise. They have pursued independent research agendas and made little effort to demonstrate the relevance or utility of their studies for technical agricultural scientists.

Some national research systems have tried to overcome these constraints by training agronomists in basic methods of socioeconomic analysis, such as survey design and partial budgeting analysis. Without leadership from a senior social scientist, however, appropriate use of these methods has been erratic and difficult to sustain (Ruano and Fumagalli, 1988; Cuellar, 1990; Soliz et al., 1990). The ISNAR study strongly suggests that this is not an effective substitute, even under resource-constrained situations, for the continued involvement of trained social scientists in OFR (Ewell, 1988).

Contribution of social scientists

The diminishing role of social sciences is a key institutional problem jeopardizing the effectiveness of many national OFR efforts. Evidence from the case studies and other experiences argues forcefully that on-going social science input is critical for maintaining dynamic and effective OFR aimed at the development of relevant technologies (Rhoades and Booth, 1982; Biggs, 1983, 1989; Horton, 1984; Maxwell, 1986: Byerlee and Tripp, 1988; Ewell, 1988; Harrington, 1989; Tripp and Woolley, 1989). Social scientists have taken lead roles in developing sound and defensible methods for selecting farmer collaborators as well as innovative mechanisms for encouraging productive farmer participation (Sutherland, 1987; Ewell, 1988; Biggs, 1989). In addition, they have helped to maintain the use of the systems perspective and the application of alternative criteria for evaluating technologies reflecting farmers' needs and priorities.

Social scientists have also been instrumental in setting OFR priorities and selecting relevant research problems, in monitoring farmers' reactions to proposed technological alternatives, in developing the required institutional links and in championing the importance of feedback from clients for relevant research (Byerlee and Tripp, 1988; Harrington, 1989). Many of the OFR programmes which lost social scientists, or had minimal social science input, stagnated, regressing to routine testing programmes.

Lessons for managers

The case studies reveal several important lessons for sustaining social science input in OFR activities.

Senior social scientist. A crucial lesson is that, at a minimum, OFR efforts should be supported by at least one senior social scientist. Experience has shown that young, inexperienced social scientists rarely have the stature or skills to maintain social science input in defining the research agenda (Collinson, 1982; Ewell, 1988). A senior social scientist is needed to interact on an equal basis with senior natural scientists, as well as to support junior scientists and reinforce their training in social science research methods. To be effective this person has to have solid field experience, a good technical knowledge of agriculture and strong leadership skills. Perhaps most importantly, he/she has to be able to use flexible and well-focused research methods that provide timely and relevant information for research planning and analysis of results (Byerlee and Tripp, 1988).

Organizational considerations. Three lessons regarding the organization of social science input in OFR should be stressed. Social scientists should not be given primary responsibility for developing OFR in national programmes. They can play an important coordinating role, but they cannot substitute for technical scientists in implementing OFR. Second, when possible, they should be based at field level, working closely with technical scientists. They easily become marginalized if isolated at headquarters. Third, in addition to agricultural economics, rural sociology, applied anthropology and geography have made valuable contributions to OFR (Rhoades, 1984; Cernea and Guggenheim, 1985; Tripp, 1985; Sutherland, 1987) by keeping client orientation and farmer participation at the centre of the agenda. These disciplines can be incorporated into OFR in a cost-effective manner. In Zambia, recognizing that they could not afford to post a rural sociologist to each of the eight regional teams, a small core group has been established at national level. Each researcher is responsible for backstopping two regional teams and goes to the field to work with them as needed (Kean and Singogo, 1988).

Developing effective links with technology transfer

For effective technology dissemination, OFR has to develop strong links with technology transfer agencies — whether extension services, NGOs, development projects or commodity organizations (Fresco, 1984; Andrew and McDermott, 1985; Cernea et al., 1985; Ewell, 1989; Norman, 1989; Merrill-Sands and Kaimowitz, 1990)

On-farm research cannot substitute for technology transfer

A key conclusion emerging from the ISNAR study is that OFR cannot substitute for technology transfer efforts. Research institutions sometimes develop OFR programmes in order to by-pass weak extension services, but this approach may have only limited impact. Although some good technologies, particularly varieties, can diffuse spontaneously from trials in farmers' fields, most cannot. Many require production and distribution of inputs, intensive training in their use, or collective action (Ewell, 1989; Merrill-Sands and Kaimowitz, 1990).

The experiences of ICTA in Guatemala provide a good example. ICTA initially pursued an independent strategy for technology transfer, assuming that farmers would spontaneously adopt technologies verified in farmers' fields (Ruano and Fumagalli, 1988). This proved successful for some improved varieties in the more favourable and uniform lowland areas, but it soon became clear that, to reach resource-poor farmers operating complex production systems in the diverse and marginal highland areas, strong and flexible links with extension had to be developed. An integrated research-extension effort, based on OFR, has met with considerable success in serving this client group in recent years (Ortíz et al., 1991).

Using on-farm research to strengthen links

OFR can be a powerful vehicle for developing links with technology transfer agencies, but the opportunity has not been fully exploited. Weak links between OFR and technology transfer persist as a chronic problem in all the cases (Ewell, 1989). Similar findings emerged from a recent review of farming systems projects supported by the United States Agency for International Development (USAID) (Frankenberger et al., 1989). What can research managers do to improve these vital links? Several relevant lessons emerged from the ISNAR study.

Recognizing limits to informal cooperation. Developing informal cooperation at the field level is not enough. In fact, relying on informal links alone can cause more harm than good. Experiences from Ecuador, Guatemala, Zambia and Zimbabwe show that when additional research tasks are simply tacked on to field agents' normal duties, these agents rarely have the time, training, mobility or motivation to perform them effectively (Gilbert et al., 1988; Kean and Singogo, 1988; Ruano and Fumagalli, 1988; Avila et al., 1989; Soliz et al., 1990). This approach is a recipe for frustration: field agents resent collaboration with research as an additional, and unrewarded, burden; and researchers quickly become disillusioned with field staff's apparent lack of interest and poor management of trials or surveys.

Building links at multiple levels of the administrative hierarchy. Formal linkage mechanisms are needed to reinforce informal collaboration. The ISNAR study indicates that links are most successful and sustainable when mechanisms are active at several administrative levels: among technicians in the field, among regional research and extension managers, and among senior managers at the highest levels. Good cooperation at field level is impossible to sustain unless senior and regional managers actively support regular opportunities for staff to meet and work together. Similarly, joint goals set at the policy level cannot be realized unless specific operational procedures and staff commitment are developed at the local levels.

Anticipating the need for links. Managers and researchers need to cultivate good working relationships with technology transfer agents early in the OFR process. This can improve research relevance by giving technology transfer agents, as clients of research, an opportunity to influence the research agenda. It can also improve efficiency in transfer because the structures and procedures will be in place when needed. More important still, developing links early gives technology transfer agents more confidence in the technologies and helps to foster the shared sense of purpose and commitment necessary for effective collaboration.

Linkage mechanisms

Four promising mechanisms for developing a partnership between OFR and technology transfer emerged from the case experiences: secondment of extension staff; coordinating committees; joint planning and review processes; and integrated field teams (Ewell, 1989).

Secondment of technology transfer staff. The secondment of technology transfer staff to provincial OFR teams has been a key linkage mechanism in Zambia (Kean and Singogo, 1988, 1990). At the field level, local agents are assigned to the teams as trials assistants. The advantages are that they live in the research areas, have the skills to communicate well with farmers and can promote informal links with other agents working in the field. Professionals from the government extension service have also been seconded to the teams to serve as research-extension liaison officers (RELOs). The responsibilities of RELOs include: revising recommendations, testing promising technologies on farm, organizing demonstrations, preparing and producing newsletters, training extension staff and strengthening feedback from extension to research. Staff secondment is difficult to manage, but it can be a powerful integrating mechanism (Baker and Norman, 1988; Bourgeois, 1990; Kean and Singogo, 1990).

Coordinating committees. Coordinating committees, often set up in response to pressures from donors or policy makers, have not had a good track record. They often exist on paper only. However, the experience from Zimbabwe and Bangladesh shows that they can be very effective linkage mechanisms when several conditions are met: they have well-defined objectives and responsibilities; they are supported by senior managers and have the power to take decisions; they are well managed; andthey reflect the interests and felt needs of both research and technology transfer staff (Abedin and Chowdury, 1989; Avila et al., 1989; Bourgeois, 1990; Merrill-Sands and Kaimowitz, 1990).

In Zimbabwe, for example, the Committee for On-Farm Research and Extension (COFRE) has improved links between research and extension considerably (Shumba and Fenner, 1989). It has played a crucial role in coordinating priority setting, planning and on-farm operations in both departments. Problem identification has improved and on-farm trials have been simplified. COFRE has stimulated extension staff participation in verification trials, accelerated the translation of research results into demonstration projects and synthesized available research results into tentative recommendations. It has also secured representation on the National Agricultural Research Council where it can influence agricultural research policy and cultivate support for OFR aimed at resource-poor farmers.

Joint planning and review processes. Again, this mechanism is often implemented in form only. In north-western Bangladesh, however, joint planning and review meetings, implemented within the Training and Visit (T&V) system, became the central mechanism for linking technology transfer, OFR and experiment station research (Abedin and Chowdury, 1989). Through these meetings, staff began to critically review methods and results together and to jointly plan research and extension strategies. The quality of OFR and demonstration activities improved and feedback was strengthened.

Integrated field teams. After 10 years of frustrated attempts to strengthen links between research and technology transfer, integrated field teams have now had considerable success in Guatemala (Ortíz et al, 1991). The key to their success has been strong and visible support from senior managers, joint training in OFR methods, availability of technologies for adaptation and testing, extension staff specialization in technology transfer and active farmer involvement which has made the teams feel accountable.

MANAGEMENT OF RESEARCH AND RESOURCES

Many lessons for increasing the effectiveness and efficiency of OFR through improved management emerged from the ISNAR study. Only a few of the most important can be highlighted in this review. These are discussed under the two broad categories of management of the OFR programme and management of resources.

Management of the on-farm research programme

Many OFR efforts have had trouble maintaining vigour and innovativeness. Without careful stewardship, they can easily succumb to the institutional norms of the 'top-down' approach to research and technology transfer and revert to conventional technology testing programmes. Managers need to be aware of this vulnerability and provide the necessary support to maintain a dynamic research programme. The most important management considerations are outlined below.

Scientific leadership

Strong scientific leadership emerged as one of the main factors determining the success of OFR efforts in national agricultural research systems. Yet, too often its importance has been overlooked. Scientific leadership in OFR has often been erratic and/or overshadowed by the more pedestrian, but pressing, demands of coordination and administration.

OFR places heavy demands on leadership. As an evolving approach, it requires methodological experimentation, creative thinking, strong team work, broad knowledge spanning various disciplines and commodities, capacity to integrate diverse types of information and data, and skills for interacting with clients. Strong leadership is particularly important for ensuring the quality and scientific credibilty of OFR. Leaders of OFR need to nurture and backstop young scientists, helping them to apply new research skills, learn from farmers and integrate knowledge from various disciplines. A strong leader, with a status equal to heads of programmes or departments, is also needed to develop effective collaboration with station-based research, to defend the legitimacy of non-traditional types of data and analysis typical of OFR and to keep the client at the centre of the research agenda. Overall, the case study experiences show that research managers who are committed to integrating OFR must give high priority to cultivating and retaining a strong scientific leader to guide its development within their systems (Merrill-Sands and McAllister, 1988; Bingen and Poats, 1990)

Focused research agenda

Setting clear priorities, based on clients' needs, and keeping the research agenda focused on them is essential for effective and efficient OFR. Priority setting is especially important under the resource-constrained situations of many national research systems. OFR has to deal with a hierarchy of priorities: the selection of geographic zones for the conduct of OFR (although this may be dictated by development priorities), the targetting of specific farming systems within those zones, and the choice of research problems to be addressed to improve the productivity and/or sustainability of those farming systems.

Holding fast to these priorities can be difficult as OFR efforts come under pressure from various stakeholders — donors, development agencies, extension services, policy makers, commodity scientists or influential farmers — to serve their specific interests. Having clear and defensible criteria and procedures for priority setting helps on-farm researchers to defend their choice of target areas, client groups and research problems. Involving stakeholders in the decisions is an even more effective mechanism.

These priority choices are also critical for the efficiency of OFR and its eventual impact. It is essential that on-farm researchers work on farming systems and research problems which affect a large number, or a significant group, of farmers. The costs of location-specific research can not be justified if results cannot be extrapolated to a wider group. In too many cases, external pressures or sloppy methods have led OFR to address problems which were too location- or client-specific to warrant the heavy investment of research resources (Ewell, 1988; Tripp et al., 1990). This has jeopardized the credibility of OFR and led some reasearch managers to conclude that it is simply too much of a luxury for their systems to afford.

Clustering trial sites

The case study experiences show that, once target areas and farming systems have been selected, clustering trials at specific sites or in specific villages can be useful (Ewell, 1988; Kean and Singogo, 1988; Avila et al., 1989; Jabber and Abedin, 1989; Kayastha et al., 1989). Clustering has several advantages. It improves efficiency in resource use as researchers and technicians spend less time in travel and demands for operating funds and logistic support are reduced. It fosters closer supervision and better management of trials. And it helps to strengthen links at the field level by promoting regular contact with farmers over long periods and providing a focal point for regular meetings with commodity scientists and extension agents.

Communication between researchers working on-farm

Because OFR is still evolving, methods and procedures are still being developed and refined. No researcher leaves university trained in the discipline of OFR and no short course can give researchers all the required skills. These skills are honed and methods are developed in the field, working closely with farmers. It is extremely important, therefore, that managers create regular opportunities for researchers and technicians working on-farm to come together to share experiences, discuss methods and approaches to working with farmers, and to critically review each other's work. Such exchanges are important in all scientific endeavours, but they are essential in areas and topics under intense development. In the case studies, strong OFR efforts all had mechanisms in place to encourage the exchange of information and experiences among researchers.

Management of resources

To link research more effectively with clients, managers have to be willing to commit the requisite resources, in terms of staff and funds, to OFR. It is not something which can be done halfway.

Staffing considerations

The case experiences confirm that, when resources are limited, it is better to sacrifice quantity of resesarch activities than quality of researchers. Experienced researchers are needed to conduct good quality OFR. Only they can deal with the complexities of field research and promote and defend the diverse kinds of information and data generated by OFR. Most importantly, experienced and respected researchers are needed for the feedback role to function effectively. Those cases where feedback was weakest were those where most on-farm researchers were junior scientists (Merrill-Sands and McAllister, 1988).

On average, 42% of the researchers in the nine research systems studied had post-graduate degrees and about 50% had 5 years or more research experience. The staffing patterns, however, varied markedly across the cases.

In five of the nine national research systems studied, on-farm researchers had lower degree levels and less research experience than the average for the research staff as a whole. OFR fared even worse when expatriate scientists were removed from the analysis (Bingen and Poats, 1990).

The pressures on managers to staff OFR with junior scientists are considerable. Managers often have trouble reassigning senior scientists to OFR efforts, especially if they are required to live in remote areas. External pressures for wide geographical coverage often lead managers to allocate new recruits to OFR simply to demonstrate that field teams have been established. Heads of research programmes or departments, seeing OFR primarily as a testing and demonstration activity, often argue hard against deploying more experienced scientists in the field. Nevertheless, the ISNAR study indicates that succumbing to such pressures can have high costs in terms of research quality and the long-term credibility of OFR. If managers simply have no other option than to allocate junior scientists to OFR, then they have to ensure that they are backstopped systematically by senior scientists who make regular visits to the field. This approach has been used very effectively in Indonesia and Nepal (Kayastha et al., 1989; Budianto et al., forthcoming).

Financial considerations

OFR is often accused of being too expensive for sustained use in national agricultural research systems. This accusation is based on perceptions, however, and not on data. No systematic study of the costs of OFR, except that of Martínez and Sain (1983), or of its costs relative to experiment station research, has ever been reported in the literature. Perceptions of high cost derive from four factors:

- much OFR, especially in earlier years, has been executed through donor projects which were usually generously funded and staffed by expensive expatriate scientists;

- OFR relies on transport and operating funds, which are often scarce resources in national agricultural research systems;

- the location-specific nature of OFR makes the ratio of costs to benefits appear high, especially when the criteria for selecting target groups and the extent of extrapolation of research results are not made explicit;

- benefits tend to be defined narrowly in terms of technologies adapted by farmers rather than in terms of less tangible impacts, such as funds saved by redirecting research to more relevant research problems through a better understanding of farmers' needs and circumstances (Biggs, 1983).

Evidence from the ISNAR study, while it cannot be considered conclusive, strongly challenges the perception of OFR as an expensive exercise (Gilbert, forthcoming). Analysis of data from eight OFR programmes in six countries shows the costs per researcher for OFR to be 60% of the costs per researcher for the research system as a whole. This finding was consistent across OFR programmes of varying scale and complexity. The data also show that OFR does not rely disproportionately upon operating funds, as is commonly asserted. While salaries and wages accounted on average for 60% of the OFR costs, operating costs, including travel expenses, accounted for only 25% of the total expenditures. This cost structure was essentially the same as that for the research system as a whole. In absolute terms, however, operational expenses per researcher for the case study research systems as a whole were 150% greater than those for the OFR programmes.

Although the data are few, they are the best currently available. Their implication that OFR does not have excessive or peculiar funding requirements bodes well for the long-term sustainability of OFR in national research systems.

While it can be argued that comparing costs of on-farm and experiment station research has serious limitations, research managers are faced regularly with tough decisions on allocating scarce resources across research activities. In such choices, OFR, prejudiced by the perception of its high cost relative to other types of research, tends to suffer (Gilbert, forthcoming). The evidence from the ISNAR study suggests that funding considerations *per se* should not determine managers' decisions about the budget allocation for OFR. Such decisions should be based firmly on research objectives and the relative balance of activities required to meet them. Moreover, managers need to bear in mind the cost of not doing OFR — the increased likelihood that research will produce knowledge and technologies which farmers will not, or cannot, adopt.

CONCLUSION

Many national research systems have launched major OFR efforts aimed at generating relevant technologies for resource-poor farmers. Institutional and policy factors, however, have often hindered the effectiveness of these efforts. Experience has shown that OFR — linking farmers, researchers and technology transfer agents — poses special organizational and managerial challenges. Institutional innovations and a strong policy commitment are needed if OFR is to succeed in meeting the needs of resource-poor farmers. Sound research methods, alone, are not enough.

Drawing on the findings of the ISNAR study on institutional factors affecting the performance and sustainability of OFR efforts in national agricultural research systems, this chapter has reviewed the key policy, organizational and managerial considerations for integrating OFR as an effective and stable component of these systems. The objective of this review is to provide research managers and on-farm researchers and advisors with concrete and practical insights for strengthening the effectiveness and efficiency of OFR and its contribution to the generation and transfer of relevant technologies.

Notes

1. In *Our Common Future* (1987), the World Commission on Environment and Development developed a useful contrast between two main types of agriculture in the developing world: 'Green Revolution agriculture' and 'resource-poor agriculture'. The latter relies on uncertain rainfall, rather than irrigation, and is found in the more marginal areas with fragile soils, such as the drylands, highlands and forests. These areas are difficult for farming, vulnerable to degradation and typically have limited infrastructure to support agricultural development. They include much of sub-Saharan Africa and the more remote areas of Latin America and Asia.

2. As the ISNAR study focused on institutional rather than methodological issues, we adopted the generic term 'on-farm client-oriented research' to cover a range of methodologies aimed at developing technologies for resource-poor farmers: 'farming systems research', 'cropping systems research', 'on-farm research with a farming systems perspective', 'farmer-back-to-farmer' and 'farmer-first-farmer-last'. All these methods share common features which define on-farm client-oriented research: they involve farmers as primary clients of research; they are designed to complement experiment station research; they emphasize diagnosis and priority setting in the context of the whole farming system; and they adapt and evaluate technologies at farm level. All are aimed at making research more responsive to clients' needs.

3. Comprehensive reports on the organization and management of OFR in each of the case study institutions are available from ISNAR and are listed in the references.

4. Tripp et al. (1990) cite this as a key issue for FSR in the 1990s.

5. The key organizational considerations and the strengths and weaknesses of various models for organizing OFR are analysed in depth in a forthcoming paper by Merrill-Sands and McAllister, 'Alternative Arrangements for Organizing On-Farm Client-Oriented Research: Comparative Strengths and Weaknesses'. See also Tripp et al. (1990).

6. Because these links are so important to the effective management of OFR, they received high priority in the ISNAR study. Papers from the study examine the management of each of these sets of links in depth, highlighting key issues in developing the links as well as innovative solutions and mechanisms developed by managers in the national agricultural research systems reviewed (Ewell, 1988; Merrill-Sands and McAllister, 1988; Merrill-Sands et al., 1989; Merrill-Sands and Kaimowitz, 1990). See also Norman (1989) and Byerlee and Tripp (1988).

7. Developing improved methods and procedures for sustained farmer participation is now the focus of numerous national and international efforts. Some of the key references are Ashby et al. (1987), Chambers and Jiggins (1987), Farrington and Martin (1988), Sagar and Farrington (1988), Biggs (1989), Chambers et al. (1989) and Ashby (1990). Norman (1989) and Tripp (1989) caution against expecting too much from farmer participation; there are limits in the efficiency and technical feasibility of farmers assuming major responsibility for adaptive research.

8. Managers stressed this concern at a recent international workshop, 'Making the Link between Agricultural Research and Technology Users', convened at ISNAR in The Netherlands, November 1989. For a summary of the deliberations, see Merrill-Sands and Kaimowitz (1990). Some preliminary data, however, suggest that a few of the more intensive approaches do not require more staff time if researchers work with farmer groups (Ashby, 1990).

The ideas and conclusions in this chapter are the product of several years of intensive research by over 35 scientists under an international study sponsored by ISNAR and funded by the Government of Italy and the Rockefeller Foundation. We are indebted to the researchers who prepared country case studies and to the members of the external advisory committee, ISNAR staff and management, and national research managers who helped shape the study, monitor its progress and review its ouput. Special thanks are extended to M. Avila, M. Collinson, E. Gilbert, H. Elliot, D. Kaimowitz, S. Kean, E. Moscardi, S. Ruano, W. Stoop and R. Tripp who made important contributions to the design and analysis of the study. We also acknowledge the useful comments made on this chapter by ISNAR colleagues T. Eponou, P. Eyzaguirre and P. Perrault. The content of the chapter is our interpretation of the important lessons emanating from the ISNAR study. The opinions presented are not necessarily those of ISNAR or of all the researchers involved in the study. An abridged version of this chapter was published in *Experimental Agriculture* October 1991.

References

Abalu, G., Mutsaers, H. and Faye, J. (eds) 1988. *Farming Systems Research in West Africa: Proc. of West African Farming Systems Research Network Workshop, Dakar, Senegal, 10-14 March 1986.*

Abedin, M. Z. and Chowdry, M. 1989. Organizational and managerial innovations for research and extension linkages in Bangladesh. Paper presented at the International Workshop on 'Making the Link between Agricultural Research and Technology Users', ISNAR, November 1989.

Aldrich, H.E. 1979. *Organizations and Environments.* Englewood Cliffs, USA: Prentice-Hall.

Andrew, C. and McDermott, K. 1985. *Farmer Linkages to Agricultural Research and Extension: A Question of Cost-Effectiveness.* Mimeograph. Gainesville, USA: Farming Systems Support Project, University of Florida.

Arnold, H. and Feldman, D. 1986. *Organizational Behavior.* New York, USA: McGraw-Hill.

Ashby, J. 1990. Small-farmer participation in the design of technologies. In Altieri, M. and Hecht, S. (eds) *Agroecology and Small Farm Development*. Boca Raton, USA: CRC Press.

Ashby, J., Quirós, C. and Rivera, Y. 1987. *Farmer Participation in On-Farm Varietal Trials*. Agricultural Administration Network Discussion Paper No. 22. London, UK: ODI.

Avila, M., Whingwiri, E. and Mombeshora, B. 1989. *Zimbabwe: A Case Study of Five On-Farm Research Programs in the Department of Research and Specialist Services, Ministry of Agriculture*. The Hague, Netherlands: ISNAR.

Baker, D. and Norman, D. 1988. A framework for assessing farming systems activities in national settings in West Africa, with special reference to Senegal, Nigeria, and Mali. In *Farming Systems in West Africa: Proc. of West African Farming Systems Research Network Workshop, Dakar, Senegal, March 1986*. Ottawa, Canada: IDRC.

Biggs, S. 1982. Generating agricultural technology: Triticale for the Himalayan Hills. *Food Policy* (7)1: 69-82.

Biggs, S. 1983. Monitoring and control in agricultural research systems: Maize in northern India. *Research Policy* 12: 37-59.

Biggs, S. 1989. *Resource-Poor Farmer Participation in Research: A Synthesis of Experiences in Nine National Agricultural Research Systems*. OFCOR Comparative Study Paper No. 3. The Hague, Netherlands: ISNAR.

Biggs, S. and Gibbon, D. 1985. *Agricultural Research to Help Poor People in Developing Countries*. Occasional Paper No. 28. Norwich, UK: School of Development Studies, University of East Anglia.

Bingen, J. and Poats, S. 1990. *Staff Management Issues in On-Farm Client-Oriented Research*. OFCOR Comparative Study Paper No. 5. The Hague, Netherlands: ISNAR.

Bourgeois, R. 1990. *Structural Linkages for Integrating Agricultural Research and Extension*. ISNAR Working Paper, No. 35. The Hague, Netherlands: ISNAR.

Budianto, J., Ismail, I., Sridodo, Sitorus, P., Tarifans, D., Mulyadi, A. and Suprat. (forthcoming). *Indonesia: Organization and Management of On-Farm Research in the Agency for Agricultural and Development*. Mimeograph.

Byerlee, D. 1988. Food for thought: Technological challenges in Asian agriculture in the 1990s. Paper prepared for the Conference of Asia and Near East Bureau's Agricultural and Rural Development Officers, USAID, February 1988, Morocco.

Byerlee, D. and Tripp, R. 1988. Strengthening linkages in agricultural research through a farming systems perspective: The role of social scientists. *Experimental Agriculture* 24: 137-51.

Byerlee, D. , Collinson, M. et al. 1980. *Planning Technologies Appropriate to Farmers: Concepts and Procedures*. El Batan, Mexico: CIMMYT.

Byrne, K. 1989. *A Review of AID Experience: Farming Systems Research and Extension Projects, 1975-87*. Center for Development Information, AID Evaluation Highlights No. 4. Washington DC, USA: USAID.

Cernea, M., Coulter, J. and Russell, J. (eds) 1985. *Research-Extension-Farmer: A Two-Way Continuum for Agricultural Development*. Washington DC, USA: World Bank.

Cernea, M. and Guggenheim, S. 1985. Is anthropology superfluous in farming systems research? *Farming Systems Research and Extension Newsletter* 4(9): 504-17.

Chambers, R. and Jiggins, J. 1987. Agricultural research for resource-poor farmers, Part I: Transfer of technology and farming systems research. *Agricultural Administration and Extension* 27(1): 35-52.

Chambers, R., Pacey, A. and Thrupp, L. (eds). 1989. *Farmer First: Farmer Innovation and Agricultural Research*. London, UK: Intermediate Technology Publications.

Chang, M. (ed). 1984. *Issues in Organization and Management of Research with a Farming Systems Perspective Aimed at Technology Generation*. Cali, Colombia/The Hague, Netherlands: CIMMYT/ISNAR.

Collinson, M. 1982. *Farming Systems Research in Eastern Africa: The Experiences of CIMMYT and Some National Agricultural Research Services: 1976-1981*. International Development Paper No. 3. East Lansing, USA: Michigan State University.

Collinson, M. 1986. *On-Farm Research and Extension Institutions*. Agricultural Administration Network Discussion Paper No. 17. London, UK: ODI.

Collinson, M. 1987. Farming systems research: Procedures for technology development. *Experimental Agriculture* 23: 365-86.

Collinson, M. 1988. The development of African farming systems: Some personal views. *Agricultural Administration and Extension* 29: 7-22.

Cuellar, M. 1990. *Panama: Un Estudio del Caso de la Organización y Manejo del Programa de Investigación en Finca de Productores en el Instituto de Investigación Agropecuaria de Panamá.* OFCOR Case Study No. 8. The Hague, Netherlands: ISNAR.

Ewell, P. 1988. *Organization and Management of Field Activities in On-Farm Research: A Review of Experience in Nine Countries.* OFCOR Comparative Study Paper No. 2. The Hague, Netherlands: ISNAR.

Ewell, P. 1989. *Linkages between On-Farm Research and Extension in Nine Countries.* OFCOR Comparative Study Paper No. 4. The Hague, Netherlands: ISNAR.

Farrington, J. and Martin, A. 1988. *Farmer Participation in Agricultural Research: A Review of Concepts and Practices.* Agricultural Administration Unit Occasional Paper No. 9. London, UK: ODI.

Farrington, J. and Biggs, S. 1990. NGOs, agricultural technology and rural poor. *Food Policy* December 1990: 479-91.

Faye, J. and Bingen, J. 1989. *Sénégal: Organisation et Gestion de la Recherche sur les Systèmes de Production, Institut Sénégalais de Recherches Agricoles.* The Hague, Netherlands: ISNAR.

Frankenberger, T., Finan, T., Dewalt, B., McArthur, H., Hudgens, R., Flora, C. and Young, N. 1989. Identification of results of farming systems research and extension activities: A synthesis. *Culture and Agriculture* 38 (Spring/ Summer 1989): 8-11.

Fresco, L. 1984. Issues in farming systems research. *Netherlands Journal of Agricultural Sciences* 32: 253-61.

Galbraith, J.R. 1977. *Organizational Design.* Reading, USA: Addison-Wesley.

Gamser, M. 1988. Innovation, technical assistance, and development: The importance of technology users. *World Development* 16(6): 711-21.

Gibbon, D., Thapa, H. and Rood, P. 1989. The development of farming systems research in Nepal: The working group as the focus of interdisciplinary activity. Paper presented at the Ninth Annual FSR/E Symposium, University of Arkansas, Fayetteville, October 1989.

Gilbert, E. 1990. *Non-Governmental Organizations and Agricultural Research: The Experience of The Gambia.* Agricultural Administration Network Paper No. 12. London, UK: ODI.

Gilbert, E. (forthcoming). *Financial Resources and Management for On-Farm Client-Oriented Research: A Review of Experiences in Nine Countries.* OFCOR Discussion Paper. The Hague, Netherlands: ISNAR.

Gilbert, E., Posner, J. and Sumberg, J. 1988. Farming systems research within a small research system: A search for appropriate models. *Agricultural Systems* 33(4): 327-46.

Hage, J. and Aiken, M. 1969. Routine technology, social strucure and organizational goals. *Administrative Science Quarterly* 14: 366-77.

Harrington, L. 1989. The role of social science in farming systems research and development. Paper presented at the Experts Symposium on Farming Systems Research and Development, Los Baños, Philippines, November 1989.

Heinemann, E. and Biggs, S. 1985. Farming systems research: An evolutionary approach to implementation. *Journal of Agricultural Economics* 36(1): 59-65.

Hildebrand, P. and F. Poey. 1985. *On-Farm Agronomic Trials in Farming Systems Research and Extension.* Boulder, USA: Lynne Rienner.

Horton, D. 1984. *Social Scientists in Agricultural Research: Lessons from the Mantaro Valley Project, Peru.* Ottawa, Canada: IDRC.

Jabbar, M. and Abedin, M.Z. 1989. *Bangladesh: The Evolution and Significance of On-Farm and Farming Systems Research in the Bangladesh Research Institute.* OFCOR Case Study No. 4. The Hague, Netherlands: ISNAR.

Kayastha, B., Mathema, S. and Rood, P. 1989. *Nepal: A Case Study of the Organization and Management of On-Farm Research.* OFCOR Case Study No. 3. The Hague, Netherlands: ISNAR.

Kean, S. and Singogo, L. 1988. *Zambia: A Case Study of the Organization and Management of the Adaptive Research Planning Team (ARPT), Ministry of Agriculture and Water Development.* OFCOR Case Study No. 1. The Hague, Netherlands: ISNAR.

Kean, S. and Singogo, L. 1990. *Research-Extension Liaison Officers in Zambia: Bridging the Gap between Research and Extension.* OFCOR Discussion Paper No. 1. The Hague, Netherlands: ISNAR.

Lawrence, P. and Lorsch, J. 1967. *Organization and Environment.* Homewood, Illinois, USA: Richard D. Irwin.

Martínez, J. and Sain, G. 1983. *The Economic Returns to Institutional Innovations in National Agricultural Research: On-Farm Research in IDIAP, Panama.* Economics Program Working Paper 04/83. Los Baños, Mexico: CIMMYT.

Martínez, J. and Arauz, J. 1984. Developing appropriate technologies through on-farm research: The lessons from Caisán, Panama. *Agricultural Administration* 17: 93-114.

Maurya, D.M., Bottrall, A. and Farrington, J. 1988. Improved livelihoods, genetic diversity and farmer participation: A strategy for rice breeding in rainfed areas of India. *Experimental Agriculture* 24: 311-20.

Maxwell, S. 1986. Farming systems research: Hitting a moving trarget. *World Development* 14(1): 65-77.

Merrill-Sands, D. 1986. *The Technology Applications Gap: Overcoming Constraints to Small-Farm Development.* FAO Research and Technology Paper No. 1. Rome, Italy: FAO.

Merrill-Sands, D. 1988. Assessing the institutional impact of on-farm client-oriented research programs: Lessons from a nine country study. In *Contributions of FSR/E towardsSustainable Agriculture.* Farming Systems Research Paper Series No. 17. Arkansas, USA: University of Arkansas/Winrock International Institute.

Merrill-Sands, D. and Kaimowitz, D. 1990. *The Technology Triangle: Linking Farmers, Technology Transfer Agents, and Researchers.* The Hague, Netherlands: ISNAR.

Merrill-Sands, D. and McAllister, J. 1988. *Strengthening the Integration of On-Farm Client-Oriented Research and Experiment Station Research in National Agricultural Research Systems: Management Lessons from Nine Country Case Studies.* OFCOR Comparative Study Paper No. 1. The Hague, Netherlands: ISNAR.

Merrill-Sands, D., Ewell, P., Biggs, S. and McAllister, J. 1989. Issues in institutionalizing on-farm client-oriented research: A review of experiences from nine national agricultural research systems. *Quarterly Journal of International Agriculture* 28 (3/4) July-December, 1989: 279-300.

Mintzberg, H. 1983. *Structure in Fives: Designing Effective Organizations.* Englewood Cliffs, USA: Prentice Hall.

Norman, D. 1980. *The Farming Systems Approach: Relevancy for Small Farmers.* Rural Development Paper No. 5. East Lansing, USA: Michigan State University.

Norman, D. 1983. *Some Problems in the Implementation of Agricultural Research Projects with a Farming Systems Perspective.* Farming Systems Support Project Paper No. 3. Gainesville, USA: University of Florida.

Norman, D. 1989. Accountability: A dilemma in farming systems research. *Culture and Agriculture* 38 (Spring/ Summer 1989).

Norman, D. and Collinson, M. 1985. Farming systems approach to research in theory and practice. In Remenyi, J. (ed) *Agricultural Systems Research for Developing Countries.* Canberra, Australia: ACIAR.

Norman, D., Baker, D., Heinrich, G. and Worman, F. 1988. Technology development and farmer groups: Experience from Botswana. *Experimental Agriculture* 24(3): 321-31.

Ortíz, R., Ruano, S., Juarez, H., Olivet, F. and Menses, A. 1991. *A New Model for Technology Transfer in Guatemala: Closing the Gap between Research and Extension.* OFCOR Discussion Paper No. 2. The Hague, Netherlands: ISNAR.

Pardy, P. and Roseboom, H. 1988. A global evaluation of national agricultural research investments. In Javier, E. (ed) *The Changing Dynamics of Global Agriculture: A Seminar/Workshop on Research Policy Implications for National Agricultural Research.* The Hague, Netherlands: ISNAR.

Peters, T. 1987. *Thriving on Chaos: Handbook for a Management Revolution.* New York, USA: Harpers & Row.

Peters, T. and Waterman, R. 1984. *In Search of Excellence: Lessons from America's Best Run Companies.* New York, USA: Warner Books.

Pingali, P.L. and Moya, P.F. 1989. *Post-Green Revolution Blues in Asian Rice Production: Farm and Experimental Station Evidence on the Dimninished Yield Gap.* Mimeograph. Los Baños, Philippines: IRRI.

Piniero, M. 1989. Generation and transfer of technology for poor, small farmers. In Kesseba, A. (ed) *Technology Systems for Small Farmers: Issues and Options.* Boulder, USA: Westview Press.

Rhoades, R. 1984. *Breaking New Ground: Agricultural Anthropology.* Lima, Peru: CIP.

Rhoades, R. and Booth, R. 1982. *Farmer-back-to-Farmer: A Model for Generating Acceptable Agricultural Technology.* Social Science Department Working Paper 1982-1. Lima, Peru: CIP.

Röling, N. 1990. The agricultural research-technology transfer interface: A knowledge systems perspective. In *Making the Link: Agricultural Research and Technology Transfer in Developing Countries.* Boulder, USA: Westview Press.

Ruano, S. and Fumagalli, A. 1988. *Guatemala: Organización y Manejo de la Investigación en Finca en el Instituto de Ciencia y Tecnologia Agrícolas (ICTA).* OFCOR Case Study No. 2. The Hague, Netherlands: ISNAR.

Sager, D. and Farrington. J. 1988. *Participatory Approaches to Technology Generation: From the Development of Methodology to Wider-Scale Implementation.* Agricultural Administration Network Paper No. 2. London, UK: ODI.

Shaner, W., Phillips, P. and Schmehl, W. 1982. *Farming Systems Research in Development: Guidelines for Developing Countries.* Boulder, USA: Westview Press.

Shumba, E. and Fenner, R. 1989. Linking research and extension through on-farm research and demonstrations: The Zimbabwe experience. Paper presented at the International Workshop on Making the Link between Agricultural Research and Technology Users, ISNAR, The Hague, Netherlands, November 1989.

Soliz, R., Espinosa, P. and Cardoso, V. 1989. *Ecuador: Un Estudio de Caso de la Organización y Manejo del Programa de Investigación en Finca de Productores (PIP) en el Instituto de Investigaciones Agropecuarias.* OFCOR Case Study No. 7. The Hague, Netherlands: ISNAR.

Sompo-Cessay, M. and Gilbert, E. 1989. Special issues for small countries in making the link between research and technology users: The case of The Gambia. Paper presented at the International Workshop on Making the Link between Agricultural Research and Technology Users, ISNAR, The Hague, Netherlands, November 1989.

Souder, W. 1980. Promoting effective R & D/marketing interface. *Research Management* 23(4): 10-15.

Stoop, W. 1987. *Issues in the Implementing Research with a Farming Systems Perspective in NARS.* Working Paper No. 6. The Hague, Netherlands: ISNAR.

Sutherland, A. 1986. *Extension Workers, Small-Scale Farmers, and Agricultural Research: A Case Study in Kabwe Rural, Central Province, Zambia.* Agricultural Administration Network Discussion Paper No. 15. London, UK: ODI.

Sutherland, A. 1987. *Sociology in Farming Systems Research.* Agricultural Administration Network, Occasional Paper No. 6. London, UK: ODI.

TAC (CGIAR). 1990. *A Possible Expansion of the CGIAR.* Mimeograph. Rome, Italy: FAO.

Thompson, J.D. 1967. *Organizations in Action: Social Science Bases of Administrative Theory.* New York, USA: McGraw-Hill.

Tripp, R. 1985. Anthropology and on-farm research. *Human Organization.* 44(2): 114-24.

Tripp, R. 1989. *Farmer Participation in Agricultural Research: New Directions or Old Problems?* Discussion Paper 256. Sussex, UK: IDS.

Tripp, R. and Woolley, J. 1989. *The Planning Stage of On-Farm Research: Identifying Factors for Experimentation.* El Batan, Mexico: CIMMYT/CIAT.

Tripp, R., Anandajayasekeram, P., Byerlee, D. and Harrington, L. 1990. Farming systems research revisited. In Eicher, C. and Staaz, J. (eds) *Agricultural Development in the Third World.* Baltimore, USA/London, UK: Johns Hopkins University Press.

Van de Ven, A. H. and Delbecq, A.L. 1974. A task contingent model of work unit strucutre. *Administrative Science Quarterly* 19: 183-97.

von Hippel, E. 1978. Users as innovators. *Technology Review* January, 1978.

Wolf, E. 1986. *Beyond the Green Revolution: New Approaches to Third World Agriculture.* World Watch Paper No. 73. Washington DC, USA: World Watch Institute.

World Commission on Environment and Development. 1987. *Our Common Future.* Oxford, UK/New York, USA: Oxford University Press.

Woodward, J. 1965. *Industrial Organization: Theory and Practice.* London, UK: Oxford University Press.

Zandstra, H., Price, E., Litsinger, J. and Morris, R. 1981. *A Methodology for On-farm Cropping Systems Research.* Los Baños, Philippines: IRRI.

Zey-Ferrell, M. 1979. *Dimensions of Organizations: Environment, Context, Structure, Process, and Performance.* Santa Monica, USA: Goodyear Publishing.

Planned Change in Farming Systems
Edited by R. Tripp
© 1991 R. Tripp
A Wiley-Sayce Co-Publication

16

Integrating On-Farm Research with Disciplinary, Commodity and Policy Research: The Potential of On-Farm Wheat Research in Pakistan

D. BYERLEE, P. HOBBS and R. TRIPP

Although on-farm research (OFR) is aimed mainly at developing improved recommendations for specific groups of farmers, the results of this type of research can often have much wider and more important implications for the mainstream commodity and disciplinary research that is usually conducted on-station. OFR programmes are in a unique position to generate farm-level information that is not generally available to on-station researchers or to policy makers. This information includes an in-depth understanding of farming systems, quantitative information on farmers' practices and their variability, and agronomic and economic responses to improved practices measured under farmers' conditions through on-farm experimentation. The results from OFR can be particularly valuable where several OFR programmes are carried out in different locations representing the major farming systems, allowing aggregation of results to the national level. Despite this potential for wider use of OFR results at the regional and national level, few research systems have been able to capitalize on the rich set of data increasingly being generated by OFR programmes.

One of the principal reasons for this under-utilization of OFR results is that those who are engaged in this type of research are usually on the periphery of their research organizations, both in terms of location and profession. They tend to be relatively junior staff, assigned to field projects at some distance from research headquarters and from policy makers. As a result, they often do not have the support of their colleagues at headquarters nor the opportunity to contribute to setting directions for research and policy. This can be true even if the on-farm researchers are part of a commodity research programme.

This chapter summarizes OFR activities carried out on wheat in Pakistan over a 5-year period. It illustrates the potential for using OFR not only to improve the efficiency and relevance of recommendations at the local level, but also to contribute to making national level research in various disciplines more responsive to the needs of farmers, as well as to providing information for policy decisions.

The results of OFR on wheat in Pakistan are of national significance because wheat is the main food grain and the major winter crop throughout the country. In 1990, Pakistan produced over 15 million t of wheat on about 7.5 million ha of land, most of it irrigated. Since 1965, it has achieved one of the most remarkable increases in wheat production in the world, as part of what is commonly called the Green Revolution — that is, the rapid adoption of semi-dwarf wheat varieties and fertilizer use, accompanied by improved supplies of irrigation water. However, in the past decade, wheat production in Pakistan appears to have entered a stage of 'post-Green Revolution blues', as yield growth has slowed and evidence accumulates that yields are low given the level of seed, fertilizer and water now being applied. Indeed, there is growing concern about the sustainability of current yields (Hobbs et al., 1988; Byerlee, 1990). In addition, increased cropping intensity in the dominant irrigated systems has posed special challenges to developing improved technology for wheat grown as part of these evolving systems.

Against this background, an OFR programme, initiated by the Pakistan Agricultural Research Council (PARC) and the International Maize and Wheat Improvement Center (CIMMYT) and carried out in several major cropping systems, has attempted to suggest new directions for improving the efficiency of wheat production in Pakistan. While this programme was initially aimed at developing improved recommendations for each of those systems, it soon became apparent that the results had much wider implications for wheat commodity and disciplinary research programmes conducted on-station and even for price and input policy with respect to wheat. We describe here how that work was organized and discuss some of the principal results and their implications for setting wheat research priorities and analysing policies at the national level. We conclude by noting some of the challenges to more effective utilization of the results of OFR at national level.

OVERVIEW OF THE ON-FARM RESEARCH PROGRAMME

Pakistan has a well-developed agricultural research system that includes a wide range of institutions. Wheat research is carried out both at the national level, by PARC, as well as at the provincial level, with each of the four provinces having a wheat or cereal crops research institute. However, OFR based on well-defined diagnosis and on-farm experimentation by a multidisciplinary team was not a part of the research tradition. In addition, and despite the trend towards wheat being grown as part of increasingly complex farming systems, the research system lacked a means to coordinate research across commodities produced as part of the system. An OFR programme was initiated in 1983 by agronomists from PARC's wheat programme, in collaboration with agronomists from the wheat programmes of provincial research institutes. Socioeconomists from PARC who were based in the provincial institutes also took part in the programme. The research focused on the major wheat growing province, Punjab, as well as North West Frontier Province (NWFP) (*see* Figure 16.1).

Wheat is grown in many different environments and conditions in Pakistan (*see* Table 16.1 *overleaf*). In order to address this diversity, it was decided to carry out work in a number of carefully delimited areas that were representative of important wheat-based systems. Among the factors which determined the choice of target research areas were the extent to which a particular system was representative and its proximity to a research station. The target areas included three major irrigated wheat systems (Punjab rice/wheat, Punjab cotton/wheat and NWFP maize/wheat) and one rainfed system (*see* Table 16.2 *overleaf*).

In the irrigated systems, wheat, the dominant food crop, is grown during the winter (*rabi*) season, after a summer (*kharif*) crop, usually a cash crop (*see* Table 16.3 *overleaf*). In the two irrigated systems in the Punjab the major *kharif* crops are rice and cotton, while in NWFP maize and sugar cane are the main *kharif* crops. In the rainfed (*barani*) system of northern Punjab, wheat is usually grown in the *rabi* season after one complete year of fallow, and a variety of fodder crops and pulses are grown in *kharif*. The number of years that wheat has been planted continuously as a *rabi* crop is also an important measure of crop rotation. In this respect, the rice/wheat area stands out with a high proportion of fields planted to wheat (and rice) every year.

Figure 16.1   Distribution of major wheat-based cropping patterns in Pakistan

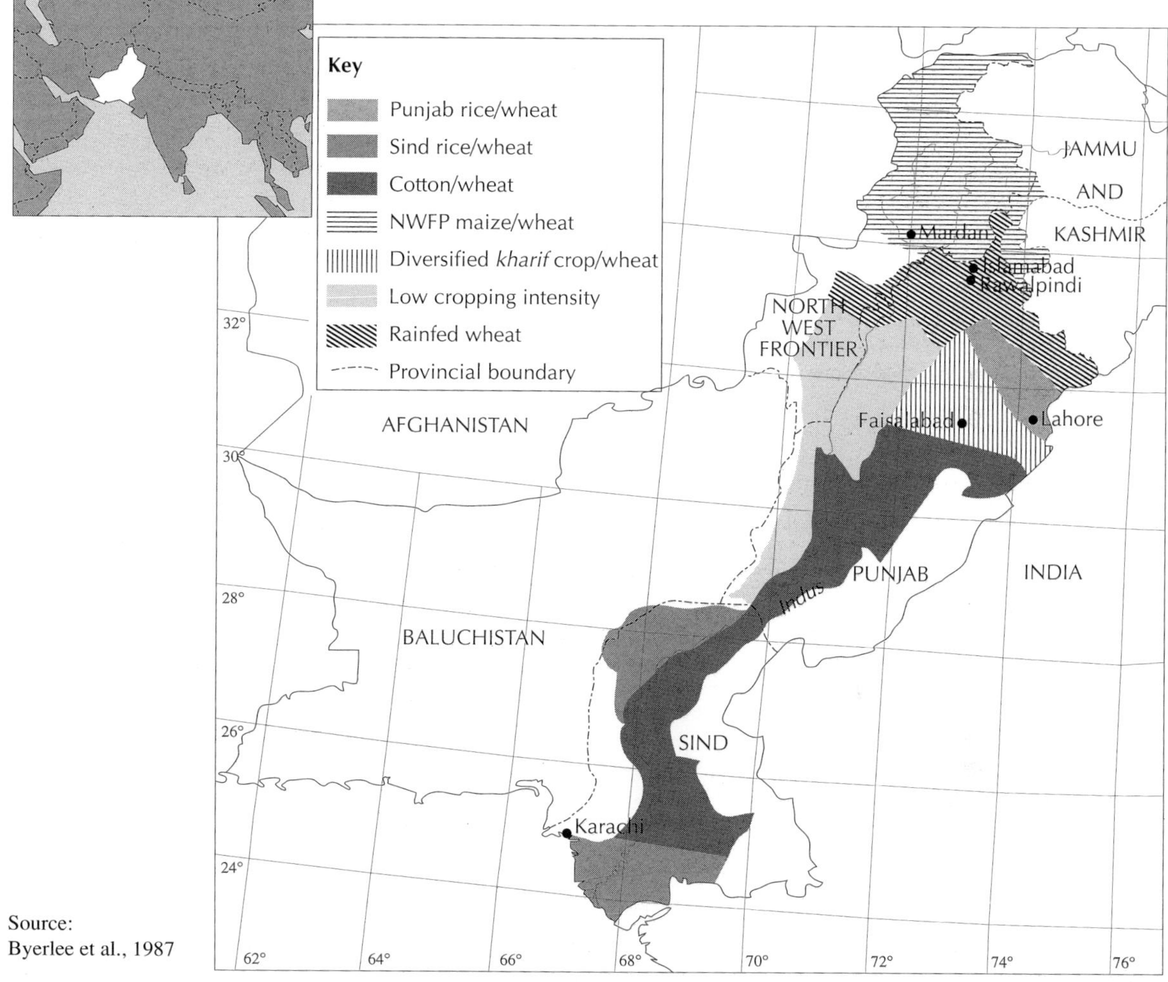

Source:
Byerlee et al., 1987

Research focused not only on demonstrating the similarities and differences between these systems but also on showing the variability within them, even in the relatively homogeneous irrigated areas. Within each area, differences in cropping patterns tend to reflect farmers' resource base and access to markets. For example, in the maize/wheat area of Mardan, sugar cane is concentrated in areas closest to sugar mills. In the cotton/wheat area, larger farmers who produce more wheat than required for home consumption tend to emphasize cotton production and leave more land fallow in the *rabi* season.

The OFR included several methods of data collection. One of the key activities in each research area was an informal diagnostic survey, in which researchers spent about a week in the field during the wheat season talking with farmers and observing their fields. Each survey was conducted by 7-15 researchers, representing a wide range of disciplines and several research institutes (Byerlee et al., 1989). The informal surveys were

Table 16.1   Area, production and yield of wheat in different cropping systems in Pakistan, 1982-83

	Area (000 ha)	Production (000 t)	Yield (t/ha)
Irrigated systems:			
Rice/wheat	1313	2318	1.77
Punjab (Basmati rice)	(1010)	(1843)	(1.83)
Sind (IR6 rice)	(303)	(475)	(1.57)
Cotton/wheat	2559	5067	1.98
Maize/wheat	129	227	1.76
Diversified *kharif* crop/wheat	1062	2131	2.00
Low cropping intensity	420	660	1.57
Rainfed systems:			
Barani areas, northern Punjab and southern NWFP	1007	975	0.97
All systems	5483	10 403	1.90

Source: Byerlee et al., 1986

Table 16.2   Resource base of farmers included in on-farm research programme, Pakistan

	Punjab rice/wheat	Punjab cotton/wheat	NWFP maize/wheat	Punjab rainfed
Average farm size (ha)	8.4	8.2	2.7	2.9
% farms less than 5 ha	56	58	85	89
% fields tenant operated	41	37	59	10
% using tractor in wheat production	82	57	81	88
% using a tubewell in wheat production	81	93	13	0

followed by formal surveys using a short questionnaire and focusing on key practices and issues that had been identified in the informal surveys. An important feature of the formal surveys was that they included a yield sample from 3-5 plots in a specific field for each farmer at the time of harvest. In addition, several other special purpose surveys were conducted during the course of the research. These surveys varied according to the cropping system and were designed to follow up on issues of special relevance to particular systems.

The survey work was followed by an extensive programme of on-farm experimentation in each research area, except for the Punjab cotton/wheat area where resource constraints prevented the initiation of the experimental programme. The experiments were conducted in farmers' fields representing the dominant crop rotations in each research area. These experiments concentrated on a few priority research themes for the target system that had been identified in the diagnostic stage.

Table 16.3   Cropping patterns of areas included in on-farm research studies

System	Location	% cropped area under wheat in *rabi* season	% cropped area under dominant crop in *kharif* season	% fields with at least 3 years' continuous wheat in *rabi* season
Rice/wheat	Gujranwala/ Sheikupura	77	64 (rice)	72
Cotton/wheat	Multan	72	63 (cotton)	51
Maize/wheat	Mardan	66	63 (maize)	27
Rainfed wheat/ fodder	Northern Punjab	94	45 (sorghum/millet)	15

Source: Byerlee et al., 1986

SELECTED RESULTS OF THE ON-FARM RESEARCH PROGRAMME

In this summary of the results of the diagnostic and experimental work, the discussion is limited to results concerning choice of variety, land preparation and planting practices, fertility management and weed control.

Variety and planting date

From the diagnostic surveys, two major results stood out at almost all the sites with respect to varieties and planting dates. The first was that although nearly all the wheat varieties grown in the study areas were semi-dwarf varieties introduced as a result of the Green Revolution (with the exception of a few local varieties grown in the rainfed area), the rate of adoption of newer varieties was very slow (*see* Table 16.4 *overleaf*). This slow rate of adoption has continued and has put many of Pakistan's wheat growing areas at risk from leaf and stripe rust attack, as the older varieties become susceptible to new races of rust.

Yield performance of the newly released varieties was not an issue, as extensive on-farm experimentation had indicated that they were superior to current farmers' varieties. Over 50 experiments in the irrigated areas showed that the new variety Pak-81 gave an average 20% increase in yield over the farmer variety. Although Pak-81 was developed for irrigated areas, it also outyielded the farmers' variety in the medium and higher rainfall parts of the rainfed area. Only in the drier parts is the availability of appropriate varieties more problematic, as farmers need varieties that can be planted deeply into residual moisture. Interviews with farmers in these areas showed they were more interested in varieties that provided more fodder and higher grain quality.

A major follow-up study was conducted to establish the reason for this slow adoption of apparently appropriate higher-yielding varieties in all but the dry rainfed area (Heisey, 1990). Although the situation is complex, it would appear that the major explanation is farmers' unfamiliarity with new varieties and their lack of knowledge about the dangers of a rust epidemic, coupled with an inefficient distribution system for seed of new varieties. Neither the extension service nor the private sector has been aggressive enough in demonstrating and promoting new varieties, and the research service has not invested sufficient resources to monitor the movement of newly released varieties.

Table 16.4   Variety use in wheat areas, Pakistan

	% of fields			
Variety classification	Punjab rice/wheat 1984[1]	Punjab cotton/wheat 1985	NWFP maize/wheat 1984	Punjab rainfed 1985
1. New recommended (Pak-81, Punjab-81, etc.)	16	15	16	55
2. Old recommended (Sonalika)	18	24	37[2]	39[3]
3. Banned (rust susceptible)	66[4]	61[5]	47[2]	—
4. Local	—	—	—	6
5. Total	100	100	100	100

Notes: 1. Year of survey.
2. Variety classification for the maize/wheat area of Mardan is very approximate, as many farmers do not recognize variety names; also, a mixture of varieties is grown in many fields.
3. Lyallpur-73.
4. Mostly Yecora.
5. Mostly WL711.
Source: Byerlee et al., 1986

The second major finding from the diagnostic surveys was the prevalence of late planting in all the irrigated systems surveyed (*see* Table 16.5). This reflected the increase in cropping intensity and the tendency for more wheat to be planted after the dominant *kharif* crop rather than after fallow, as had traditionally been practised. Survey data have confirmed this increasing tendency to plant late (*see* Figure 16.2).

Wheat yields in Pakistan are very sensitive to date of planting because of the effect of high temperatures in the flowering and grain-filling stages. Both on-station and on-farm experiments indicated that each day's

Table 16.5   Distribution of planting dates of wheat in the irrigated areas of Pakistan

Planting date	Punjab rice/wheat (%)	Punjab cotton/wheat (%)	NWFP maize/wheat (%)
Recommended time (before Dec.1)	60	29	66
Somewhat late (Dec.1-15)	30	29	19
Very late (after Dec.15)	10	41	15

Source: Byerlee et al., 1990

Figure 16.2   Percentage of wheat planted late (after 1 December) in the Punjab of Pakistan

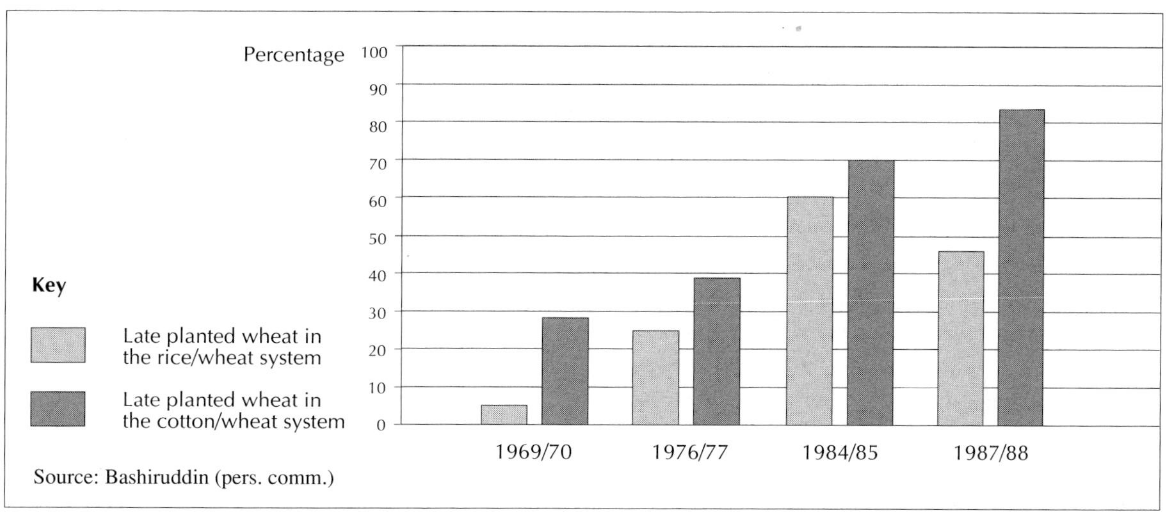

Figure 16.3   Effect of planting date on the yield of Pak-81 and Sonalika wheat varieties at two locations in the Punjab (average from 1981 to 1984)

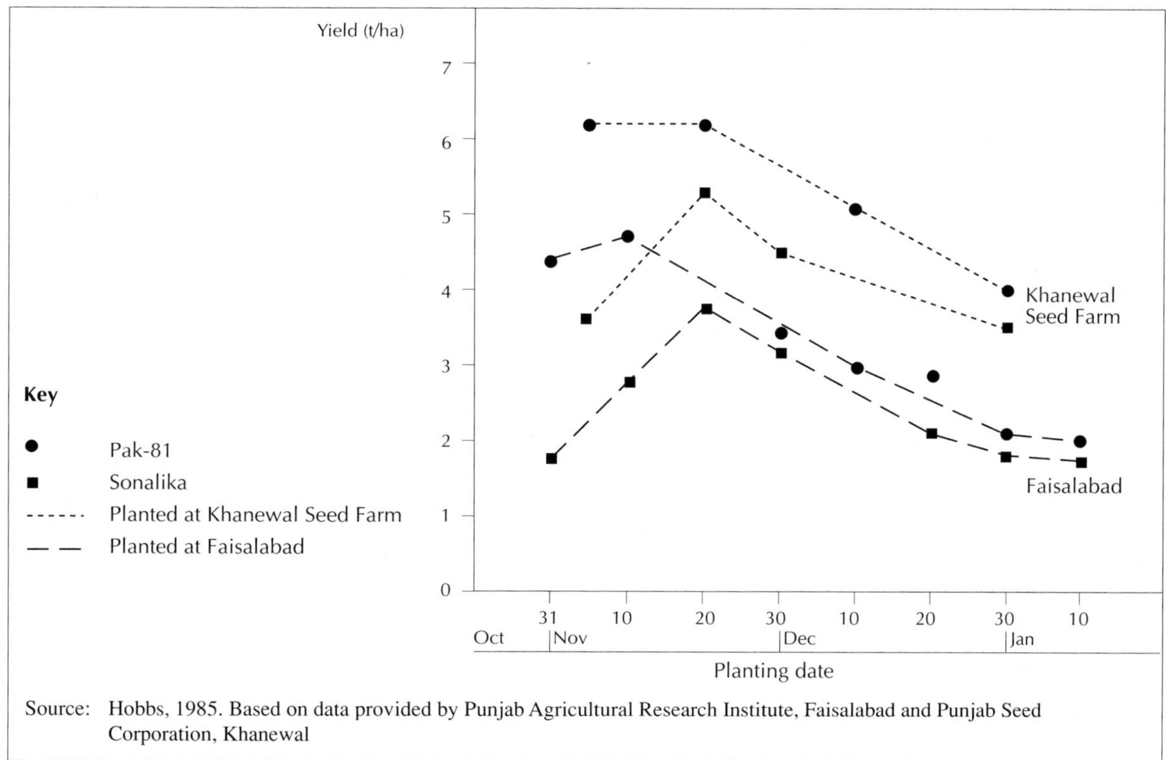

delay in planting after mid-November decreased yields by 1% (Hobbs, 1985; Aslam et al., 1989). However, conflicts with the harvest of the previous crop often push wheat planting well beyond the optimal period, especially in the cotton/wheat system. The wheat planting date is also affected by the management and maturity date of the previous *kharif* crop. However, simple economic calculations show that, at present prices, the returns to one additional cotton picking are more than double the value of wheat lost to delayed planting, so farmers are behaving quite rationally in this respect (Byerlee et al., 1987).

The diagnostic survey data were important in establishing a priority for wheat varieties that performed well at late planting. Wheat breeders have traditionally given priority to evaluating yields at 'optimum' planting dates. Some research was in progress on developing early maturing varieties that were assumed to be most appropriate for late planting. However, an analysis of available experimental data on variety by date of planting showed that the recently released Pak-81 variety, although later maturing than the variety Sonalika recommended for late planting, yielded better at all planting dates (Hobbs, 1985) (*see* Figure 16.3). Also, farmers were found to be using one variety at all planting dates, rather than specific varieties for optimal and late planting.

On the other hand, in the rainfed areas, farmers would benefit from a wheat variety that could be planted earlier in the season, in order to take advantage of the monsoon rains. In drier areas, wheat must be sown deeply into residual moisture at planting time, and the available improved varieties have difficulty emerging vigorously from this depth.

Land preparation and planting practices

The land preparation and planting practices in the four wheat-based systems are shown in Table 16.6. In the irrigated areas, the turnaround time between crops is affected by the intensity of land preparation. This is particularly crucial in the rice/wheat area, where a seedbed for wheat must be prepared on the puddled, compacted soils left from the previous rice crop. Although the survey showed that farmers prepared land more intensively in this area, the resulting seedbed was often inadequate to achieve a good stand of wheat. Farmers sometimes took 3 weeks or longer to prepare the seedbed, and thus sacrificed considerable wheat yields because of the delay in planting. Experiments using a specially designed drill showed that direct drilling of wheat with zero tillage reduced turnaround time, with no loss in wheat yields (Aslam et al., 1989). In other areas, farmers have shown considerable imagination in managing turnaround time between crops. In parts of NWFP, for instance, they have developed an innovative method of relay cropping sugar cane into wheat.

Table 16.6   Land preparation and planting practices in wheat, Pakistan

	Punjab rice/wheat	Punjab cotton/wheat	NWFP maize/wheat	Punjab rainfed
Average number of tillage operations	6.1	4.8	3.3	8.2
Average seed rate	98	110	113	94
% broadcast seed	98	98	92	3

Source: Byerlee et al., 1986

Although research and extension had recommended and widely demonstrated line planting of wheat in the irrigated areas, the vast majority of farmers broadcast wheat seed. Extension loses credibility by persisting with the recommendation to plant in lines, especially when farmers usually do not have the time, draught animals or equipment to follow the recommendation and when experiments show little or no yield advantage over broadcasting.

The survey showed that land preparation and planting practices were very different in the rainfed area, where limited soil moisture near the surface restricts possibilities for broadcasting wheat seed. Instead, most farmers drilled their wheat using an animal-drawn planter or a tractor-drawn seeder. Turnaround time was not such a factor here, except perhaps on the more intensively cultivated *lepara* land (*see below*) near the village. Most of the rainfed wheat was planted on land that had been fallowed during the previous *kharif* season, where farmers ploughed following rain to conserve moisture and control weeds. This accounted for the high number of tillage operations observed. Nevertheless, most ploughing was done with a shallow-tyned cultivator, and repeated passes with a tractor contributed to soil compaction. On-farm experiments at more than 50 locations over 5 years showed a significant yield response of over 25% to the use of a mouldboard plough (Razzaq et al., 1990a). This difference can be attributed to several factors, especially the breaking of the soil compact layer to allow better rooting and water infiltration.

Fertilizer use

Almost all the farmers in the wheat systems studied used chemical fertilizer, but practices and problems varied considerably from one farmer to another and according to field conditions. As expected, in the rainfed area fertilizer use tended to be greater in the higher rainfall zone. But a more important factor in determining fertilizer use in rainfed areas, and in shaping future research, was the local land type classification. Farmers distinguished two types of land — *lepara* fields that are located close to the village and receive farmyard manure, and *mera* fields that are farther from the village and are not regularly manured. The use of farmyard manure was important for maintaining soil fertility and for improving soil structure and moisture retention capacity. *Lepara* land accounted for an average of only 27% of total farm area, but over half of the crop production. On-farm experiments showed that it was necessary to derive separate fertilizer recommendations for these two land types, particularly with regard to phosphorus.

In irrigated areas, few farmers (except those with very little land in the NWFP maize/wheat system) applied farmyard manure to their wheat crop, although almost all of them had considerable experience with chemical fertilizers.[1] Recommendations have traditionally given one fertilizer level for all irrigated wheat areas in Pakistan, but farmers have adjusted their fertilizer doses to reflect differences in cropping patterns, access to irrigation water and availability of resources. On-farm experimentation confirmed the rationality of many of the choices made by farmers and pointed the way towards the development more efficient fertilizer recommendations.

Experiments carried out in both the rainfed and irrigated areas emphasized differential response to fertilizer by crop rotation and land type, and produced recommendations for more homogeneous groups of farmers based on realistic estimates of farmers' total costs of applying fertilizer and acceptable rates of return on capital. Figure 16.4 (*overleaf*) shows estimated fertilizer responses for wheat after maize, sugar cane and rice.

The conventional wisdom in Pakistan has been that nitrogen and phosphorus should be combined in a ratio of 2:1, and indeed government policy reflected the objective of moving to this target ratio by placing a subsidy

Figure 16.4   Fertilizer response curves for wheat in different cropping patterns

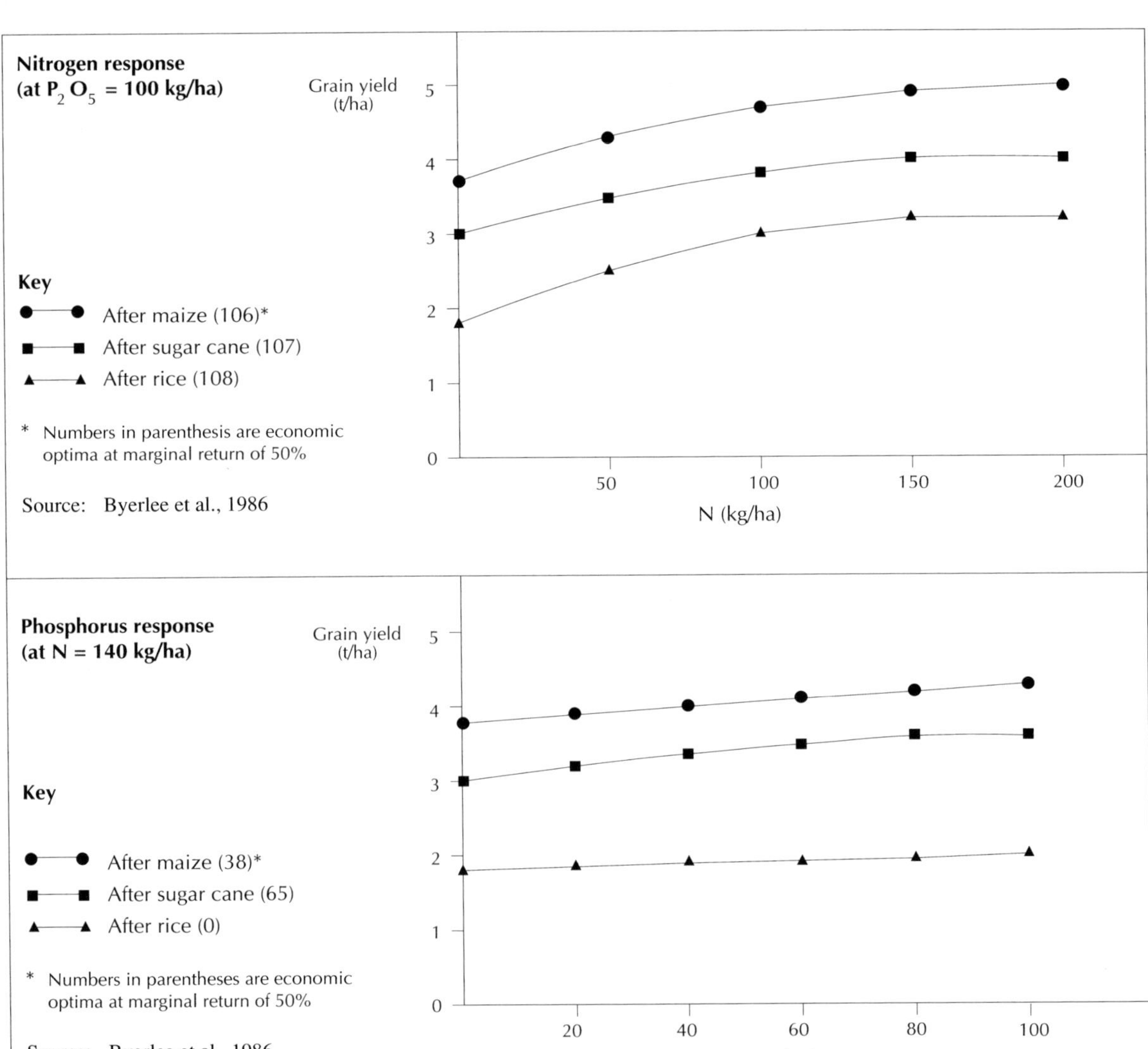

on phosphatic fertilizers. Data from on-farm experiments showed little justification for this rule, and enabled more efficient fertilizer recommendations to be tailored to a variety of farming conditions. The research showed, for instance, that in some cases (for example, in the rice/wheat area) fertilizer use would be more productive if more nitrogen were applied relative to phosphorus, while in other cases the nitrogen : phosphorus ratio should be lowered. In the long run, however, more efficient fertilizer use will depend as much on complementary improvements in irrigation practices, weed control and stand establishment, as on fertilizer management *per se*.

Weed control

As expected, the importance and type of weed problems in wheat varied considerably among the systems studied. In the rainfed area, weeds did not appear to be a major problem. In addition, most of the wheat was intercropped with mustard, which was then harvested on a continuous basis and used for fodder. Although this may appear an irrational practice, experiments showed that mixing wheat and mustard in the same field was quite profitable (Hobbs et al., 1985).

In the irrigated areas, handweeding was practised in only a few fields, and very few farmers used herbicide on wheat. Handweeding was declining because of the increasing cost of labour, but farmers used several other weed control techniques, including crop rotation and pre-planting irrigation and cultivation. Rotation, particularly with the fodder crop Egyptian clover or *berseem* (*Trifolium alexandrinum*), was a common way of controlling grassy weeds, but was profitable only for farmers who had enough animals or were close enough to city markets to justify the fodder produced. Indeed, the diagnostic surveys in the three irrigated areas showed a strong correlation between grassy weed infestation and the continuous planting of wheat in the same field for 3 or more years (without rotation).

OFR identified two particularly important issues related to weed control. In the maize/wheat area, research showed that herbicides applied to control broadleaf weeds were relatively cheap and quite effective (*see* Table 16.7) and that farmers were very open to learning how to use them. In fact, some farmers had taken the lead in experimenting on their own with efficient methods for broadcasting herbicides mixed with soil or sand. The principal constraints appeared to be an inadequate extension effort and apathy on the part of private chemical suppliers. In the rice/wheat area, grassy weeds, particularly *Phalaris minor*, had become a serious problem, especially in fields sown continuously to wheat. Experimentation showed good returns to using herbicides for

Table 16.7   Economics of herbicide application in two irrigated wheat areas

	NWFP maize/wheat (20 sites)		Punjab rice/wheat (14 sites)	
	Broadleaf herbicide (Buctril M)	**Grassy weed herbicide (Dicuran MA)**[1]	**Broadleaf herbicide (Buctril M)**	**Grassy weed herbicide (Dicuran MA)**
Increase in yield (kg/ha)[2]	550[3]	700	0	1000
Gross benefit (Rs/ha)[4]	775	990	—	1420
Cost of herbicide (Rs/ha)[5]	240	525	—	575
Net benefit (Rs/ha)	535	465	—	845
Marginal return (%)	220	89	-ve	152

Notes: 1. Dicuran MA also controls many broadleaf weeds.
2. Experimental yield adjusted downward by 10%.
3. US$ 1 = Rs 17.
4. Net wheat price Rs 1.4/kg after subtracting harvesting, threshing and marketing costs.
5. Doses were 1.4 l/ha for Buctril M and 2.5 kg/ha for Dicuran MA. Costs include Rs 75/ha for application.
Source: Aslam et al., 1989

grassy weeds in this system, although the relatively high cash outlay associated with this practice suggested that further research was required on more efficient application methods, dosages and products.

IMPLICATIONS FOR COMMODITY AND DISCIPLINARY RESEARCH

The OFR results reflected a wide range of wheat production practices and problems in Pakistan. The contrasts and similarities between areas, and the ingenuity and dynamism of Pakistan's farmers, emerged clearly from the data. The challenge is how to integrate this rich set of data from several key systems in order to improve the overall research and policy strategy for wheat at the national level. Although the data derived from location-specific studies on wheat, the implications go well beyond recommendations for a particular location or for a single commodity. Possible courses of action for plant breeding, crop management research, extension, and socioeconomic research and policy analysis are discussed here. In some cases, these actions have been implemented, but in others the links between OFR and commodity or disciplinary research and policy are not well developed enough to utilize the OFR results effectively.

Plant breeding

A number of potential priorities for wheat breeding can be identified from the OFR data (*see* Table 16.8). Wheat breeding in Pakistan is carried out mainly at the provincial level in close coordination with PARC, which organizes national yield trials and sets up the committees for varietal evaluation and release. In the interests of efficiency, crop breeding programmes try to develop varieties that are as widely adaptable as possible. This effort has been particularly successful in Pakistan, as shown by the remarkably consistent performance of varieties such as Pak-81 across a wide range of environments. However, the threat of a rust epidemic continues to hang over the country unless farmers take up these new varieties more quickly than they have in the past. Although much of the responsibility seems to lie with extension in this case, a problem uncovered in interviews with farmers was that the breeding programmes often used similar names for their new varieties, which confused the farmers. As a result of this finding, new varieties now carry distinctive local names.

The OFR programme also identified the need for wheat breeders to more explicitly address the needs of Pakistan's rapidly evolving farming systems. Varieties were traditionally evaluated in variety by planting date trials in order to establish the optimum planting date for a given variety (assuming that the farmer had flexibility in planting time). Moreover, separate programmes were developed for optimum and late planting, with the emphasis on the former. Largely as a result of the OFR work, breeders now give attention to finding varieties for specified planting dates, dictated by the cropping system in which wheat is grown. In addition, instead of separate programmes for 'normal' and late planting, breeders now evaluate their materials over a wide range of planting dates, seeking those varieties that perform well at both normal and late planting, regardless of maturity period. A national-level variety by planting date trial has been implemented for this purpose. More opportunities need to be established for breeders to get early feedback from farmers regarding other characteristics of new varieties, such as chapati qualities, forage quality and shattering, that often slow the adoption of new varieties. This can be managed through more informal meetings between farmers and breeders, combined with occasional surveys and follow-ups of on-farm experiments.

The OFR work also highlighted the need for close communication between wheat breeders and breeders for other crops, especially cotton and rice, grown in rotation with wheat in irrigated cropping patterns. The

Table 16.8   Priorities for wheat breeding, Pakistan

Priority	Rice/wheat	Cotton/wheat	Maize/wheat	Rainfed
Stripe rust resistance	*		*	*
Leaf rust resistance	*	*	*	*
Varieties suitable for late planting	*	*	*	
Varieties with better emergence from deep planting				*
Varieties suitable for early planting				*
Better contact with other commodity programmes	*	*	*	

conflict between cotton harvesting and wheat planting can be resolved, for example, by using earlier varieties of cotton or varieties of wheat suitable for late planting. Likewise, changes in maturity of cotton or rice varieties, or even changes in the management or the price of these crops, can have important implications for wheat breeders (Byerlee et al., 1987). In some cases these conflicts may best be resolved by substituting other crops, such as oilseeds, for late planted wheat if development of appropriate varieties (especially early maturing ones) of these crops appears more likely than development of wheat varieties for late planting (Tetlay et al., 1990). Unfortunately, mechanisms for close coordination between commodity breeding programmes for crops grown in the same cropping system are virtually non-existent.[2] Indeed, several hundred kilometres separate the main cotton and rice breeding institutes from the wheat breeding institute.

Long-term crop management research

Despite the potential for further gains though breeding, it is obvious that much of the challenge of increasing the productivity of Pakistan's wheat-based farming systems must be taken up by crop management research, especially on issues requiring a longer time period to assess the implications for crop management and productivity.

Cropping intensity is increasing in all the systems studied, and researchers need to explore ways to help farmers gain experience with new options. In the rainfed areas, most wheat is still grown after fallow because of a perceived need for moisture conservation. New cropping patterns need to be designed so that farmers can take better advantage of available land and moisture. In the irrigated systems, where cropping intensities tend to be higher, there is much scope for exploring alternative crops and management practices. However, it will be necessary to examine the sustainability of these more intensive systems, paying particular attention to factors such as soil fertility (especially micronutrients), pest populations (for example, soil-borne pathogens), and irrigation water management (*see* Table 16.9 *overleaf*).

These issues require a fresh outlook on agronomic research and planning in order to deploy research resources in the most efficient manner on those issues most relevant to the future stability of major farming systems. Some agronomists may have to specialize in more strategic research to do justice to issues related to the sustainability of intensive cropping systems.

Table 16.9   Long-term priorities for crop management research, Pakistan

Priority	Rice/wheat	Cotton/wheat	Maize/wheat	Rainfed
Increase cropping intensity	*	*	*	*
Increase crop diversification	*	*		
Effects of long-term rotations on fertility and pests	*	*	*	*
Micronutrients	*	*	*	*
Irrigation water efficiency	*	*	*	
Effects of reduced tillage	*			*
Direct drilling of wheat	*	*		
Alternative tillage methods and soil structure improvement	*			*
Efficient herbicide application	*	*	*	
Soil health problems	*			*
Farmyard manure management and conservation			*	*
Fodder management and conservation			*	*

The future of effective agronomic research in Pakistan also depends upon the development of effective collaboration between commodity programmes and research institutes, and better integration of the various disciplines related to crop management, such as soil science, irrigation, water management, soil fertility, tillage and weed science. These disciplines have traditionally worked in isolation from each other, usually in separate disciplinary research institutes and often at different locations.

Most of the long-term issues in crop management research will be specific to a particular system, such as tillage in rice/wheat, direct drilling of wheat following cotton, or the management of mixed cropping in the rainfed areas. Some research may also be relevant across systems, such as the management of crop residues and organic manures in order to stabilize or build soil organic matter. This work will require the attention of multi-disciplinary teams that have one foot in the field and the other at the research station.

A good example of interdisciplinary collaboration stimulated by the results of adaptive research is provided by Inayatullah et al. (1989). In the rice/wheat system, after zero tillage on wheat had been shown to be feasible and cost effective, there was some concern that leaving rice stubble in the field would increase the problem of rice stem-borer damage in the following rice crop. Accordingly, the rice entomologist organized field surveys during the wheat season to examine the population of stem-borers in zero-tilled and conventionally tilled fields. On the basis of this work, he was able to devise a strategy for stem-borer control that allowed for zero-tillage.

This type of strategic crop management research suggests not only a need for better integration of crop management disciplines but also a longer-term perspective in research. Both long-term experiments in representative sites on experiment stations, as well as long-term monitoring of farmers' fields and practices, may be appropriate. The organization of long-term research that cuts across commodities and disciplines is especially challenging. In addition, while much of this research will be done initially under fairly controlled conditions, it should evolve as quickly as possible to adaptive research with the full participation of farmers.

In rainfed areas, an even more difficult challenge facing crop management research is the issue of fodder management and conservation and its interaction with wheat management. More efficient ways of producing and conserving fodder need to be explored (Byerlee et al., in press). This may require experimentation with speciality fodder crops and a closer examination of the trade-offs in intercropping fodder and grain in maize and wheat production. It may even involve OFR with animal production and collaboration with livestock scientists. In Pakistan there has been very little contact between crop and livestock scientists, despite the great importance of animals in many of the country's farming systems. The recent establishment by PARC of a coordinated farming systems programme has helped initiate better communication between crop and livestock scientists.

Extension services

The OFR programme has already contributed to strong adaptive research efforts in several areas. It has become obvious in the course of this research that an important priority is to improve the capacity of the extension services to transfer the results of the adaptive research. The extension services are organized at the provincial level and have recently adopted the Training and Visit (T&V) extension method. Nonetheless, coordination between research and extension, especially at the local level, remains weak.

There are a number of immediate opportunities for extension to make an impact (*see* Table 16.10). One of the most pressing of these is to organize better demonstrations of new wheat varieties and to make farmers more aware of the dangers of rust epidemics and the risks of a breakdown in rust resistance of wheat varieties. In those areas where chemical weed control has potential, the extension services could stimulate the private sector to acquaint farmers with products and equipment, teach safe application methods and help organize the availability of sprayers. In one case study in a Punjab village in the rice/wheat system where herbicide use had been widely adopted, it was found that the factors determining adoption were the presence of an extension agent who had made available a backpack sprayer for rental to farmers and a local farmer who then developed his

Table 16.10   Short-term priorities for adaptive research and extension, Pakistan

Priority	Rice/wheat	Cotton/wheat	Maize/wheat	Rainfed
Demonstration of rust-resistant varieties	*	*	*	*
Adjust fertilizer dose to higher N:P ratio	*			*[1]
Adjust fertilizer dose to lower N:P ratio		*	*	
Broadleaf weed control			*	
Grassy weed control	*			
Deep tillage			*	*

Note: 1. On *lepara* land.

own rental service (Ahmad et al., 1988). As the OFR programme moves into issues such as alternative tillage methods, long-term strategies to maintain fertility and integrated pest control, the extension services must play an increasingly important role in providing farmers with the information and skills to use these management-intensive technologies more effectively, as well as in organizing or encouraging the availability of inputs and machinery needed to implement the technologies where appropriate.

Given the large number of systems and the diversity within them, it is unreasonable to expect that the OFR teams currently in operation will be sufficient to provide the location-specific information required for further increases in the productivity of these systems. Increasingly, the extension services will have to carry the responsibility for adaptive research, with orientation and training from the research system.

Socioeconomic research and policy analysis

An important achievement of the OFR programme has been the strengthening of socioeconomic research units in the provincial research institutes. To improve collaboration with crop scientists, socioeconomists need to develop expertise in particular research areas (such as soil fertility) and systems (such as rice/wheat). Some of the priority issues for socioeconomists will require virtually their full-time presence in the field, often in association with an adaptive research and extension effort (*see* Table 16.11); other issues will require close interaction with research directors and policy makers. Although there will inevitably be some specialization among socioeconomists, leaders of socioeconomic research units should have a staff development strategy that allows researchers to develop confidence and familiarity with both the farmer's field and the policy maker's office.

A wide range of issues arising from the OFR programme require socioeconomic research. Some of them involve analysing trends and changes in national policy and interpreting the implications to crop researchers. These include analysing fertilizer response data, in the light of the expected removal of subsidies on phosphatic fertilizer, and studying the comparative advantage of oilseed crops with respect to wheat, in view of the growing demand for vegetable oil and government emphasis on oilseed production. Changes in the demand for and the importance of livestock products also has implications for setting research priorities in several of the wheat-

Table 16.11   Priorities for socioeconomic research, Pakistan

Priorities	Rice/wheat	Cotton/wheat	Maize/wheat	Rainfed
Seed distribution systems	*	*	*	*
Fertilizer subsidy	*	*	*	*
Comparative advantage of oilseeds	*	*		*
Costs of harvesting, labour displacement	*	*	*	*
Irrigation policy	*	*	*	
Demand for livestock			*	*
Monitoring technological change	*	*	*	*
Community decision-making in land use				*

Figure 16.5   Analysis of varietal development, seed system and information transfer in understanding slow wheat varietal replacement in Pakistan

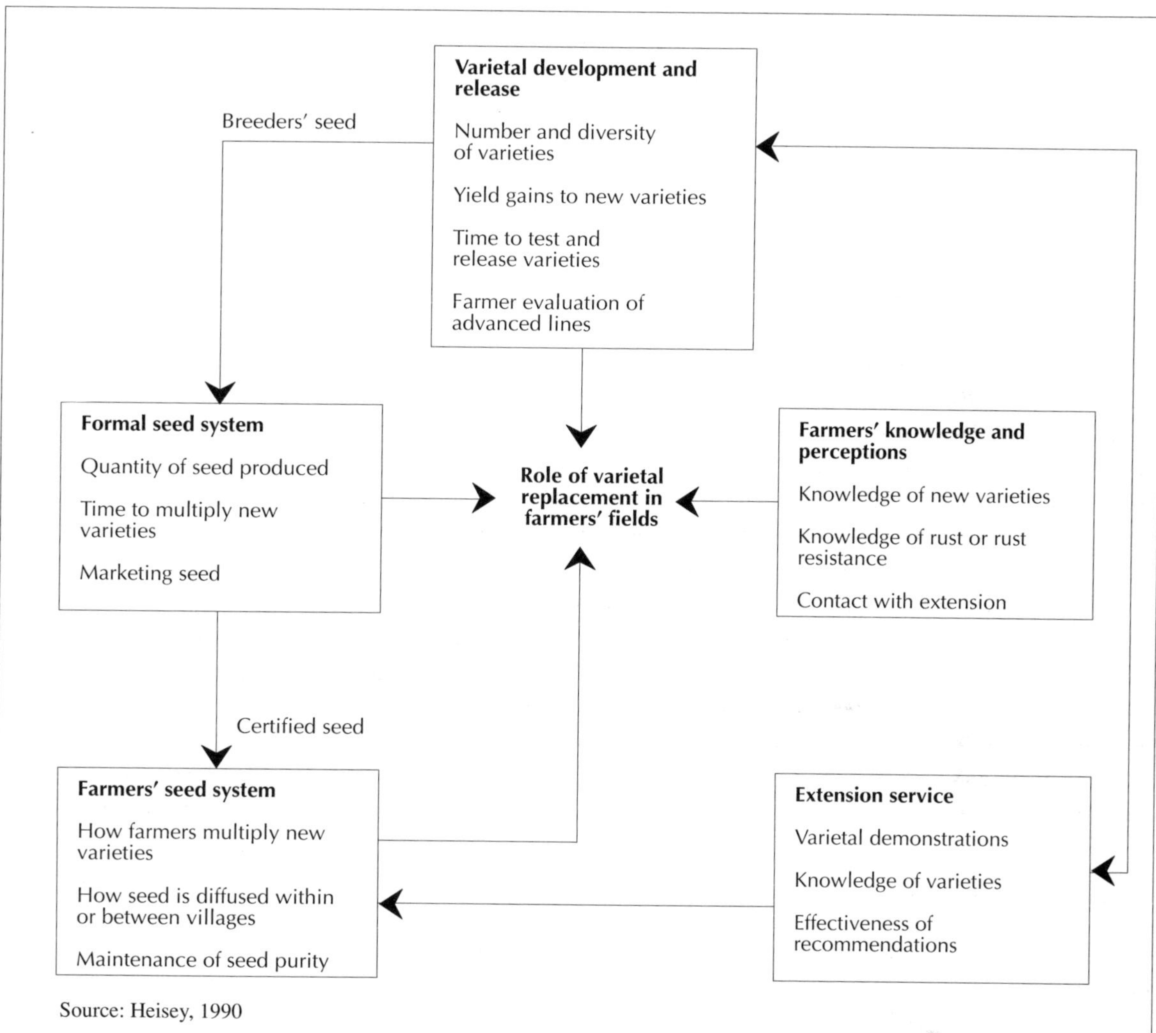

based farming systems. A particularly complicated issue, related to wheat harvesting, is the trend toward mechanized harvesting and its implications for costs of production, labour supply and equity (Smale, 1987). In rainfed areas, intensification of cropping may require extensive research at the community level, on such issues as the nature of community decision-making regarding land use, fallowing and livestock grazing.

Socioeconomists should also take responsibility for developing an effective system for monitoring technological change and evaluating research impact. The OFR studies filled in some large gaps in researchers' knowledge about how new technologies are spreading (for example, mechanization) or stagnating (for example, wheat varieties). Such knowledge is crucial for research decision-making and for justifying research pay-offs to those who fund agricultural research. The Agricultural Economic Research Units (AERUs) have already initiated monitoring by conducting annual surveys of wheat, rice and cotton varieties. Recently, these surveys have expanded to include key management practices such as planting date and the use of fertilizers and herbicides. Given the size of these surveys (1000 farmers), socioeconomists must ask whether they should continue this work, or work with the crop reporting service responsible for crop statistics to collect the appropriate data and make it available in a timely manner to agricultural research institutes.

Clearly, much of the OFR work has important implications for policy, especially in improving the efficiency of input delivery. A major challenge for socioeconomists working at research institutes is how to make this information available to policy makers. The case study on the slow adoption of wheat varieties was aimed specifically at identifying the policy changes needed to accelerate varietal adoption. This study included analyses of the farmers' information system, the formal seed distribution system, the informal seed distribution system and the wheat breeding system (*see* Figure 16.5), and produced a number of policy recommendations, ranging from the procedure involved in naming varieties to commissions given to private vendors of certified seed (Heisey, 1990). Although meetings were held to discuss the results of the findings, the large number of agencies involved from the research, seed and extension system made it difficult to effectively reach all the relevant decision makers.

Another policy implication arising from the OFR programme concerns the subsidy on phosphatic fertilizer. The research results suggested that in some systems phosphate was a limiting nutrient, while in others, such as the rice/wheat system, farmers may have already exceeded the optimum dose. Thus, while the subsidy may have been appropriate in some situations, in others it may have led to inefficiency. The results were considered in the negotiations between donor agencies and the government that eventually led to the removal of the subsidy.

An important condition for fuller use of OFR results in policy analysis is the need for results to be synthesized and reported in a timely manner. In the Pakistan OFR on wheat, diagnostic surveys were usually reported within 6 months of completion. The results of the on-farm experiments have now been synthesized over 5 years at each site (Aslam et al., 1989; Razzaq et al., 1990a; Razzaq et al., 1990b). While reporting results does not guarantee that policy makers will read them, it is an important first step toward facilitating their use in policy decisions.

CONCLUSION

The main conclusion arising from this review of the OFR programme in Pakistan's wheat-based systems is that further progress demands better integration between disciplines, commodities and institutions, as well as integration between strategic research, adaptive research and extension. Without this integration there is little possibility of effectively addressing the issues identified in the OFR programme or of exploiting the potential for using OFR to set research priorities and guide policy decisions at the national level.

Take the case of the rice/wheat system. Crop breeding research needs to develop alternative varieties of rice and wheat that are able to minimize conflicts in the planting time of wheat. Extension and the private sector need to ensure that farmers know about these varieties and can judge their performance. Research is needed on alternative tillage and planting methods for both rice and wheat, and long-term rotation trials on reduced tillage, oilseeds and organic manures should be initiated to analyse the sustainability of these new systems. Many of these trials will initially be conducted on the research station through close collaboration between the rice and wheat commodity programmes, but they need to be based on an understanding of farmers' existing cropping systems and the problems inherent in those systems. There is thus an inevitable link to the adaptive research on fertilizer use or weed control currently in progress in the field.

The common thread through this complex research programme is the ability to understand and work under farmers' conditions. OFR provides the perspective and the methods to do this. The farming systems perspective gives researchers a framework for dealing with agricultural systems that are becoming increasingly complex. The perspective itself is not enough, however. High-quality research will be required, and it must be founded in a research strategy that goes beyond narrow disciplines or commodity interests. If the methods of OFR are understood and applied *throughout the research organization*, there is a chance of giving small farmers a 'voice' at all levels (Norman, 1980) and making sure that that voice is heard wherever research decisions are made. But simply working with farmers in their fields will not be enough either. Progress will still depend on providing better education and political organization to the farming population so that farmers can exert pressure more directly on research and extension organizations to meet their needs.

A strategy that demands that a considerable proportion of research be carried out in farmers' fields and requires coordination and communication among a variety of researchers and institutions is potentially expensive and beyond normal management capacity (a review of the management factors in OFR is provided in Chapter 15). It should be kept in mind that the research described here relates only to a few representative and important wheat-based systems in Pakistan. Extending this strategy to include all the major wheat growing areas, or to include other farming systems, represents a tremendous challenge.

Selecting priority farming systems and target groups for a research effort is a difficult task. Locations for strategic research must be representative of the conditions of important farming systems. Adaptive research must be organized so that it reaches large numbers of farmers. This is especially difficult in the rainfed systems of Pakistan that exhibit varying conditions and problems within a small area. Where possible, the selection of areas for strategic research and adaptive OFR should be coordinated to take advantage of the inevitable links between the two.

Staffing assignments represent an additional challenge. Most scientists would profit from the opportunity to move back and forth between farm-level and experiment station assignments. Adaptive research may at times demand full-time field presence, but there should be opportunities to rotate in and out of such assignments. Initially, it may be advisable to establish a small working group for an important cropping system with representatives of each major commodity programme. This group would be charged with designing strategic research and on-farm adaptive research that address important constraints in the system.

Lastly, the issue of better communication among all the agents involved in technology development and transfer is a fundamental issue which must be addressed if this strategy is to function. Rather than establishing new departments or programmes, a series of opportunities must be presented to researchers, extension agents and farmers in order to facilitate more effective communication between these parties. Working groups (whose lifetime is a function of their productivity), joint planning meetings and joint field activities are all possibilities. In the Pakistan case, where OFR was organized by wheat researchers, communication of the OFR results was most effective with on-station wheat researchers, especially breeders. This was facilitated by travelling seminars during which on-station and on-farm wheat researchers reviewed and discussed each other's work. Communication with other commodity programmes was more difficult, as was communication with the wide array of disciplinary research programmes. Communicating results to policy makers is undoubtedly the most challenging task, if only because of the tenuous links between researchers and policy makers.

Although it is easy to design communication systems built around planning meetings and workshops, perhaps the most effective technique for improving communication is through the informal diagnostic survey where teams representing different commodities and disciplines talk informally to farmers and observe their fields, and then meet daily to synthesize findings and come to a common understanding of problems and opportunities. This type of diagnostic survey has become well established in OFR programmes. The challenge now is to extend the method to setting priorities for on-station research, including varietal development and long-term crop management research, and even to guide policy making.

Notes

1. The use of farmyard manure has decline in irrigated areas, with tractors replacing draught animals and with competing non-farm uses of farmyard manure (for example, as cooking fuel).

2. Despite this, there has been some recent success in developing earlier maturing rice and cotton varieties (Akhtar et al., in press; Sharif et al., in press).

This chapter draws on the work of many scientists who participated in the joint PARC/CIMMYT research effort, including Munir Ahmed, Zulfiqar Ahmed, Ramzan Ahktar, Mohammed Aslam, Mohammed Azeem, Naeem Hashmi, Sajidin Hussain, Baktmand Khan, Bakht Roidad Khan, Abdul Majid, Mohammed Munir, Abdul Razzaq, Mohammed Shafique, Mohammed Sharif, A.D. Sheikh, and Khaleel Tetlay.

References

Ahmad, Z., Sharif, M., Longmire, J. and Tetlay, K.A. 1988. *Weed Management Strategies for Wheat in the Irrigated Punjab: Farmers' Knowledge, Adoption and Economics.* PARC/CIMMYT Paper No. 88-3. Islamabad, Pakistan: PARC.

Akhtar, M.R., Byerlee, D. and Tetlay, K. (in press). Developing appropriate cropping patterns for the cotton/wheat system of the southern Punjab: Plant breeding and policy perspectives. In Byerlee, D. and Hussain, T. (eds) *Farming Systems of Pakistan. Diagnosing Priorities for Agricultural Research.* Lahore, Pakistan: Vanguard Publishing.

Aslam, M., Majid, A., Hobbs, P.R., Hashmi, N.I. and Byerlee, D. 1989. *Wheat in the Rice-Wheat Cropping System of the Punjab: A Synthesis of On-Farm Research Results, 1984-1988.* PARC/CIMMYT Paper No. 89-3. Islamabad, Pakistan: PARC.

Byerlee, D. 1990. Technical change, productivity, and sustainability in irrigated wheat systems of Asia: Emerging issues. Paper presented at the Symposium, Post-Green Revolution Asian Agriculture: Prospects for Sustaining Productivity Growth, at the Annual Meetings of the American Agricultural Economics Association, August 1990, Vancouver, Canada.

Byerlee, D., Hobbs, P., Khan, B.R., Majid, A., Aktar, M.R. and Hashmi, N.I. 1986. *Increasing Wheat Productivity in the Context of Pakistan's Irrigated Cropping Systems: A View from the Farmer's Field.* Research Report 86-7. Islamabad, Pakistan: PARC.

Byerlee, D., Akhtar, M.R. and Hobbs, P. 1987. Reconciling conflicts in sequential double cropping patterns through plant breeding: The example of cotton and wheat in Pakistan's Punjab. *Agricultural Systems* 24: 291-304.

Byerlee, D., Heisey, P. and Hobbs, P. 1989. Diagnosing research priorities for small farmers: Experiences from on-farm research in Pakistan. *Quarterly Journal of International Agriculture* 28(3/4): 254-65.

Byerlee, D., Sheikh, A.D. and Azeem, M. (in press). Food, fodder, and fallow: Analytics of the *barani* farming systems of northern Punjab. In Byerlee, D. and Hussain, T. (eds) *Farming Systems of Pakistan. Diagnosing Priorities for Agricultural Research.* Lahore, Pakistan: Vanguard Publishing.

Heisey, P.W. (ed). 1990. *Accelerating the Transfer of Wheat Breeding Gains to Farmers: A Study of the Dynamics of Varietal Replacement in Pakistan*. CIMMYT Research Report No. 1. El Batan, Mexico: CIMMYT.

Hobbs, P.R. 1985. Agronomic practices and problems for wheat following cotton and rice in Pakistan. In *Wheats for More Tropical Environments: Proc. of International Symposium*. El Batan, Mexico: CIMMYT.

Hobbs, P.R., Razzaq, A., Hashmi, N.I., Munir, M. and Khan, B.R. 1985. The effect of mustard grown as a mixed intercrop on the yield of wheat. *Pakistan Journal of Agricultural Research* 6(4): 241-47.

Hobbs, P., Hashmi, N.I. and Ahmed, N.U. 1988. Rice-wheat cropping systems in irrigated rice environments. Paper presented at the International Rice Research Conference, November 1988. Los Baños, Philippines: IRRI.

Inayatullah, C., Ul-Haq, E., Ul-Mohsin, A., Rehman, A. and Hobbs, P.R. 1989. *Management of Rice Stem-Borers and the Feasibility of Adopting No-Tillage in Wheat*. Islamabad, Pakistan: Entomological Research Laboratories, PARC.

Norman, D.W. 1980. *The Farming Systems Approach: Relevancy for the Small Farmer*. Rural Development Paper No. 5. East Lansing, USA: Department of Agricultural Economics, Michigan State University.

Razzaq, A., Khan, B.R., Khan, B., Hobbs, P.R., Hashmi, N.I. and Heisey, P. 1990a. *Irrigated Wheat in North West Frontier Province: A Synthesis of On-Farm Research Results 1983-86*. PARC/CIMMYT Paper 90-1. Islamabad, Pakistan: PARC.

Razzaq, A., Hashmi, N.I., Khan, B., Khan, B.R. and Hobbs, P.R. 1990b. *Wheat in Barani Areas of Northern Punjab: A Synthesis of On-Farm Research Results, 1982 to 1988*. PARC/CIMMYT Paper 90-2. Islamabad, Pakistan: PARC.

Sharif, M., Longmire, J., Shafique, M. and Ahmad, Z. (in press). System implications of varietal adoption: Basmati-385 in the rice-wheat zone of the Punjab. In Byerlee, D. and Hussain, T. (eds) *Farming Systems of Pakistan. Diagnosing Priorities for Agricultural Research*. Lahore, Pakistan: Vanguard Publishing.

Smale, M. 1987. *Wheat Harvest Technology in Punjab's Rice-Wheat Zone: Combines, Laborers and the Cost of Harvest Delay*. PARC/CIMMYT Paper No. 87-23. Islamabad, Pakistan: PARC.

Tetlay, K., Byerlee, B. and Ahmad, Z. 1990. Role of tractors, tubewells and plant breeding in increasing cropping intensity in Pakistan's Punjab. *Agricultural Economics* 4: 13-25.

Acronyms

AARD   Agency for Agricultural Research and Development (Indonesia)

ACIAR   Australian Centre for International Agricultural Research

ADD   Agricultural Development Division (Malawi)

AERUs   Agricultural Economic Research Units (Pakistan)

AGRITEX   Agricultural Extension (Zimbabwe)

ARFSN   Asian Rice Farming Systems Network

ARPT   Adaptive Research Planning Team (Zambia)

BAEx   Bureau of Agricultural Extension (Philippines)

BARC   Bangladesh Agricultural Research Council

BARI   Bangladesh Agricultural Research Institute

BDA   Banco de Desarrollo Agropecuaría (Panama)

BPI   Bureau of Plant Industry (Philippines)

CCRI   Cereal Crops Research Institute (Pakistan)

CGIAR   Consultative Group on International Agricultural Research

CIAT   Centro Internacional de Agricultura Tropical (International Center for Tropical Agriculture)

CIDA   Canadian International Development Agency

CIMMYT   Centro Internacional de Mejoramiento de Maíz y Trigo (International Maize and Wheat Improvement Center)

CIP   Centro Internacional de la Papa (International Potato Center)

COFRE   Committee for On-Farm Research and Extension (Zimbabwe)

CRI   Crops Research Institute (Ghana)

CRIFC   Central Research Institute for Food Crops (Indonesia)

CTA   Technical Centre for Agricultural and Rural Development (Lomé Convention)

DR&SS   Department of Research and Specialist Services (Zimbabwe)

ENDEMA   Empresa Nacional de Maquinaria Agrícola (Panama)

FAO   Food and Agriculture Organization of the United Nations

FITT   Farmer Innovation and Technology Testing programme (The Gambia)

FSR   farming systems research

FSR&DD   Farming Systems Research and Development Division (Nepal)

GGDP   Ghana Grains Development Project

GLDB   Grains and Legumes Development Board (Ghana)

IARCs   international agricultural research centres

IBRD   International Bank for Reconstruction and Development

ICRA   International Centre for Development-Oriented Research in Agriculture (Netherlands)

ICRAF   International Council for Research in Agroforestry

ICRISAT   International Crops Research Institute for the Semi-Arid Tropics

ICTA   Instituto de Cienca y Tecnología Agrícolas (Guatemala)

IDIAP   Instituto de Investigaciones Agropecuarias de Panama

IDRC   International Development Research Centre

IDS   Institute of Development Studies (UK)

IFPRI   International Food Policy Research Institute

IICA   Instituto Interamericano de Cooperacíon para la Agricultura

IITA   International Institute of Tropical Agriculture

ILCA   International Livestock Centre for Africa

INADES   Institut Africain de Développement Economique et Social

INGER   International Network for Genetic Evaluation of Rice

INIAP   Instituto Nacional de Investigaciones Agropecuarias (Ecuador)

INIPA   Instituto Nacional de Investigación Pecuaria e Agricola (Peru)

INTSORMIL   International Sorghum and Millet Program

IPRA	Participatory Research in Agriculture (CIAT)
IRAZ	Institut de Recherche Agronomique et Zootechique (Great Lakes Region, Africa)
IRRI	International Rice Research Institute
ISAR	Institut des Sciences Agronomiques du Rwanda
ISNAR	International Service for National Agricultural Research
ISRA	Institut Sénégalais de Recherches Agricoles
LAC	Lumle Agricultural Centre (Nepal)
MARIF	Malang Research Institute for Food Crops (Indonesia)
MAWD	Ministry of Agriculture and Water Development (Zambia)
MIDA	Ministry of Agricultural Development (Panama)
MINAGRI	Ministry of Agriculture (Rwanda)
MLARR	Ministry of Lands, Agriculture and Rural Resettlement (Zimbabwe)
MOA	Ministry of Agriculture (Ghana)
NFTA	Nitrogen-Fixing Trees Association
NGOs	non-governmental organizations
NWFP	North West Frontier Province, Pakistan
ODI	Overseas Development Institute (UK)
OFCOR	on-farm client-oriented research
OFR	on-farm research
PAC	Pakhribas Agricultural Centre (Nepal)
PAP	Projet Agropastoral de Nyabisindu (Rwanda)
PARC	Pakistan Agricultural Research Council
PCARRD	Philippine Council for Agriculture and Resources Research and Development
PCCMA	Proyecto Colaborativo Centroamericano para el Mejoramiento de Cultivos Alimenticios
PUDOC	Centre for Agricultural Publishing and Documentation (The Netherlands)
RADIP	Rainfed Agricultural Development (Iloilo) Project (Philippines)
RELOs	research-extension liaison officers
SADCC	Southern Africa Development Coordination Conference
SDC	Swiss Development Corporation
SERED	Socio-Economic Research and Extension Division (Nepal)
SPAAR	Special Program for African Agricultural Research
T&V	Training and Visit extension method
TAC	Technical Advisory Committee (CGIAR)
UNDP	United Nations Development Programme
UPLB	University of the Philippines, Los Baños
USAID	United States Agency for International Development
WCED	World Commission on Environment and Development

Index

AARD (Indonesia) 143, 292
adaptation, farmer *see* farmer adaptation
adaptive research *see* research (adaptive)
ADD (Malawi) 267
adoption *see* surveys (adoption); technology adoption;
  technology diffusion
AERUs (Pakistan) 333
Africa 5, 6, 19, 66, 90, 99, 257-70, 283, 289
  see also individual countries
agricultural economists *see* economic analysis; social
  science; socioeconomic research
agricultural intensification 12, 41, 45, 89, 109, 117, 287,
  318, 329, 333
AGRITEX (Zimbabwe) 267
agroforestry 52, 53, 59, 283
  see also alley farming; trees
alley farming 85-107, 283
  see also agroforestry; trees
animal damage 52, 68, 69, 70, 72, 88, 91, 127
animal traction *see* draught power
anthracnose 40, 59
anthropology *see* social science; sociocultural issues
applied research *see* research (applied)
ARFSN (Asia) 7
ARPT (Zambia) 264, 268, 277
ascochyta blight 40
Asia 135, 140, 267, 287, 289
  see also individual countries
autonomous changes, farmer *see* farmer innovation
avocado 40, 56

BAEx (Philippines) 115, 116, 136
bananas 40, 41, 43, 46, 58, 86
Bangladesh 303, 307
BARC (Bangladesh) 292
BARI (Bangladesh) 291, 292
barley 252, 253
BDA (Panama) 216, 217, 218, 219, 227, 229, 230,
beans 39-61, 191-212, 216, 218, 220, 224, 226, 251,
  253, 261, 262, 279, 284
    bush bean 39-61, 195, 198, 199
    climbing bean 39-61, 195, 198, 199, 252, 253
    snap bean 39, 275
    varieties 43, 50, 51, 191, 192, 194, 195, 197, 198,
    199, 251

see also diseases and disease control; pests and pest
  control; planting practices
Botswana 257, 280
BPI (Philippines) 115, 116, 117
broad bean 252
browse 88, 91, 92, 97, 98, 104
buffaloes *see* draught power
Burma 114
Burundi 41, 45

Calliandra spp. 52, 53, 59
Cameroon 99, 280
Caqueza project (Colombia) 6
career advancement *see* extension (staff issues); on-farm
  research (staff issues)
cassava 40, 41, 56, 65, 66, 80, 86, 87, 88, 90, 91, 94, 97,
  98, 103, 143, 144, 145, 146, 276
cattle 52, 88, 101, 146, 173, 174, 216
  see also draught power
CCRI (Pakistan) 18, 170, 171
CIAT 17, 18, 41, 42, 48, 51, 56, 61, 192, 194, 281, 284
CIDA (Canada) 63
CIMMYT 18, 19, 63, 66, 144, 148, 182, 194, 215, 216,
  217, 228, 230, 252-54, 318
citrus *see* fruit trees
CIP 12, 19, 233-44
climate 48, 65
  see also drought; rainfall
clover, Egyptian 172, 173, 175, 184, 327
co-operatives *see* farmers' organizations
cobrot 261
cocoa 5, 86, 90, 99
cocoyam 86
coffee 49
COFRE (Zimbabwe) 267, 307
collaborative research *see* on-farm research
  (complementarity); research (interdisciplinary);
  research-extension links/collaboration
Colombia 6, 243, 275, 278, 279, 284
commercial agriculture 80, 242, 263, 288, 289, 293
committees *see* COFRE; on-farm research (monitoring);
  research-extension links/collaboration
consumer preferences 23, 27, 39, 51, 58, 59, 80, 146,
  160, 185, 232, 233, 253, 258, 279, 328
consultative participation *see* farmer participation (types)

cooking quality 8, 25, 27, 31, 43, 78, 233, 253, 328
Costa Rica 215, 251
costs of on-farm research *see* donor agencies; on-farm
    research (costs)
cotton 261, 318, 319, 320, 321, 323, 324, 328, 329, 330,
    334
cowpea 56, 63, 86, 90, 117
credit availability 10, 33, 130-32, 134-35, 139, 215,
    216, 217, 218, 219, 227, 229, 230, 252, 269, 288
CRI (Ghana) 63, 64, 66, 68
CRIFC (Indonesia) 143, 292
crop management 208-09, 329-31
    see also crops
crop rotation 8, 28, 29, 30, 65, 184, 216, 252, 253, 264,
    318, 325, 327, 334
cropping systems research 11, 12, 19, 21, 109-10, 114,
    117, 136, 138, 140, 292
crops *see* bananas; barley; beans; cassava; cocoa;
    cocoyam; coffee; cotton; cowpea; groundnut; maize;
    millet; oil palm; oilseed; okra; peas; pigeonpea;
    potato; rice; sorghum; sugar cane; sunflower;
    tomato; wheat; yam
cultural issues *see* social science; sociocultural issues

data collection *see* experiments; surveys
decentralization 10, 64, 281, 289, 296
    see also research (management)
decision-making participation, farmer *see* farmer
    participation (types)
demonstrations, on-farm 10, 24, 25, 26, 33, 34, 64,
    73-76, 77, 93, 94, 101, 136, 227, 237, 244, 253, 265,
    266, 267, 307, 325, 331, 333
    see also 'verification/demonstrations'
density, plant *see* plant density
'design and diagnosis' technique 86
development agencies *see* donor agencies; NGOs;
    private sector
diagnosis *see* 'design and diagnosis' technique; 'rapid
    rural appraisal' technique; surveys (diagnostic)
diffused light storage 231-44, 272
diffusion, technology *see* technology diffusion
diseases and disease control 40, 44, 45, 46, 59, 66, 83,
    175, 191, 196-97, 209
Dominican Republic 274
donor agencies 4, 6, 7, 248, 251, 257, 269, 289, 294,
    295, 298, 310
double cropping 115, 135, 145, 274
DR&SS (Zimbabwe) 267, 293
draught power 8, 114, 127, 135, 173-74, 176, 259, 325
drought 39, 72, 75, 115, 119, 123, 196-97, 226, 253

economic analysis 9, 25, 26, 27, 53-54, 74-76, 92, 96,
    127-29, 159-60, 163, 179-80, 181, 186-87, 210-211,
    220, 223, 224, 227, 229, 239, 304, 324, 327, 332
    see also labour; input costs; social science; socio-
    economic research
Ecuador 252-54, 283, 293, 294, 306
education *see* farmer education; farmer knowledge;
    training (farmers)
ENDEMA (Panama) 216
evaluation *see* farmer evaluation; on-farm research
    (impact, monitoring); surveys
expatriate staff 144, 262, 295, 304, 310
experiment station research *see* experiments (experi-
    ment station); on-farm research (complementarity)
experiments 23-27, 136, 262
    clustering 309
    experiment station 25, 26, 90, 95, 98, 100, 101, 136-
    37, 152, 153, 182, 185, 233-35, 250, 264, 279, 322
    management 24, 45, 66, 92-95, 118-28, 138-39,
    156, 158, 172, 181-82, 183, 187, 249, 279, 280
    non-experimental variables 24, 25, 26, 156, 172,
    219, 220, 277
    on-farm 9, 23-27, 64, 82-83, 92-98, 137, 147,
    181-87, 219-27, 235-38
    see also 'design and diagnosis' technique; farmer
    participation; 'key site' model; 'minus-one' trial
    designs; on-farm research (implementation);
    research
extension 33, 49, 56, 77, 93, 99, 130-32, 133, 135, 138,
    166, 187-88, 251-52, 261, 264-67, 281, 282, 305-07,
    327, 331-32, 333
    agents 33, 34-35, 58, 64, 73, 93, 94, 114, 135, 166,
    240, 242, 264, 273, 277, 280
    materials 57, 268, 307
    packages/messages 13, 30, 114, 115, 169, 264-65,
    266, 274
    rapport with farmers 48, 49, 55, 244, 282, 307
    staff issues 55, 106, 242, 252, 266, 306-07
    see also research-extension links/collaboration;
    training (extension staff); 'transfer of technology'
    model; T&V extension method

fallow 65, 71, 86, 87, 88, 89, 90, 91, 92, 96, 101, 103,
    318, 319, 325
FAO 7, 70, 238
farm conditions
    income 88, 113-14, 173, 193, 212
    mechanization 65, 79, 101, 106, 127, 135, 173-74,
    217, 229, 284, 333
    size 44, 46, 48-49, 57, 65, 86, 105, 113, 134-35,
    145, 165, 172, 174, 193, 216, 228, 239, 320

see also input costs; input supply; labour;
land tenure; markets and marketing; women
farmers
farm-level prices *see* markets and marketing; prices,
agricultural product
farmer adaptation 5, 6, 31-32, 43, 54, 91-95, 98, 100,
139, 227, 240-41, 273-80
farmer adoption *see* technology adoption
farmer education 102, 103, 130-32, 281
see also farmer knowledge; training (farmers)
farmer evaluation of problems and priorities 21, 44, 46,
47, 89, 170, 206, 218, 233, 253, 266, 277, 278, 280,
288, 335
of technologies 27, 51, 54, 68, 73, 94, 105-06, 160,
185, 188, 212, 238, 241, 253, 269, 274, 275-76, 277,
279, 303, 304
farmer innovation 5, 30, 43, 56, 139, 203-07, 212, 240,
275, 278, 279-80, 281, 324, 327
farmer knowledge 21, 43, 47, 60, 85-86, 130-32, 133,
135, 171, 191, 232, 233, 239-40, 242, 280-81, 282,
283, 333
farmer participation 9, 23, 25-26, 31-32, 35, 55, 73, 99,
115, 138, 182, 185, 187, 212, 227, 243, 249, 266,
273-84, 301-03, 335
meetings 50-51, 55, 93, 94, 135, 266, 277, 281
selection of participants 48, 49, 50, 86, 93, 94, 129,
217, 262, 280, 302-03, 304
types of participation 276-84, 302
see also experiments; farmers' organizations;
on-farm research (adjustments; implementation);
surveys; women farmers
farmer-back-to-farmer 12, 169, 233, 243, 244
farmers' organizations 92, 102, 216, 238, 255, 279, 281,
282, 283-84, 335
see also farmer participation
farmers, women *see* women farmers
farming systems 7, 20, 46, 85, 113-14, 143, 171-74,
191, 212, 232, 250, 317, 329
interactions 8, 27, 162
parameters 27-29
farming systems research 3-12, 48, 85, 135-36, 138,
143, 247, 248, 252
'farming systems research and extension' 276
feed gardens 91, 100, 101, 105
fertilizer use and application 45, 54, 70-72, 82-83, 90,
91, 104, 119, 128, 152, 155-56, 158, 159, 175, 186,
187, 196, 202, 206, 218, 219, 224, 226, 253, 259-60,
261, 265, 268, 278, 325-26, 334
nitrogen 52, 70, 71, 72, 82, 88, 92, 119, 152, 155,
156, 158, 175, 182, 186, 198, 219, 221-22, 226, 268,
325, 326

organic 45, 52, 54, 89, 91, 100, 175, 187, 325, 330,
334
phosphorus 54, 70, 71, 82, 119, 149, 150-52, 158,
175, 180, 182, 186, 187, 198, 221-22, 268, 325, 326
potassium 82, 152, 226
see also input costs; input supply; soil (fertility)
field days 26, 33, 34, 64, 73, 95, 115, 166, 227, 240,
277, 282, 283
firewood 104-05
FITT (The Gambia) 279, 280
fodder, animal *see* livestock (fodder)
food consumption *see* consumer preferences; cooking
quality; food supply and demand
food supply and demand 8, 27, 40, 51, 58, 59, 140, 238,
260, 269, 287
forage 30, 88, 89, 91, 97, 100, 105, 328
Ford Foundation 192
forest zone 64, 66, 67, 70, 80, 86, 87-88, 92, 95, 104
fruit trees 146, 280
FSR&DD (Nepal) 292
funding *see* donor agencies; NGOs; on-farm research
(costs); private sector

Gambia, The 279
GGDP (Ghana) 63, 64, 65, 66, 73, 77, 80, 81, 82-83
Ghana 5, 63-83
GLDB (Ghana) 34, 63, 68, 73
Gliricidia spp. 90, 91, 92, 94, 95, 97, 98, 100, 101, 104,
105
goats 52, 85, 88, 91, 92, 97, 98, 101
government policies 33-35, 49, 100, 215, 232, 248, 251,
252, 293, 325, 332
see also prices, agricultural product
grasses 47, 52, 88, 91, 96, 98, 101, 278
Green Revolution 5, 6, 109, 265, 287, 288, 318, 321
Grevillea spp. 52
groundnut 143
group meetings *see* farmer participation (meetings)
Guatemala 6, 7, 243, 282, 294, 296, 305, 306, 307

harvest density *see* plant density
health *see* food supply and demand; population, human
hedgerows *see* agroforestry
Helminthosporium spp. 226
Honduras 280
humid zone 85-107
'hunger periods' *see* food supply and demand

IARCs 6, 7, 12, 136, 137, 138, 243, 248, 289
see also CIAT, CIMMYT, CIP, ICRAF, ICRISAT,
IITA, ILCA, IRRI, ISNAR

ICRAF 86
ICRISAT 12
ICTA (Guatemala) 291, 296, 305
IDIAP (Panama) 215, 216, 217, 220, 228, 229, 230, 291
IDRC 7, 238, 239
IITA 63, 89, 90
ILCA 18, 19, 85-107
implementation of on-farm research *see* on-farm
    research (implementation)
INADES 55, 56
income, farm *see* farm conditions (income)
India 279
Indonesia 143-67, 278, 310
INGER 137
INIAP (Ecuador) 252-54, 291
INIPA (Peru) 34, 192, 194, 237, 240, 242
innovation, farmer *see* farmer innovation
inoculation *see Rhizobium* inoculation
input costs 75, 128, 159, 175, 186, 196, 236, 325
input supply 33, 46, 80, 252, 268-69, 283, 305, 327,
    332
insect control *see* pests and pest control
institutionalization of on-farm research *see* on-farm
    research (institutionalization)
Integrated Pest Control Program for Tropical Asian
    Rice 282
integration of on-farm research *see* on-farm research
    (integration)
intensification *see* agricultural intensification
intercropping 4, 8, 28, 29, 40, 46, 58, 59, 65, 79, 80, 86,
    192, 194, 260, 327
interdisciplinary research *see* on-farm research
    (complementarity); research (interdisciplinary)
interviews, farmer *see* surveys
IPRA 281, 284
IRRI 12, 18, 19, 109-140
irrigation 115, 140, 144, 169, 172, 318
ISAR (Rwanda) 18, 41, 42, 47, 48, 51, 52, 54, 55, 56,
    59, 61
ISNAR 12, 290-311
ISRA (Senegal) 291
Italy 290

job definition *see* extension (staff issues); on-farm
    research (staff issues)
joint activities *see* on-farm research (complementarity);
    research (interdisciplinary); research-extension
    links/collaboration

Kabsaka project (Philippines) 115, 116, 117, 129, 133,
    134, 139

'key site' model 138
knowledge levels *see* farmer knowledge

labour 8, 29, 45, 52, 71, 79, 130, 233
    availability 29, 46, 50, 73, 130-32, 133, 134
    costs 92, 114, 134, 163, 218, 327, 333
    hire 52, 94, 96, 106, 134, 196, 219
    off-farm 39, 49, 50, 88, 146, 173, 193, 195
    requirements 47, 54, 69, 72, 95, 96, 106, 134, 227,
    250, 261, 265
    see also economic analysis; farm conditions; input
    costs
LAC (Nepal) 298
land preparation 65, 83, 96, 115, 116, 118, 136, 175,
    176, 217, 224, 226, 259, 260, 324-25
land pressure *see* agricultural intensification; farm
    conditions; food supply and demand; population,
    human
land tenure 75, 87, 98-99, 102, 103, 106, 113-14, 130-
    32, 135, 172
Latin America 6, 66, 90, 194, 212, 218, 289
    see also individual countries
leadership *see* research (management)
leaf blight 226
Leucaena spp. 52, 53, 59, 90, 91, 92, 95, 97, 98, 100,
    101, 104, 105
linkage mechanisms *see* on-farm research
    (complementarity); research (interdisciplinary);
    research-extension links/collaboration
livestock 8, 22, 28, 29, 85, 86, 88-89, 90, 91, 92, 106,
    146, 169, 173, 277, 331, 332
    fodder 8, 27, 28, 29, 30, 52, 85, 88, 90, 92, 94, 97,
    98, 100-101, 104, 106, 148, 150, 170, 172-73, 176,
    180, 181, 327, 331
    see also draught power; feed gardens
location specificity *see* on-farm research (location
    specificity)
lodging 69, 78, 148, 151, 152, 155, 175, 176, 184, 217,
    218

maize 41, 52, 53, 58, 63-83, 86, 87, 90, 91, 96, 114,
    143-67, 169-89, 192, 195, 202, 216-30, 252-54,
    257-70, 318
    hybrids 6, 146, 161, 162, 258, 259, 260, 265, 269
    varieties 66-68, 148, 160, 161, 162, 165, 174-75,
    178, 179, 180, 182, 185-86, 218, 220, 253-54, 258,
    260, 262, 269
    see also diseases and disease control; pests and pest
    control; planting practices
Malawi 257, 258, 260, 261, 264, 267, 269
Mali 252

management, crop *see* crop management

management, research *see* experiments (management); research (management)

marginal environments 39, 294, 299, 305

MARIF (Indonesia) 143-67

markets and marketing 5, 30, 53-54, 80, 146, 204, 207, 241, 252, 253, 269

maturity, plant *see* plant maturity

mechanization *see* farm conditions (mechanization)

'menu' approach 279-80, 283-84

methodology, on-farm research *see* on-farm research (implementation)

Mexico 7, 51, 194, 210

millet 56

ministries responsible for agriculture 40, 42, 48, 55, 57, 63, 70, 86, 215, 251, 269, 291, 292

'minus-one' trial designs 44

monocropping 76, 79, 80

Mosca minadora 238

mucuna 280

mulch 89, 90, 96-97, 100, 104, 105

multidisciplinary approach *see* research (inter-disciplinary)

mungbean 117, 128, 129, 143

mustard 327

Myanmar *see* Burma

national agricultural research systems 41, 63, 117, 215, 236, 247, 248, 251, 257, 283, 288-311, 318

national research programmes 136, 138, 143, 230, 243, 267, 279

Nepal 282, 294, 298, 303, 310

newsletters *see* extension (materials)

NGOs 4, 231, 252, 254, 255, 279, 283, 293, 305

Nicaragua 210

Nigeria 6, 85-107

nitrogen *see* fertilizer use and application

nutrition, human *see* food supply and demand; population, human

off-farm activities *see* labour (off-farm)

oil palm 86, 275

oilseed 173, 329, 332, 334

okra 86

on-farm client-oriented research 12, 311

on-farm experiments *see* experiments

on-farm research 12, 13, 17-36, 140, 247, 276, 288, 300

adjustments of priorities and objectives 46, 47, 98, 100, 101, 126, 148, 150-51, 181, 186, 224, 226, 229, 233, 277-78, 281, 294, 301, 325, 326, 328

complementarity/links with experiment station research 26, 61, 138, 172, 185, 192, 211-12, 230, 250, 253, 263-64, 295-301, 306, 308, 330, 334

costs and funding 11, 36, 48, 54-56, 117, 210, 229, 248, 283-84, 293, 295, 297, 305, 308-09, 310-11

impact and impact evaluation 13, 61, 98, 167, 202-07, 208-11, 212, 228, 230, 243, 249, 254, 257, 260-61, 262-69, 290, 293, 294, 333

implementation and methodology 9-11, 48, 54-55, 81, 92-98, 138, 144, 192, 194-95, 212, 230, 235, 248, 249, 250, 262-63, 264, 267, 289, 293, 296-98, 300, 309, 335

institutionalization 247, 248-49, 257, 289-90, 295, 299

integration 248, 250-51, 254, 257, 263-64, 270, 287-311, 317-36

location specificity 10, 11, 17-20, 42, 81, 250, 251, 308, 310, 328, 332

monitoring and follow-up 23, 60, 68, 94, 212, 240, 249, 280, 294, 303, 333

planning 9, 13, 20, 45-48, 56, 81, 147-52, 172, 181-82, 194, 219, 262, 264, 278, 298, 301, 303, 307

site selection 19-20, 48, 86, 93, 111-13, 138, 182-83, 192, 219-20, 262, 335

staff issues 55, 99, 248, 249, 262, 289, 293, 295-98, 305, 306-07, 309-10, 332, 335

see also farmer participation; farmer-back-to-farmer; farming systems research; on-farm client-oriented research; 'rapid rural appraisal' technique; research; research-extension links/collaboration

on-station research *see* experiments (experiment station)

on-the-shelf technology 81, 138, 154, 250, 263, 297

PAC (Nepal) 292

Pakistan 169-89, 317-36

Panama 215-30, 294

PAP (Rwanda) 48, 50, 52, 55, 56

PARC (Pakistan) 171, 318, 328, 331

participation, farmer *see* farmer participation

peas 40, 41, 252, 253

pepper 86

Peru 191-212, 231-44, 277

pests and pest control 8, 44, 68, 78, 83, 94, 115, 119, 150, 152, 153, 154, 155, 197, 201, 207, 218, 238, 275, 282

Phalaris minor 327

Philippines 109-140, 240, 242, 243, 278, 280, 283

phosphorus *see* fertilizer use and application

pigeonpea 86

pigs 233

planning *see* on-farm research (planning); research
    (management)
plant breeding *see* crops; research; varieties
plant density 59, 68-70, 78-79, 82-83, 100, 148, 149,
    150, 151, 155, 170, 176, 177, 178, 180, 183, 184,
    186, 196-97, 219, 220-23, 259
plant diseases *see* diseases and disease control
plant maturity 44, 51, 59, 109, 115, 174, 324, 328
    see also crops; planting practices
planting practices 59, 68-70, 78-79, 81, 90, 91, 100,
    103, 114, 115, 116, 118, 119, 148, 155, 162, 164,
    173, 186, 195, 196, 204, 206, 217-19, 258, 322,
    324-25, 328
    see also crops; fertilizer use and application; land
    preparation; plant density
policy environment *see* government policies; research
    (policy)
population, human 5, 86, 111
    growth 39, 44, 65, 88, 287
    health status 45, 58
population, plant *see* plant density
post-harvest technology 231-44
    see also consumer preferences; storage
potassium *see* fertilizer use and application
potato tuber moth 32, 243
potato 40, 41, 56, 192, 196, 231-44, 252, 253, 279
    varieties 237, 240
    see also diffused light storage
poultry 146
prices, agricultural product 53-54, 74, 75, 78, 92, 146,
    174, 232, 237, 238, 241, 324
    see also economic analysis; markets and marketing
priorities *see* farmer evaluation; research (priorities)
private sector 4, 5, 81, 146, 255, 283, 288, 321, 327,
    331, 334
problem-solving approach 9, 13, 20-21, 243, 263, 265,
    267, 281, 308
PROGETAPPS (Guatemala) 282
Project ATA-272 (The Netherlands) 144, 148
Puebla project (Mexico) 6

questionnaires *see* surveys

RADIP (Philippines) 117, 118
rainfall 39, 46, 48, 57, 65, 75, 86, 111-12, 118, 145,
    192, 216, 226, 253, 324
rainfed agriculture 12, 109-40
'rapid rural appraisal' technique 9
recommendations *see* research (recommendations);
    farmer adaptation
RELOs (Zambia) 268, 307

research
    adaptive 7, 9, 12, 19, 21, 33, 35, 41, 114, 136, 137,
    138, 143, 144, 267, 288, 299
    applied 6, 114-15, 116, 136, 143, 299
    commodity 11, 13, 42, 48, 81, 136, 257, 263-64,
    289, 297, 317-36
    disciplinary 143, 263-64, 269, 289, 294, 296, 297,
    298, 317-36
    interdisciplinary 9-10, 35, 41, 85, 233, 244, 263-64,
    296, 299, 330, 331
    management 55, 248, 262, 276, 287-90, 293-94,
    298, 299-311
    policy 33-35, 248, 251-52, 269, 287-95, 317-36
    priorities 9, 13, 20, 30, 152, 212, 219, 233, 249,
    264, 277, 294, 300, 303, 307, 308, 328-34
    recommendations 30, 59, 73, 74, 126, 138, 139,
    162, 169, 187, 229, 258, 268, 273, 274, 275, 276,
    307, 325, 326
    see also cropping systems research; experiments;
    farming systems research; national research
    programmes; on-farm research; socioeconomic
    research
research-extension links/collaboration 33, 48, 54, 63-64,
    93, 116, 138, 166, 243, 251-52, 263, 264, 266, 267,
    268, 281, 305-07, 331
Rhizobium inoculation 197, 198, 201, 202
rice 90, 109-40, 145, 146, 173, 318, 324
    varieties 109, 114, 115, 119, 120
    see also diseases and disease control; pests and pest
    control; planting practices
risk 8, 27, 46, 74, 75, 115, 128, 134, 135, 180, 240, 241,
    288
Rockefeller Foundation 7
root rot 45, 197, 204, 206, 207
rust 321, 322, 328, 329, 331, 333
Rwanda 39-61, 279

savanna zones 64, 86, 87-88, 95, 96, 104
SDC (Switzerland) 41
seed 46, 162, 197, 198, 201, 218, 236, 240, 241, 242
    production 100, 187-88, 201, 236-37, 275, 283
    supply 33, 34, 46, 47, 50, 60, 68, 78, 160, 174, 189,
    236, 239, 253, 321, 333, 334
    see also planting practices
'seel' operation 175, 176, 177, 178, 186
selection, farmer *see* farmer participation (selection of
    participants)
Senegal 294, 298
SERED (Nepal) 292
Sesbania spp. 52, 53, 59
shaftal see clover

share cropping 75, 113, 173
sheep 85, 88, 91, 92
shootfly 150-51, 152, 156, 162
site selection *see* on-farm research (site selection)
slash-and-burn agriculture 88
social science 10, 233, 243, 278, 303-05
sociocultural issues 98, 232, 242, 282, 283
    see also farmer evaluation
socioeconomic research 139, 304, 332-33
    see also economic analysis; social science
soil
    erosion 52, 217, 219, 224, 278, 281
    fertility 8, 43, 44, 45, 48, 71, 88, 90, 148, 155-60,
    193, 216, 329
    types 65, 111, 113, 117, 128, 144, 216
sorghum 41, 261
soybean 40, 41, 143
Sri Lanka 240, 243
staff issues *see* expatriate staff; extension (staff issues);
    on-farm research (staff issues); training
stem-borer 330
storage 8, 68, 78, 80, 231-44, 258, 260
    see also diffused light storage; post-harvest
    technology
subsidies 75, 81, 146, 154, 159, 325-26, 332, 334
sugar cane 144, 145, 318, 319, 324, 325
sulphur 149, 152
sunflower 261
surveys
    adoption 13, 76, 80-81, 101, 129, 136, 139, 162,
    202, 228, 238, 249, 294
    diagnostic 21-23, 42-45, 65, 85, 86, 89, 147-52,
    171-81, 194-201, 204, 212, 217, 232-33, 249, 262,
    304, 319-20, 335-36
    see also 'design and diagnosis' technique; 'rapid
    rural appraisal' technique
sustainability 4, 44, 89, 90, 140, 264, 308, 329, 334
Swaziland 257, 258, 260, 261, 266, 268, 269
sweet potato 40, 41, 56, 57, 58-59, 143, 280

T&V extension method 4, 252, 266, 268, 307, 331
Tanzania 45
target group 215, 249
technological change *see* farmer innovation; on-farm
    research (impact)
technology adaptation *see* farmer adaptation; on-farm
    research; research
technology adoption 6, 13, 56-61, 76-81, 99, 101-06,
    115, 129-135, 139, 162-65, 187-88, 204, 207-12,
    228-29, 238-42, 260-62, 263, 273-80, 283, 305,
    321

adoption sequence 13, 24, 30-32, 79-80, 162, 208,
    219, 229
    see also surveys; technology diffusion
technology diffusion 6, 56-61, 115, 134, 135, 203, 238,
    239, 240, 275-76, 284
    see also technology adoption
technology packages *see* extension (packages)
technology transfer *see* extension; 'transfer of
    technology' model
terminology *see* farmings systems research
theories, farming systems *see* farming systems research
tillage *see* land preparation
Togo 99
tomato 86, 173
training
    extension staff 50, 64, 115, 230, 236, 242, 262-63,
    264, 266, 267, 268, 283, 307, 332
    farmers 50, 102, 115, 240, 281-82, 283, 335
    research staff 191-92, 194, 230, 262-63, 267, 269,
    304, 305, 307
    see also T&V extension method
'transfer of technology' model 264, 265, 276
transition zone 63-83
trees 52, 88, 90, 91, 97, 99, 100, 101, 106, 278
    see also agroforestry; alley farming; *Calliandra*
    spp.; firewood; *Gliricidia* spp.; *Grevillea* spp.;
    Leucaena spp.; *Sesbania* spp.
trials *see* experiments
tsetse fly 88
tuber crops 40, 45, 103, 104, 232
    see also cassava; cocoyam; potato; sweet potato;
    yam

Uganda 39
UNDP 70
USAID 7, 248, 306
universities 7, 114, 136, 202, 215

varieties
    improved 6, 43, 51, 66, 68, 69, 78, 160, 161, 174,
    182, 191, 233, 252, 258, 321
    local 66, 68, 148, 149, 174, 186, 218, 240, 258,
    279
    see also crops; experiments
vegetables 6, 173, 174, 175
'verification/demonstrations' 73-76, 77
water resources *see* irrigation; rainfall; rainfed
    agriculture
weeds and weed control 6, 8, 72-73, 82-83, 94, 96, 106,
    114, 119, 128, 134, 136, 176, 206, 218, 219, 220-23,
    226, 260, 265, 269, 278, 280, 327-28

wheat 56, 172, 173, 175, 187, 252, 287, 317-36
    varieties 318, 321-24, 328, 333, 334
    see also diseases and disease control; pests and pest
    control; planting practices
white fly 275
white grubs 226, 227
women farmers 33, 86, 87, 88, 94-95, 99, 102, 106,
    279
World Bank 117, 252, 298

yam 86, 87, 88, 90, 94, 104

Zaïre 39, 41, 45
Zambia 250, 257, 258, 260, 261, 262, 264, 265, 266,
    267, 268, 269, 277, 294, 301, 303, 304, 305, 306
Zimbabwe 257, 258, 259, 260, 261, 265, 266, 267,
    269, 293, 294, 298, 303, 306, 307
zinc 82, 119, 144